Science and Technology
of Magnetic Oxides

MATERIALS RESEARCH SOCIETY
SYMPOSIUM PROCEEDINGS VOLUME 494

Science and Technology of Magnetic Oxides

Symposium held December 1–4, 1997, Boston, Massachusetts, U.S.A.

EDITORS:

Michael F. Hundley
Los Alamos National Laboratory
Los Alamos, New Mexico, U.S.A.

Janice H. Nickel
Hewlett-Packard Laboratories
Palo Alto, California, U.S.A.

Ramamoorthy Ramesh
University of Maryland
College Park, Maryland, U.S.A.

Yoshinori Tokura
University of Tokyo
Tokyo, Japan

Materials Research Society
Warrendale, Pennsylvania

CAMBRIDGE
UNIVERSITY PRESS

University Printing House, Cambridge CB2 8BS, United Kingdom

One Liberty Plaza, 20th Floor, New York, NY 10006, USA

477 Williamstown Road, Port Melbourne, VIC 3207, Australia

314-321, 3rd Floor, Plot 3, Splendor Forum, Jasola District Centre, New Delhi - 110025, India

79 Anson Road, #06-04/06, Singapore 079906

Cambridge University Press is part of the University of Cambridge.

It furthers the University's mission by disseminating knowledge in the pursuit of education, learning and research at the highest international levels of excellence.

www.cambridge.org
Information on this title: www.cambridge.org/9781558993990

Materials Research Society
506 Keystone Drive, Warrendale, PA 15086
http://www.mrs.org

© Materials Research Society 1998

First published 1998
First paperback edition 2013

Single article reprints from this publication are available through University Microfilms Inc., 300 North Zeeb Road, Ann Arbor, MI 48106

CODEN: MRSPDH

A catalogue record for this publication is available from the British Library

ISBN 978-1-558-99399-0 Hardback
ISBN 978-1-107-41349-8 Paperback

CONTENTS

*Invited Paper

v

*Invited Paper

*Invited Paper

*Invited Paper

*Invited Paper

PREFACE

This proceedings volume contains papers presented at the "Metallic Magnetic Oxides" symposium (Symposium V) held in Boston, Massachusetts, December 1-4, 1997 as part of the 1997 MRS Fall Meeting. The considerable degree of interest in metallic magnetic oxides was demonstrated by the attendance at the symposium sessions as well as by the 82 papers presented during the four-day symposium.

Research into the science and technology of magnetic oxides has undergone a renaissance during the past seven years. In large measure this stems from the rediscovery of the colossal magnetoresistance associated with the ferromagnetic-order-induced, metal-insulator transition exhibited by the doped lanthanum manganites. These are not "new" materials. Indeed, pioneering work was carried out by Jonker, Van Santen, and Volger at the Dutch Phillips Research Laboratory in the 1950s. Research today is focused both on improving our understanding of the phenomena exhibited by these compounds and on developing technological applications that utilize their extremely magnetic-field-dependent conductivity near room temperature.

With the development of advanced oxide thin-film growth techniques in recent years it has become possible to produce novel materials with exciting electronic and magnetic properties which may be candidates for future device applications. One key class of these materials is the metallic magnetic oxides. This symposium focused on colossal magnetoresistance (CMR) materials, including manganites and cobalites. Transport and magnetic properties and their dependence on stress, growth conditions, stoichiometry, and elemental composition are now being explored quite extensively. These new and exciting results are driving an effort to explain the underlying physical mechanisms responsible for the remarkable electrical properties exhibited by these compounds. The large magnetic field required to obtain the CMR effect has been perceived as a technological roadblock for commercialization of this phenomenon. This has motivated research aimed both at reducing the intrinsic field dependence as well as at developing novel device structures that will reduce the magnetic field required to realize the CMR effect. Technologically useful devices utilizing these compounds will undoubtedly involve multilayer, spin-valve or tunneling-junction heterostructures. Extremely impressive low field effects have indeed been observed recently at low temperatures in CMR heterostructure devices. Due to the strong interplay between spin, charge, and lattice degrees of freedom in these compounds, the magnetic and transport properties of CMR systems are extremely stress dependent. As such, CMR heterostructures will most likely involve other metallic or insulating oxide materials. Hence, CMR device research must involve other metallic magnetic oxide systems as well. Other compounds of interest include half-metallic ferromagnets, yttrium garnet materials, ferrites, spinels, and vanadates. In addition to their consideration for magnetic recording applications, these systems are also under consideration for more generic magnetic sensing uses, microwave, bolometric, and other high-frequency applications.

The research on metallic magnetic oxides presented in this proceeding volume is composed of both device-related technology work and basic research studies focusing on the novel phenomena exhibited by these systems. Device-related research is presented that examines the fabrication and properties of CMR-based spin valves, tunnel junctions, and bolometers grown via MBE, pulsed-laser deposition, and sputtering techniques. Hybrid CMR/high-T_c devices are also discussed. These devices are characterized via magnetization, magnetotransport, and microstructural microscopy measurements. Extensive research is also presented that examines the underlying properties from which the CMR effect originates. Progress in elucidating the influence of strain on the magnetic and electronic properties of CMR compounds is reported from both experimental and theoretical viewpoints. Advances in our understanding of local structure effects are presented which clarify the nature of the charge transport process in CMR manganites below T_c. Optical and Raman spectroscopies, spin-dynamic measurements, results from isotope-effect experiments, magnetostriction, and thermal expansion measurements are also presented that extend our understanding of the way in which the spin, charge, and lattice act in unison to produce the novel properties that CMR materials exhibit.

The contents of this proceedings volume represent the latest research concerning the science and technology of magnetic oxides performed at academic, government, and industrial laboratories world wide.

Michael F. Hundley
Janice H. Nickel
Ramamoorthy Ramesh
Yoshinori Tokura

January 1998

ACKNOWLEDGMENTS

The organizers wish to thank all those who participated in the 1997 MRS symposium "Metallic Magnetic Oxides". We would especially like to thank the invited speakers for their presentations; each added significantly to the symposium, and, as a whole, the invited talks formed the foundation for a very successful symposium. The invited speakers include:

C.H. Booth	A.J. Millis
Alexander Bratkovsky	J.J. Neumeier
S-W. Cheong	M. Rajeswari
L.F. Cohen	Yuri Suzuki
David Emin	Hitoshi Tabata
J.B. Goodenough	T. Venkatesan
Tsuyoshi Kimura	X-D. Xiang
V. Kiryukhin	Gang Xiao

We also thank the session chairs for their assistance in orchestrating the sessions and the associated discussions. We extend our appreciation to all of the participants who took the time to prepare a manuscript for this proceeding volume. We are also grateful to those who promptly and thoroughly reviewed the proceedings manuscripts.

The symposium organizers wish to thank the following organizations for their generous financial support, which enabled us to present the "Metallic Magnetic Oxides" symposium:

Hewlett-Packard Corporation
Joint Research Center for Atom Technology
Lake Shore Cryotronics, Inc.
Los Alamos National Laboratory

Our thanks go to the Materials Research Society, its staff, and the 1997 MRS Fall Meeting chairs, for a highly successful meeting. We also gratefully acknowledge the assistance of Pamela Rockage at Los Alamos, as well as the MRS publications staff, in assembling these proceedings.

MATERIALS RESEARCH SOCIETY SYMPOSIUM PROCEEDINGS

MATERIALS RESEARCH SOCIETY SYMPOSIUM PROCEEDINGS

Part I

Materials Processing of Metallic Magnetic Oxides

DRY AND WET ETCH PROCESSES FOR NiMnSb, LaCaMnO$_3$ AND RELATED MATERIALS

J. Hong*, J. J. Wang*, E. S. Lambers*, J. A. Caballero*, J. R. Childress*, S. J. Pearton*, K. H. Dahmen**, S. Von Molnar***, F. J. Cadieu**** and F. Sharifi*****
*Department of Materials Science and Engineering, University of Florida, Gainesville, FL
**Department of Chemistry/MARTECH, Florida State University, Tallahassee, FL
***Department of Physics/MARTECH, Florida State University, Tallahassee, FL
****Physics Department, Queens College of CUNY, Flushing, NY
*****Department of Physics, University of Florida, Gainesville, FL

ABSTRACT

A variety of plasma etching chemistries were examined for patterning NiMnSb Heusler thin films and associated Al$_2$O$_3$ barrier layers. Chemistries based on SF$_6$ and Cl$_2$ were all found to provide faster etch rates than pure Ar sputtering. In all cases the etch rates were strongly dependent on both the ion flux and ion energy. Selectivities of $\geq$20 for NiMnSb over Al$_2$O$_3$ were obtained in SF$_6$-based discharges, while selectivities $\leq$5 were typical in Cl$_2$ and CH$_4$/H$_2$ plasma chemistries. Wet etch solutions of HF/H$_2$O and HNO$_3$/H$_2$SO$_4$/H$_2$O were found to provide reaction-limited etching of NiMnSb that was either non-selective or selective, respectively, to Al$_2$O$_3$. In addition we have developed dry etch processes based on Cl$_2$/Ar at high ion densities for patterning of LaCaMnO$_3$ (and SmCo permanent magnet biasing films) for magnetic sensor devices. Highly anisotropic features are produced in both materials, with smooth surface morphologies. In all cases, SiO$_2$ or other dielectric materials must be used for masking since photoresist does not retain its geometrical integrity upon exposure to the high ion density plasma.

INTRODUCTION

Ferromagnetic thin films and multilayers are currently being used in various magnetic recording, magnetic sensor and non-volatile memory applications. Interest in these materials for microelectronic applications has increased dramatically since the discovery of giant magnetoresistance (GMR) in multilayers comprised of alternating ultrathin (10-50Å) ferromagnetic/noble metal layers[1] and more recently the study of La-manganite perovskite collosal magnetoresitive (CMR) materials.[2] The Heusler alloy NiMnSb is a strong candidate for useful half-metallic behavior, which shows metallic for one spin type and insulating (or semiconducting) for the other, due to its high Curie temperature (720K).[3] Recently, significant experimental effort has been expanded to deposit high-quality thin-films of NiMnSb for magnetoresistive applications.[4-6] The spin filtering effect of NiMnSb thin layers will be maximized when the current flows normal to the layer plane, either resistively or by tunneling through an oxide barrier such as Al$_2$O$_3$. On the other hand, La-manganite materials requires above magnetic fields of about 1 Tesla to achieve most sensitive field-induced resistivity transition. Thus the necessary bias field is too large to be produced by an electrical current within the device, as is done for typical low-field magnetoresistive sensors. Consequently it may be necessary to provide a fixed, built-in bias field within the device, from a hard magnet materials such as SmCo. The fabrication of small, high quality etched patterns is particularly important to the potential applications of all these materials. In this paper, we report on selective wet and dry etch processes for NiMnSb and Al$_2$O$_3$ structures, and on the Cl$_2$-based plasma etching of LaCaMnO$_3$ and SmCo.

3

EXPERIMENT

5000Å-thick NiMnSb thin films were deposited by magnetron sputtering with 2.5mTorr of Ar pressure, 30W of rf power onto glass substrate at the temperature of 350°.[6] Al_2O_3 films were deposited with 2.5mTorr Ar pressure and 140W of rf power onto Si wafer substrates held at room temperature. Thin films of $La_xCa_{1-x}MnO_3$ with x=0.41 were prepared on MnO(001) and $Al_2O_3(0001)$ single crystal substrates by liquid delivery metalorganic chemical vapor deposition (LD-MOCVD). The SmCo-based films were directly crystallized by RF diode sputtering (100mTorr Ar) onto moderately heated (375-425°C) polycrystalline aluminum oxide substrates. The samples had a nominal composition of Sm 13%, Co 58%, Fe 20%, Cu 7%, Zr 2% and were directly crystallized upon deposition into the disordered TbCu7 type crystal structure.

The samples were masked with either photoresist, SiN_x, SiO_2 or apezon wax for etching experiments. Dry etching was performed in a Plasma Therm SLR 770 system.[7] The plasma is generated in a low profile Astex 4400 Electron Cyclotron Resonance (ECR) source operating at 2.45GHz and input powers from 0-1000W. The He back side cooled sample chuck is separately biased with 13.56MHz rf power from 50-450W, with induced dc self-bias of -90 to -1000V depending on the gas chemistry and microwave source power. Etch depths were obtained by stylus profilometry after removal of the mask materials in either acetone (photoresist) or trichloroethylene (wax). The etched surface morphologies were examined by Scanning Electron Microscopy (SEM) and Atomic Force Microscopy (AFM) in the tapping mode.

RESULTS AND DISCUSSION

The sputter and etch rates for NiMnSb and Al_2O_3 in ECR Ar, $CH_4/H_2/Ar$, Cl_2/Ar discharges are shown in Figure 1, for a fixed rf chuck power (250W for Ar, $CH_4/H_2/Ar$ and 150W for Cl_2/Ar) and pressure (1.5mTorr), as a function of microwave source power. As this source power is increased, the ion density in the discharge is increased (from $\sim 10^9$ cm^{-3} at 0W to $\sim 3 \times 10^{11}$ cm^{-3} at 1000W), and even though this suppresses the dc self-bias in the sample chuck (i.e. the acceleration potential for ions incident in the sample), the etch rates of both materials increase due to the higher ion flux. For microwave power source powers up to ~400W, there is no measurable sputtering of the Al_2O_3, and thus at lower powers the Al_2O_3 can act as an etch stop when removing overlying NiMnSb films. In $CH_4/H_2/Ar$ discharges under similar conditions of rf power, pressure and total gas flow rate, the etch rates of both materials are almost independent of microwave source power, and somewhat lower than with pure Ar. This may be due to shielding of the surface by polymer deposition at high source power, as is found for $CH_4/H_2/Ar$ ECR etching of semiconductors.[7] Faster etch rates were obtained for both materials in Cl_2/Ar discharges. The enhancement in NiMnSb etch rates relative to pure Ar under the same conditions ranged from ~10% at low microwave source powers to ~30% at 1000W, even at lower ion energies. The etch rates for NiMnSb were up to a factor of two higher than for Al_2O_3 at high source powers. While etch products such as $SbCl_5$ and $AlCl_3$ are quite volatile, nickel and manganese chlorides have relatively low vapor pressures and require ion assistance to promote their desorption. The advantage of the high ion fluxes under ECR conditions is two-fold.[8] First, in strongly bonded materials such as Al_2O_3, one of the rate-limiting steps will be the ability to initially break bonds in order to allow the etch products to form. Therefore, at constant etch yield (i.e. atoms of the substrate removed per unit incident ion), a higher ion flux will produce a higher etch rate. The second advantage of the ECR discharges is that the high ion flux more effectively assists etch product desorption. Under more conventional reactive ion etching conditions this ion-assisted desorption is inefficient, allowing a thick selvedge or reaction layer of the involatile etch products to form on the sample surface. This layer shields the surface from further interaction with the plasma and etching stops. The selectivity for etching NiMnSb over Al_2O_3 is ≤ 2 in Cl_2/Ar over the microwave source power range 300-800W.

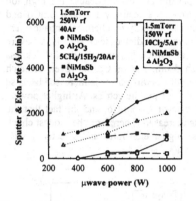 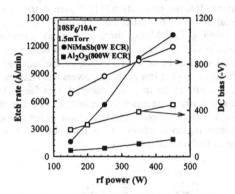

Figure 1. Etch rates of NiMnSb and Al_2O_3 as a function of microwave source power in 1.5mTorr 250W rf chuck power discharges of Ar or $CH_4/H_2/Ar$ and 150W rf chuck power discharges of Cl_2/Ar.

Figure 2. Etch rates of NiMnSb and Al_2O_3 as a function of rf chuck power in 1.5mTorr discharges of SF_6/Ar at either 800W (Al_2O_3) or 0W (NiMnSb) microwave source power.

The most efficient etching of NiMnSb was found with the SF_6/Ar plasma chemistry. In fact the etch rates were $\geq 1.6\mu m\ min^{-1}$ even for the lowest microwave source power at which ECR discharges were stable, namely 400W. The etch rates were impossible to accurately quantify at high powers because the entire NiMnSb film disappeared in ≤ 15secs under these conditions. By contrast, the etch rates of Al_2O_3 are ≤ 1200Å min^{-1} over the entire range of source powers, leading to selectivities of NiMnSb over Al_2O_3 of ≥ 20. This is not too surprising given that AlF_3 is significantly less volatile than $AlCl_3$, reducing the etch rate of Al_2O_3 in fluorine-based plasma chemistries relative to that in chlorine-based chemistries.

Besides ion flux, the other critical parameter in etching strongly bonded materials is ion energy.[8] This is controlled by the rf power applied to the sample chuck. At fixed microwave source power, the ion energy will be increased in a roughly linear fashion by increasing the chuck power.[9,10] Figure 2 shows the etch rates of both materials in SF_6/Ar discharges as a function of rf chuck power. In this case we used a relatively high microwave source power for Al_2O_3 etching, but did not power the ECR source for the NiMnSb experiments because the rates were unmeasurably high as discussed earlier. At higher dc self-biases the etch rates are expected to increase because of more efficient bond breaking initially in the materials at higher incident ion energies, and the higher sputter yields of the etch products. Figure 3 shows the sputter rates in pure Ar discharges and etch rates in Cl_2/Ar discharges of $LaCaMnO_3$ and SmCo as a function of rf chuck power. Note that the results for Cl_2/Ar basically follow those for pure sputtering (Ar), indicating that the La, Ca and Mn chlorides are not particularly volatile even at the high ion fluxes. As a comparison, there was a substantial degree of chemical enhancement observed for the etching of SmCo in Cl_2/Ar chemistries. The etch rate is approximately a factor of 10 to 12 higher than pure Ar up to dc self-biases of $\sim$-217V; at higher biases the etch rate with Cl_2/Ar saturates and then decreases. Up to a particular ion energy, the etch rate is increased by the higher sputtering efficiency that more efficiently desorbs the etch products. However, above this energy (in these experiments ~250eV) the ions are able to desorb the chlorine radicals before they are able to react with the SmCo and hence the etch rates decreases.

Another important consideration is selection of a mask material for the etching process. Photoresist is typically not suitable for high density plasma processes because the high ion currents lead to reticulation.[11] Figure 4 shows Cl_2/Ar etch selectivity for both SmCo and $LaCaMnO_3$ over the dielectric SiO_2 and SiN_x. Since there is basically no chemical enhancement for etching $LaCaMnO_3$, there is also no selectivity over the dielectrics. This is a severe limitation if one needed to pattern deep features into $LaCaMnO_3$ because the mask thickness would need to be at least as thick as the required etch depth. For SmCo, however, the etch selectivity is ~4 at low rf chuck powers and increases initially as this power is increased because the etch rate of the magnetic material rises faster than that of the dielectrics. At higher powers the selectivity decreases because of the fall-off in etch rate of the SmCo, and the fact that the dielectric etch rate continues to increase as ion energy is increased. Therefore, the modest chuck self-bias region is advantageous from the viewpoint of higher etch rates and selectivity with respect to the mask materials.

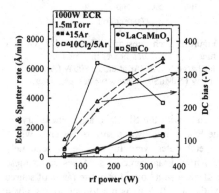 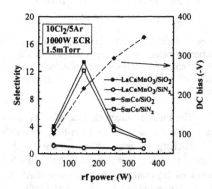

Figure 3. Etch rates of $LaCaMnO_3$ and SmCo in 1.5mTorr, 1000W microwave power discharges of pure Ar or Cl_2/Ar as a function of rf chuck power.

Figure 4. Selectivity for etching SmCo films or $LaCaMnO_3$ over either SiO_2 or SiN_x mask materials, as a function of rf chuck power.

Both NiMnSb and Al_2O_3 were found to exhibit a linear dependence of etch depth on time in HF/H_2O solutions (Figure 5), and there was no measurable effect of solution agitation on etch rate. These are both characteristics of reaction-limited etching, where the critical parameter is temperature. Note in Figure 5 that the selectivity for NiMnSb is ~3:1 over Al_2O_3 based on the fact that the etch rates are equal at compositions of $1HF : 2H_2O$ (NiMnSb) and pure HF (Al_2O_3). As mentioned previously an oxide forms readily on NiMnSb and thus HF solution would be expected to etch this material by removal of this oxide.

To obtain more highly selective etching of the NiMnSb, we examined the $HNO_3 : H_2SO_4 : H_2O$ system at 25°C. The solution composition dependence of etch rates is shown in Figure 6. The etch rates of NiMnSb become extremely rapid as HNO_3 concentration is increased and thus dilution with H_2SO_4 or H_2O, or both, is necessary to achieve controllable rates (a few thousand angstroms per minute). The $HNO_3 : H_2SO_4 : H_2O$ system is completely selective over Al_2O_3 and thus thin layers of the latter can be employed as etch-stops. It is also desirable for practical applications to determine the nature of the etch reaction. Figure 7 shows an Arrhenius plot of the

NiMnSb etch rate in a 1:3 solution HNO_3 : H_2SO_4. The activation energy of 16.7 ± 3.4kcal/mole is typical of reaction-limited etching.[12] This is the preferred situation for obtaining reproducible pattern transfer, since the important parameter, namely temperature, can be well-controlled. If the etching is diffusion-limited, then the etch rate is dependent on solution agitation which is more difficult to reproduce, and moreover the etch depth proceeds as the square root of immersion time in the solution rather than the more straightforward linear dependence of reaction-limited etching.

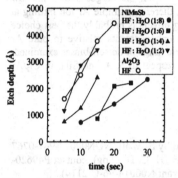

Figure 5. Etch depth versus time of NiMnSb and Al_2O_3 in HF/H2O solutions at 25°C.

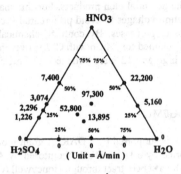

Figure 6. Etch rates (in Å/min) of NiMnSb as a function of solution formulation in $HNO_3/H_2SO_4/H_2O$ at 25°C.

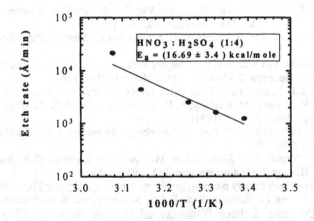

Figure 7. Arrhenius plot of NiMnSb etch rate in $1HNO_3/3H_2SO_4$ solutions.

CONCLUSIONS

Several plasma chemistries (Cl_2/Ar, CH_4/H_2/Ar and SF_6/Ar) were investigated for the etching of NiMnSb/Al_2O_3 structures under ECR conditions. The latter plasma chemistries produces the highest etch rates for NiMnSb and consequently the highest etch selectivity over Al_2O_3. Non-selective etching of both materials is obtained over a wide range of conditions in the other plasma chemistries. These processes will be useful in fabrication of small geometry spin-valve and GMR devices, especially at sub-micron dimensions. For larger area structures, pattern transfer can be obtained using two different wet etch processes, namely HF/H_2O for non-selective etching of $NiMnSb/Al_2O_3$ and $HNO_3/H_2SO_4/H_2O$ for selective etching of NiMnSb over Al_2O_3.

LaCaMnO$_3$ showed no chemical enhancement with Cl_2/Ar discharges due to the low volatilities of the potential etch products. For this material, therefore, simple Ar ion milling at modest acceleration voltages to avoid preferential sputtering effects is probably the best choice for pattern transfer processes. By contrast, chemical etch enhancements relative to pure Ar sputtering were obtained for SmCo with Cl_2/Ar over the whole range of dc self-biases examined. Selectivities as high as ~12 were obtained for SmCo with respect to SiO_2 and SiN_x in Cl_2/Ar discharges.

ACKNOWLEDGEMENT

The work is supported by a DARPA grant monitored by S. Wolf, NON00014-961-0767 ,partially supported by a DOD MURI monitored by AFOSR (H. C. DeLong), contact F49620-96-1-0026, and by a subcontract through Honeywell (ONR grant N00014-96-C-2114).

REFERENCES

1. M. N. Baibich, J. M. Broto, A. Fert, F. Nguyen Van Dau, F. Petroff, P. Etienne, G. Creuzet, A. Friederich and J. Chazelas, Phys. Rev. Lett. **66**, 2472 (1988).
2. R. von Helmolt, J. Wecker, B. Holzapfel, L. Schultz and K. Samwer, Phys, Rev. Lett. **71**, 2331 (1993).
3. M. J. Otto, R. A. M. van Woerden, P. J. van der Valk, J. Wijngaard, C. F. van Bruggen, C. Hass, and K. H. J. Buschow, J. Phys. : Condens. Matter **1**, 2341 (1989).
4. J. S . Moodera, L.R. Kinder, T.M. Wong and R. Meservey, Phys. Rev. Lett. **74**, 3273 (1995).
5. J. F. Bobo, P. R. Johnson, M. Kautzky, F. B. Mancoff, E. Tuncel, R. L. White and B. M. Clemens, J. Appl. Phys., **81** 4164 (1997).
6. J. A. Caballero, F. Petroff, Y. D. Park, A. Cabbibo, R. Morel and J. R. Childress, J. Appl. Phys, **81** 2740 (1997).
7. J. W. Lee, S. J. Pearton, C. J. Santana, J. R. Mileham, E. S. Lambers, C. R. Abernathy, F. Ren and W. S. Hobson, J. Electrochem. Soc. **143** 1093 (1996).
8. for a discussion of high density plasmas, see for example High Density Plasma Sources, ed. O. A. Popov (Noyes Publications, Park Ridge, NJ 1996), M. A. Lieberman and A. J. Lichtenburg, Principles of Plasma Discharges and Materials Processing (Wiley and Suns, NY 1994) ; J. Asmussen, J. Vac. Sci. Technol. A **7** 883 (1989).
9. F. F. Chen, Chapter 1 in High Density Plasma Sources, ed. O. A. Popov (Noyes Publications, Park Ridge NJ 1996).
10. R. J. Davis and E. D. Wolf, J. Vac. Sci. Technol. B8 1798 (1990).
11. J. W. Lee and S. J. Pearton, Semicond. Sci. Technol. **11** 812 (1996).
12. S. S. Tan and A. G. Milnes, Solid State Electron **38** 17 (1995).

EPITAXIAL GROWTH MECHANISM AND PHYSICAL PROPERTIES OF ULTRA THIN FILMS OF $La_{0.6}Sr_{0.4}MnO_3$

Yoshinori Konishi[1], Masahiro Kasai[1], Masashi Kawasaki[1,2], and Yoshinori Tokura[1,3]

1 Joint Research Center for Atom Technology (JRCAT), Tsukuba 305, Japan

2 Department of Innovative and Engineered Materials, Tokyo Inst. of Technology, Yokohama 226, Japan

3 Department of Applied Physics, The University of Tokyo, Tokyo 113, Japan

ABSTRACT

Thin films of $La_{1-x}Sr_xMnO_3$ ($x=0.4$) were fabricated using pulsed laser deposition (PLD) methods. The surface morphology of the films was sensitively affected by oxygen pressure during deposition. At high oxygen pressure (~150 mTorr), randomly aligned grains were nucleated on the epitaxial film. When the pressure was reduced to 100 mTorr, the epitaxial film had very smooth surface. Under this condition, the thickness dependence of resistivity and magnetization were analyzed. Even 6 nm thick film showed ferromagnetic metallic behavior. The AFM images of ultra thin films deposited on wet-etched $SrTiO_3$ showed atomically flat terraces and 0.4 nm steps. The film growth mode can be tuned to either layer by layer or step flow by the deposition temperature.

INTRODUCTION

Perovskite-type oxides have such versatile properties as superconductivity, ferroelectricity, and colossal magnetoresistivity [1-5]. Epitaxial multilayers composed of these oxides should explore new functional devices. Actually, trilayer tunnel junctions [6], current injection transistors [7], and superlattices [8] having manganite thin films as one of the electroactive components have been fabricated to demonstrate its capability to be utilized in future electronics. However, the surface and interface structures have been neither understood nor controlled in an atomic scale as have been done in semiconducting devices. In order to elucidate novel phenomena and utilize them in devices, each oxide layer must have not only smooth surface but also physical properties as expected. It is of great importance for this purpose to understand and control the epitaxy dynamics on an atomic scale.

In this study, we optimized the epitaxial growth conditions for perovskite-type manganese oxides $La_{1-x}Sr_xMnO_3$ ($x=0.4$) and measured physical properties of ultra thin films. We have also investigated the epitaxial growth behaviors on wet-etched $SrTiO_3$ substrates. These substrates enabled us to control the growth mode in an atomic scale.

EXPERIMENT

Thin films were fabricated using high vacuum PLD apparatus with an ArF eximer laser (193 nm) [9]. Background pressure was typically 2×10^{-9} Torr. A sintered

9

pellet of $La_{0.6}Sr_{0.4}MnO_3$ (LSMO) was used as a target. We used two types of (100) $SrTiO_3$ (STO) single crystals for substrates. One is as-polished and the other is wet-etched substrates [10]. Oxygen pressure and substrate temperature were changed to optimize the epitaxial growth condition. After the deposition, films were slowly cooled down to room temperature in 760 Torr oxygen atmosphere for an hour. Magnetization of the films was measured using a superconducting quantum interference device (SQUID) magnetometer and resistivity was measured by conventional four probe method.

RESULTS

The surface morphology and crystal orientation of the films were sensitively affected by oxygen pressure. Scanning electron microscope (SEM) pictures and reflection high energy electron diffraction (RHEED) patterns shown in Fig.1 clearly indicate the oxygen pressure dependence of the surface morphology of the films deposited on as-polished substrates. Under high oxygen pressure condition, randomly aligned grains were nucleated on the epitaxial film, resulting in RHEED pattern having both streaks from the underlying epitaxial film and rings associated with the grains. When the pressure was reduced to 100 mTorr, the epitaxial films had very smooth surface as indicated by fine streaks in RHEED pattern as well as SEM picture.

We now show the thickness dependence of resistivity and magnetization. The films are deposited at the optimum condition, i.e. 100 mTorr. The results are summarized in Fig.2. For the films thicker than 20 nm, the residual resistivity was as low as $1\text{-}2\times10^{-4}$ Ωcm, the Curie temperature (T_C) reached 350 K and magnetization saturated at 590 emu/cc (3.5 μ_B / Mn atom). These values are comparable to those of single crystals. For the films thinner than 12 nm, the residual resistivity increased and saturated magnetization and T_C decreased as decreasing the film thickness. However, ferromagnetic metallic behavior was observed for a film as thin as 6 nm. The origin of the degraded ferromagnetic and metallic properties was not clear yet. The lattice constant of STO is 3.905 Å and longer than LSMO by 0.9%. Therefore, it is plausible that the tensile strain caused by the coherent growth at the interface induced the degraded properties. Fig.3 shows the θ-2θ X-ray diffraction peaks of the films. As can be seen, the lattice constant was reduced for thinner films. Although it was difficult to detect the peaks for the films thinner than 4 nm, this trend agrees with the tensile strain at the interface.

For further investigation of the growth dynamics, we have fabricated thin films on wet-etched STO substrates which have atomically flat terraces and 0.4 nm steps[10]. Fig.4 shows the AFM images of about 10 nm thick film. When the substrate temperature was 750 °C, the surface morphology was rather rough but each grain had same height (~0.4 nm) agreeing with the size of the unit cell. The original step structure on the substrate was preserved as shown in the line profile. When the substrate temperature was increased to 820 °C, atomically smooth terraces and 0.4 nm steps were seen on the film. This fact clearly indicates the latter film grew in a step flow mode [11]. We illustrate the epitaxial growth of manganite thin film as shown in Fig.5. At relatively low substrate temperature, the migrating atoms on the terraces do not have enough kinetic energy to

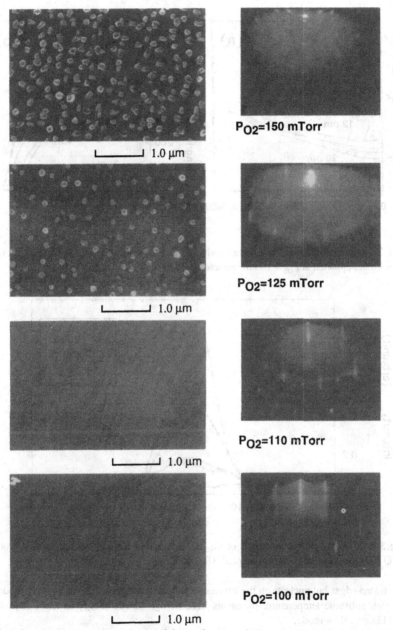

Fig.1 Oxygen pressure dependence of the surface morphology tor the 200nm thick films deposited at 750℃. SEM images (left) and RHEED patterns (right).

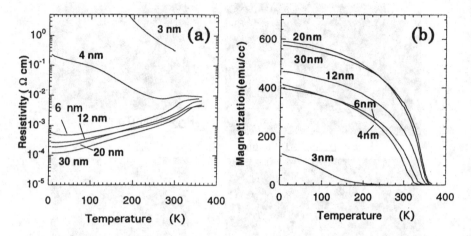

Fig.2 Temperature dependence of resistivity (a) and magnetization (b) of $La_{0.6}Sr_{0.4}MnO_3$ thin films deposited at $P_{O2} = 100$ mTorr and $T_s = 750$ °C.

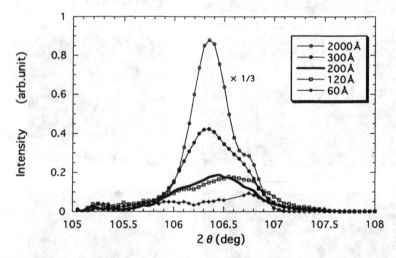

Fig.3 θ-2θ X-ray diffraction peaks of $La_{0.6}Sr_{0.4}MnO_3$ thin films. The contribution from STO (004) was subtracted from original XRD patterns.

reach step edges but nucleate on the terraces so that the films grow in layer by layer mode. At high substrate temperature, the atoms have enough kinetic energy to reach steps to yield in step flow mode.

(a)   **(b)**

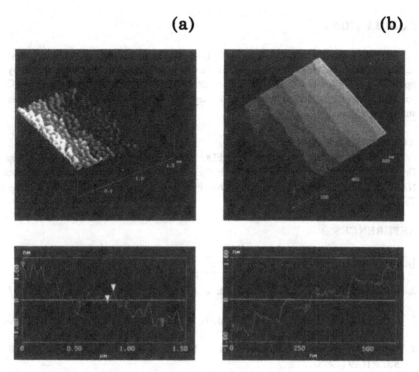

Fig.4   AFM images of about 10nm thick films deposited on wet-etched STO substrates. The substrate temperature was 750℃ (a) and 820℃ (b).

EPITAXIAL GROWTH MECHANISM

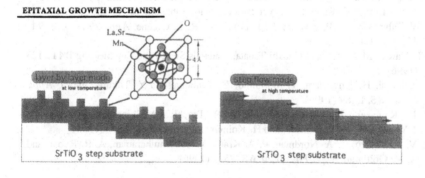

Fig.5 Schematic drawing of epitaxial growth mechanism of manganite thin films.

CONCLUSIONS

We have fabricated $La_{1-x}Sr_xMnO_3$ (x=0.4) thin films by PLD. Epitaxial films having excellent electro-magnetic properties with fairly smooth surface could be grown by optimizing the conditions. The growth mode of ultra thin film on wet-etched STO substrates could be tuned to either layer by layer or step flow modes by changing the substrates temperature.

ACKNOWLEDGMENTS

This work, partly supported by NEDO (New Energy and Industrial Development Organization), was performed in the JRCAT under the joint research agreement between the NAIR (National Institute for Advanced Interdisciplinary Research) and the ATP (Angstrom Technology Partnership).

REFERENCES

1. Y. Tokura, A. Urushibara, Y. Moritomo, T. Arima, A. Asamitu, A. Kido, and H. Furukawa, *J. Phys. Soc. Jpn.*, **63**, 3931 (1994).
2. A. Asamitu, Y. Moritomo, Y. Tomioka, T. Arima, and Y. Tokura, *Nature* **373**, 407 (1995).
3. Y. Tomioka, A. Asamitu, Y. Moritomo, H. Kuwahara, and Y. Tokura, *Phys. Rev. Lett.* **74**, 5108 (1995).
4. Y. Tomioka, A. Asamitu, H. Kuwahara, Y. Moritomo, and Y. Tokura, *Phys. Rev. B* **53**, R1689 (1996).
5. H. Kuwahara, Y. Tomioka, A. Asamitu, Y. Moritomo, and Y. Tokura, *Science* **270**, 961 (1995).
6. J. Z. Sun, W. J. Gallagher, P. R. Duncombe, L. Krusin-Elbaum, R. A. Altman, A. Gupta, Yu Lu, G. Q. Gong, and Gang Xiao, *Appl. Phys. Lett.* **69**, 3266 (1996)
7. Z. W. Dong, R. Ramesh, T. Venkatesan, Mark Johnson, Z. Y. Chan, S. P. Pai, V. Talyansky, R. P. Sharma, C. J. Lobb, and R. L. Greene, *Appl. Phys. Lett.* **71**, 1718 (1997)
8. T. Kawai, M. Kanai, and Hitoshi Tabata, Materials Sience and Engimeering **B41**, 123 (1996)
9. M. Kasai, H. Kuwahara, Y. Moritomo, Y. Tomioka, and Y. Tokura, *Jpn. J. Appl. Phys.*, **35**, L489 (1996).
10. M. Kawasaki, K. Takahashi, T. Maeda, R. Tsuchiya, M. Shinohara, O. Ishiyama, T. Yonezawa, M. Yoshimoto, and H. Koinuma, *Science*, **266**, 1540 (1994)
11. V. A. Vas'ko, C. A. Nordman, P. A. Kraus, V. S. Achutharaman, A. R. Ruosi, and A. M. Goldman, *Appl. Phys. Lett.* **68**, 2571 (1996)

THIN FILM GROWTH AND MAGNETOTRANSPORT STUDY OF (La, Sr)MnO₃

Takashi Manako, Takeshi Obata, Yuichi Shimakawa, and Yoshimi Kubo
Fundamental Research Laboratories, NEC Corporation,
34, Miyukigaoka, Tsukuba, Ibaraki 305, Japan, manako@sci.cl.nec.co.jp

ABSTRACT

Thin-film samples of $La_{1-x}Sr_xMnO_3$ (x = 0.20 - 0.30) were grown by pulsed-laser deposition using various target compositions, substrate materials, growth temperatures, oxygen partial pressures, and laser-pulse repetition rates. The crystal structure and the transport and magnetic properties of these films were then examined. Of the growth conditions, the oxygen partial pressure (Po_2) had the greatest influence on the electrical and magnetic properties. Films grown under a low Po_2 had a low ferromagnetic transition temperature (T_c) and a wide resistive transition width. None of the heat treatments done after growth improved these films' quality. The film morphology was significantly affected by the substrate material. Our x-ray diffraction analysis and AFM measurements revealed that the films deposited on both MgO (100) and LaAlO₃ (100) were epitaxially grown but contained defect structures. In contrast, grain-free thin films were epitaxially grown on the SrTiO₃ (100) substrates. The surface roughness of films grown on SrTiO₃ was less than 0.3 nm, even for films up to 150 nm thick. Under optimized growth conditions, as-deposited films for x ≥ 0.2 showed a sharp transition in resistivity at T_c. Magnetoresistance at far below T_c was as low as that reported for single-crystal sample. Since large magnetoresistance was often observed in polycrystalline samples and believed to be a grain boundary effect, these results indicate the high quality of the films grown on the SrTiO₃ substrates.

INTRODUCTION

The rediscovery of perovskite-type manganites [1,2] that show a colossal magnetoresistance (CMR) has given rise to considerable enthusiasm in both scientific and technological fields. In particular, recent observation of a huge tunnel magnetoresistance (TMR) in thin-film trilayer junction [3] has provided a new way of applying CMR manganites in a low-field magnetic sensor. An approach using thin film junction, e. g., TMR, is one of the most promising methods to overcome the hurdle of the high applied-field and narrow working-temperature region required for CMR.

In this approach, microscopic control of the thin-film surface is very important. For example, the surface roughness of the films used for a tunnel junction must be less than the thickness of the barrier layer (~ 1 nm) over the entire area of the sensor. This is very difficult to achieve given the present level of oxide-film deposition technology. Moreover, the thin film must have good magnetotransport properties, particularly a high spin-polarization ratio which is the origin of the large TMR ratio of these materials. If we can satisfy both of these requirements, extremely sensitive magnetic sensors will be attainable.

In this work, (La, Sr)MnO₃ (LSMO) thin films were grown under various conditions, and the dependence of the film structure, and the magnetic and transport properties on growth conditions was examined with the aim of developing practical multilayer-junction devices.

EXPERIMENTAL

Thin-film samples were prepared by pulsed-laser deposition (PLD) using a KrF excimer laser (λ = 248 nm). Single-phase $La_{1-x}Sr_xMnO_3$ (x = 0.20, 0.25, 0.30) pellets with a density of more

15

than 90% were used as targets for the film growth. Thin films were deposited on $SrTiO_3$ (100), MgO (100) and $LaAlO_3$ (100) substrates under the growth conditions listed in Table 1. The film thickness was typically 150 nm. After the growth, the samples were cooled in 1 atm. oxygen. The structure of the films was examined by a $\theta - 2\theta$ measurement and a pole figure measurement of the x-ray diffraction. The surface image of the films was observed with an atomic-force microscope (AFM). The temperature and field dependence of the dc magnetization was measured with a SQUID magnetometer. The resistivity of the films was measured as a function of the temperature and magnetic field using a standard dc four-probe technique.

Table 1 Growth conditions for the (La, Sr)MnO$_3$ films

Laser power	550 mJ – 950 mJ per pulse
Pulse rate	1 Hz , 3 Hz
Substrate temperature	600 – 700°C

RESULTS AND DISCUSSION

Film structure

The θ - 2θ x-ray diffraction measurement revealed that all the films deposited on the $SrTiO_3$ (STO), $LaAlO_3$ (LAO), and MgO were perfectly c-axis oriented. However, the films on LAO and MgO had weak peak intensities and broad peak shapes compared to the film on STO, suggesting poor crystallinity in these films. These differences in film structure can be clearly seen in the (102) pole figures shown in Fig. 1. Although each sample had a fourfold symmetry, which indicates epitaxial growth on each substrate, the peak shape and width was rather different for the three samples. In the case of the film on STO, the peak sharpness was almost the same as that of the STO substrate, indicating good film crystallinity. The AFM image of this film (Fig. 2(a)) also suggests that this film had a grain-free single-crystal structure with surface roughness of less than 0.3 nm. This level of flatness is close to the detection limit of AFM and should be sufficient for the use in a multi-layer device. On the other hand, the film on LAO shows asymmetrical broadening in the pole figure. For this film, a modulated surface structure with a height difference of 1 nm appeared in the AFM image (Fig. 2(b)). This structure can be explained in terms of the twinning of LAO substrate. Since LAO has a phase transition from a high-temperature cubic phase to a room-temperature rhombohedral phase below the film-growth temperature, part of the LSMO film grown on the LAO is forced to align to the twinned structure of the substrate during cooling. The surface roughness of 1 nm is not negligible in comparison with the barrier thickness of the tunnel junction (~ 2-5 nm). Therefore, an LAO substrate is not a good candidate for our purpose even if the twinning of the substrate does not cause a strain that could degrade the electrical properties of the film. The films on MgO had a granular morphology with a surface roughness of up to 2 nm (Fig. 2 (c)). This grain structure was also observed in scanning-electron-microscope (SEM) images, whereas the films on STO and LAO did not show a clear structure under the same magnification. The peak broadening of the films on MgO is almost symmetrical in the pole figure. However weak peaks that were misaligned by 45 degrees were observed. The direction of the main peak was parallel to the MgO (100) direction, so this film included a small amount of epitaxial grain that was aligned 45 degrees off direction to the MgO (100) direction. These results,

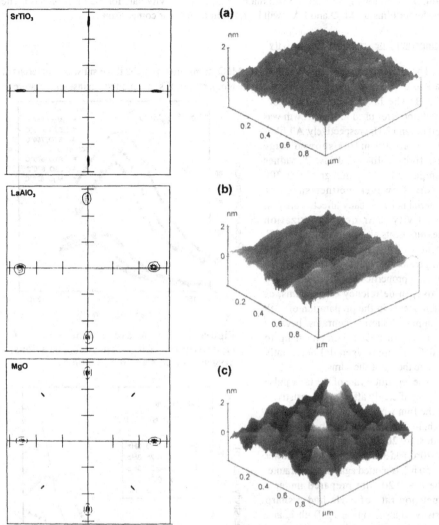

Figure 1 Pole figures of the LSMO (102) diffraction

Figure 2 AFM images of the LSMO films deposited on (a) SrTiO$_3$, (b) LaAlO$_3$, and (c) MgO substrates. Film thickness was about 150 nm.

- the granular structure and the coexistence of two alignment directions of the grain-, are probably caused by the large lattice mismatch between the LSMO and the MgO, this was the largest mismatch among the three substrates.

All these results suggest that STO is the best choice as the substrate material for the LSMO film. In the following section, we will mainly discuss resistivity data for the films on STO. The results for films on MgO and LAO will be referred to only for comparison.

Temperature dependence of resistivity

The resistivities of as-deposited $La_{0.8}Sr_{0.2}MnO_3$ films grown on the three substrate materials at various oxygen pressures (PO_2), and substrate temperatures (Ts) are plotted against temperature in Fig. 3. The laser power and pulse-repetition rates used for the growth was 950 mJ and 3 Hz, respectively. All films grown at 200 mTorr showed large resistivity values and low T_c values compared to the films grown at 300 mTorr. However, neither substrate material nor Ts greatly affected the film resistivity. Our dc magnetization measurements also indicated that the T_c value did not depend on Ts in the temperature range of 600°C to 700°C. the film properties cannot be attributed to oxygen deficiency of the films, as often occurs in the preparation of high T_c superconducting cuprates, because post-growth heat treatment at up to 700°C in pure oxygen did not greatly change the T_c of the films.

The repetition rate of the laser pulse also significantly affected the resistivity of the film when the Sr content, x, was high. In Fig. 4, the resistivity of samples with x = 0.20, 0.25 and 0.30, deposited at 670°C and 300 mTorr with laser power of 950 mJ, is plotted against temperature. The x = 0.20 film prepared under a repetition rate of 3 Hz had a sharp resistive transition near T_c. Both T_c and the absolute value of the resistivity were almost equal to those observed in single-crystal sample [4]. In contrast, the films prepared from the targets with a higher Sr content (x = 0.25 and x = 0.30) showed a lower T_c and higher resistivity than does a single-crystal sample. As shown in Fig. 4, these values approached the single-crystal values [4] when the

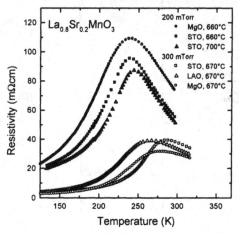

Figure 3 Temperature dependence of resistivity for films deposited at various oxygen pressures and temperatures.

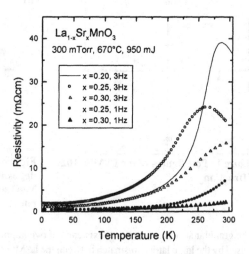

Figure 4 Temperature dependence of resistivity for $La_{1-x}Sr_xMnO_3$ films (x = 0.20, 0.25, 0.30) deposited at pulse rates of 1 Hz and 3 Hz.

repetition rate was reduced to 1 Hz. It is interesting (and strange) that the electrical properties were so highly influenced by the repetition rates, even though 1000 ms (1 Hz) and 300 ms (3 Hz) are very long periods of time compared to the actual laser pulse width which is on the order of nanoseconds. These results might indicate that the surface construction process of the film needs a rather long time, which is on the order of seconds. Further investigation, including a structural analysis, is now in progress.

Field dependence of resistivity

As we mentioned, the electrical properties of the film, T_c and absolute value of resistivity, were not greatly affected by the substrate material despite the large difference in film morphology. However, the magnetoresistance (MR) at a low temperature below T_c is fairly sensitive to the substrate material. As shown in Fig. 5, the film on STO showed a very low MR up to 60000 Gauss. This agrees with the results of the crystal structure analysis, because such behavior is very similar to that of single crystal sample.

On the other hand, the resistivity of films on LAO and MgO showed a strong dependence on the field up to a high field. A possible reason of this high MR is the grain-boundary effect. For the polycrystalline samples, two components of the MR have been reported [5]. The first, a very abrupt resistivity change observed in the low-field between each grain of the polycrystalline. The other, which remains in a very high field with a gentler slope than the first component, has not been explained. If we assume that the high-field MR in the films on MgO and LAO has the same origin as that observed in polycrystalline samples, a problem remains: why was the first type of MR not observed in these films? At present, there is no clear answer to this question because the origin of high-field MR itself is not yet understood. The grain boundaries in these films differ from those in polycrystalline samples in that each grain is not randomly oriented as in polycrystalline. This difference might account for the different MR behavior.

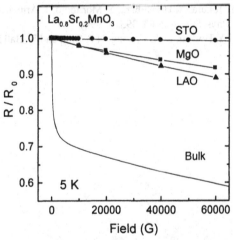

Figure 5   MR at 5 K for La$_{1-x}$Sr$_x$MnO$_3$ films deposited on STO, LAO and MgO. The data for a polycrystalline sample is also plotted for comparison.

CONCLUSION

Thin-film samples of La$_{1-x}$Sr$_x$MnO$_3$ (x = 0.20 - 0.30) were grown by pulsed laser deposition. Our x-ray diffraction analysis and AFM observation revealed that the film deposited on STO is an epitaxially grown single crystal with an extremely flat surface (surface roughness of less than 0.3

nm). Films are also epitaxially grown on both MgO and LAO, but these films had different defect structures. For example, the film on MgO had a granular structure, and the film on LAO had a twinned structure. The surface roughness, which is caused by these defect structures, was much greater for the films on MgO and LAO than for the film on STO. Therefore, STO is the best substrate choice for a multilayer junction device such as a TMR device. Of the growth conditions, the oxygen partial pressure (P_{O_2}) had the most significant influence on the electrical and magnetic properties of the films. A low P_{O_2} does not cause an oxygen deficiency in the film, but leads to other qualitative changes in the bulk nature, and these cannot be eliminated by annealing. Under optimized growth conditions, the structural, magnetic and electric quality of as-deposited films with x ≥ 0.2 was close to that of a single crystal.

References

1. K. Chahara, T. Ohno, M. Kasai, and Y. Kozono, Appl. Phys. Lett, **63**, p. 1990 (1993),
2. R. von Helmolt, J. Wecker, B. Holzapfel, L. Schultz, and K. Samwer, Phys. Rev. Lett., **71**, 2331 (1993).
3. Yu Lu,. X.W. Li, G.Q. Gong, Gang Xiao, A. Gupta, P. Lecoeur, J.Z. Sun, Y.Y. Wang, and V.P. Dravid, Phys. Rev. B, **54**, 8357 (1996).
4. Y. Tokura, A. Urushibara, Y. Moritomo, T. Arima, A. Asamitsu, G. Kido, and N. Furukawa, J. Phys. Soc. Jpn. **63**, 3931 (1994).
5. H.Y. Hwang, S-W. Cheong, N.P. Ong, and B Batlogg, Phys. Rev. Lett., **77**, 2041 (1996).

CRYSTALLINITY AND MAGNETORESISTANCE IN CALCIUM DOPED LANTHANUM MANGANITES

E.S. Gillman and K.H. Dahmen
Departments of Chemistry and Physics,
Center for Materials Research and Technology – MARTECH,
Florida State University,
Tallahassee, FL 32306-3006 USA

Abstract

Thin films of calcium doped lanthanum manganites $La_{1-x}Ca_xMnO_3$ (LCMO) with $x \sim 0.41$ have been prepared on $LaAlO_3(001)$ (LAO) Y-stablized $ZrO_2(001)$ (YSZ), and $Al_2O_3(0001)$ (SAP) substrates by liquid delivery metal-organic chemical vapor deposition (LD-MOCVD). The films on YSZ and SAP substrates have a textured, polycrystalline morphology with a preferred orientation of (110). The films on LAO show a single-crystalline morphology and a (100) orientation. Transport measurements show the polycrystalline films have a resistance peak approximately 60K lower than the films on LAO and, in general, have a much higher overall resistance. The magnetoresistance (MR) ratio ($[R(H) - R(0)]/R(H)$) is sharply peaked near the maximum in resistance for the films on LAO, while the polycrystalline films show a noticeable absence of this sharply peaked behavior and a flat, rather large ($\sim 100\%$) MR ratio over a large temperature range. These results will be discussed in terms of grain boundary scattering, crystallite size, and magnetization.

INTRODUCTION

$LaMnO_3$ is an anti-ferromagnetic (AFM) insulator. Partial substitution of the La^{+3} ions with a divalent cation, M^{2+} results in alkaline earth doped lanthanum manganites of composition $La_{1-x}M_xMnO_3$. A charge carrier doping $0.2 < x < 0.5$ results in the occurrence of "colossal MR" near the Curie temperature (T_c)[1, 2, 3, 4]. In this charge carrier doping range the material is a paramagnetic insulator above T_c and a ferromagnetic (FM) metal below T_c. Traditionally, the theory of double exchange has been used to explain this phenomenon[5, 6, 7]. However, recent data suggests that this can not totally account for this phenomenon[8].

These materials are the subject of considerable scientific investigation and technological interest due to their extraordinary electronic and magnetic properties[9, 10, 11, 12]. Thin films of perovskite materials have potential applications as electrochemical and magnetic sensors and as possible replacements for GMR read heads in the magnetic recording industry. Continued development of this technology requires that thin films of these materials be economically developed so that their magnetic and electronic properties are optimized. In this paper we report on the dependence of transport and structural properties of LCMO films grown by LD-MOCVD.

EXPERIMENTAL PROCEDURE

Thin films of Ca^{2+} doped lanthanum manganites with $x \sim 0.41$ were prepared on SAP, LAO and YSZ substrates under identical reaction conditions[13, 14, 15]. A NZ-Applied Technologies liquid delivery vaporization system was used to deliver the 2,2,6,6-tetramethyl-3,5-heptanedionato (TMHD) precursors; $La(TMHD)_3$, $Sr(TMHD)_2$, $Mn(TMHD)_3$ and// $Ca(TMHD)_2$, which were diluted in 25 ml of freshly distilled solvent (diglyme) and vaporized (T = 250°C) at a rate of 1.66 $\mu L/min$. The gaseous mixture was introduced into an EMCORE reactor where the substrates were held at a temperature of 700°C during the experiment. Nitrogen (100 sccm) was used as a carrier gas. O_2(600 sccm) and N_2O (500 sccm) served as as oxidants. The reactor pressure was maintained at 5 Torr. After deposition, the films were slowly cooled under reaction conditions until the the susceptor temperature was below 100°C when the substrates were removed from the susceptor. The deposition rate was ~ 15 Å/min.

To establish the morphology and in-plane epitaxy of the films, Θ-2Θ x-ray diffraction (Siemens D-500) and pole figure (Phillips X-Pert) measurements were performed. Electrical resistance and MR were determined as a function of temperature and magnetic field using a four-point technique in a superconducting magnet with a maximum applied field of 6T.

RESULTS

X-ray measurements (Figure 1) show that there are two preferred modes of growth onto the substrates used. Films on LAO are single-crystalline and a have a (100) orientation. Only a small amount of background from other phases is evident on magnification. Films on SAP and YSZ are polycrystalline. In these films multiple phases are clearly evident, but a preferred orientation in the (110) direction is indicated from the data. Pole figures (not shown) support in-plane epitaxy for films on LAO, but not for the films on YSZ and SAP. The results of the x-ray measurements are summarized in Table I.

Table I: X-Ray diffraction patterns for LCMO ($x = 0.41$) thin films grown by LD-MOCVD.

Substrate	Growth Direction	FWHM of Rocking Curve About (004) (degrees)	FWHM of Rocking Curve About (022) (degrees)	In-Plane Epitaxy
SAP	(011)	-	~ 3.0	no
YSZ	(011)	-	~ 2.5	no
LAO	(001)	~ 0.2	~ 0.5	yes

The MR ratio , defined as $\Delta R/R(H) = (R(H) - R(0))/R(H)$, was determined for the films from four-point resistivity measurements. Results are shown for the films in Figure 2. Films on YSZ and SAP show a nearly constant MR ratio below T_C that decreases in magnitude linearly in vicinity of T_C. In these films the peak in MR and resistance, T_{MR}, which occurs at $\sim 190K$ is broad. Well below T_C these films still show signifigant resistivity. Films on LAO show a sharply peaked MR ratio that closely corresponds to $T_{MR} \sim 250K$. In these films resistivity drops rapidly below T_C to very low values.

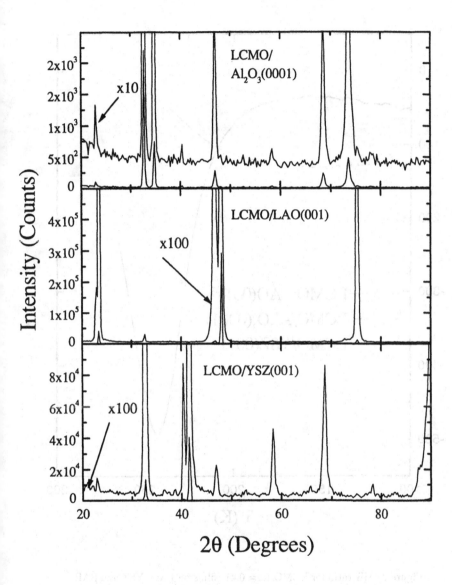

Figure 1: $\Theta/2\Theta$ x-ray diffraction patterns for LCMO ($x = 0.41$) on SAP, YSZ and LAO.

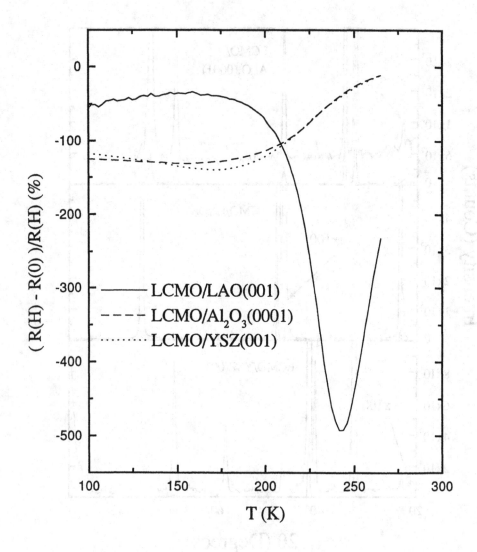

Figure 2: MR ratio for LCMO ($x = 0.41$) films on LAO, YSZ and SAP.

DISCUSSION

The degree of crystallinity can be related to the lattice matching of the substrate and the LCMO films. Films on LAO have 2.1% lattice mismatch. These films tend to grow highly ordered and have sharply peaked MR and MR ratio. YSZ and SAP have 6.8% and 22% lattice mismatches, respectively. These films have a broad MR and flat MR ratios below T_C. The resistivity of the the SAP films was approximately an order of magnitude greater than that of the films on YSZ, both at zero and in applied field. This can be attributed to the degree of crystallinity of the two films. However, films on LAO had resistivities that are signifigantly lower than films on YSZ and SAP at all temperatures.

Similar low temperature MR has recently been observed in both Sr and Ca doped manganites[16, 17, 18, 19]. The proposed explanation for the low temperature MR in polycrystalline films is disorder-induced canting of the Mn spins in the grain boundary region and spin-polarized tunneling at the grain boundaries. In the highly oriented films on LAO, at least below T_C, MR is generally acknowledged to be dominated by Zener double exchange[5, 7].

CONCLUSION

The ability to control the response to a magnetic field over a large temperature range is important for continued technological development of the lanthanum manganites. The results of these experiments demonstrate that it is possible to have a fairly large response to a magnetic field ($\sim$ 125%) over a large temperature range, as opposed to the sharply peaked, though greater, response seen in single-crystal films over a very narrow temperature range. Furthermore, these experiments demonstrate that it may be possible to manipulate texture in these films to achieve the desired physical properties which may make them useful materials for micro-electronic devices in the 21st century.

To summarize, single-crystalline films on LAO are characterized by a sharply peaked MR and MR ratio. These films have low resistances, especially well below the peak in MR. Polycrystalline films on YSZ and SAP are characterized by a broad MR and relatively flat MR ratio over a large temperature range. In these films the transport is dominated by a grain-boundary scattering process and the degree of crystallinity can correlated with magnitude of the resistance.

ACKNOWLEDGMENTS

The authors acknowledge support under grant ONR-N00014-96-1-0767. We gratefully acknowledge the contributions of Tom Fellers, Eric Lochner, Hamid Garmestani, S. Watts, X. Yu, S. Wirth, J. J. Heremans, and Mark Weaver.

REFERENCES

1. H. van Santen and G. H. Jonker, Physica **16**, 599 (1950).

2. H. van Santen and G. H. Jonker, Physica **16**, 337 (1950).

3. E. O. Wollan and W. C. Koehler, Phys. Rev. **100**, 545 (1955).

4. H. Y. Hwang *et al.*, Phys. Rev. Lett. **75**, 914 (1995).

5. C. Zener, Phys. Rev. **81**, 440 (1951).

6. P. W. Anderson and H. Hasegawa, Phys. Rev. **100**, 675 (1995).

7. P. E. deGennes, Phys. Rev. **118**, 141 (1960).

8. J. S. Zhou, W. Archibald, and J. B. Goodeneough, Nature **381**, 770 (1996).

9. C. H. Chen and S.-W. Cheong, Phys. Rev. Lett. **76**, 4042 (1996).

10. S. Jin *et al.*, Science **264**, 413 (1994).

11. R. von Helmolt *et al.*, Phys. Rev. Lett. **71**, 2331 (1993).

12. G. J. Snyder *et al.*, Phys. Rev. B **53**, 1 (1996).

13. K. H. Dahmen and M. Carris, J. Alloys and Compounds **251**, 270 (1997).

14. K. H. Dahmen and M. Carris, Chemical Vap. Deposition **3**, 27 (1997).

15. J. J. Heremans *et al.*, J. Appl. Phys. **81**, 4967 (1997).

16. A. Gupta *et al.*, Phys. Rev. B **54**, R15629 (1996).

17. H. Y. Hwang, S. W. Cheong, N. P. Ong, and B. Batlogg, Phys. Rev. Lett. **77**, 2041 (1996).

18. K. Steenbeck *et al.* (unpublished).

19. R. Shreekala *et al.*, Appl. Phys. Let. **71**, 282 (1997).

MICROSTRUCTURAL ASPECTS OF NANOCRYSTALLINE LiZn FERRITES DENSIFIED WITH CHEMICALLY DERIVED ADDITIVES

Yong S. Cho, Vernon L. Burdick and Vasantha R. W. Amarakoon,
New York State College of Ceramics at Alfred University, Alfred, NY14802

Elijah Underhill and Leo Brissette
Electromagnetic Science (EMS) Technologies Inc., Norcross, GA 30092

ABSTRACT

Densification behavior and microstructural characteristics of nanocrystalline LiZn ferrites with chemically derived additives were investigated. Nanocrystalline $Li_{0.3}Zn_{0.4}Fe_{2.3}O_4$ powders having a $\approx$ 15 nm size were prepared at a low temperature of 450°C by a chemical synthesis using a combustible polyacrylic acid (PAA). Small amounts of Si, Ca and Mn were incorporated into the nanocrystalline ferrites via sol-gel reactions utilizing tetraethyl orthosilicate, calcium isopropoxide and manganese acetate. This process was believed to give a homogeneous distribution of the additives over the nanocrystalline ferrites. A uniform microstructure was obtained without any evidence of exaggerated grain growth after sintering at 1100°C. Saturation magnetization and coercive force were found to increase with the chemical additives. The results were compared with those of the same composition, but processed by the conventional batch-mixing of corresponding oxide additives.

INTRODUCTION

Obtaining fine oxide powders has been a long-term trend in ceramic industries. Ultrafine powders with nanometer-sized particles can significantly improve sintering rates and thus can decrease sintering temperatures. Other than novel chemical processing for preparing nanocrystalline oxide powders, some additional approaches such as isostatic-pressing[1], hot-pressing[2], sinter-forging[3-4] and fast-firing[5-6] have been required to obtain promising microstructures and densities. These approaches focus primarily on reducing the detrimental effects of agglomerates and pores which are present in the starting nanocrystalline powders and pressed pellets. Agglomerates can lead to abnormal grain growth because a different sintering rate exists in the agglomerates. An external pressure from sinter-forging or hot-pressing is known to be helpful in obtaining homogeneous distribution of grain size in the resultant samples.

In this work, a new chemical approach is introduced to inhibit the abnormal grain growth of nanosized powders. Several additives includng Ca, Si and Mn were incorporated into the nanocrystalline LiZn ferrite powder using sol-gel reactions. Homogeneous distribution of the gel additives is expected to contribute to rearranging the nonosized ferrite particles during the initial stage of sintering. Calcium and silicon are known to inhibit grain growth with some benefits to electrical and magnetic properties[7]. Finally, magnetic properties of the sintered samples containing the additives will be correlated with observed microstructural characteristics.

EXPERIMENT

A 0.5M aqueous solution of nitrates, $LiNO_3$, $Zn(NO_3)_2 \cdot 6H_2O$ and $Fe(NO_3)_3 \cdot 9H_2O$ corresponding to a composition of LiZn ferrite ($Li_{0.3}Zn_{0.4}Fe_{2.3}O_4$) was prepared for the synthesis of nanocrystalline ferrites. The solution was mixed with a combustible 50 wt% aqueous polyacrylic acid (PAA) solution which is commercially available. An atomic ratio, 0.5 of carboxyl ion (of the PAA) to cations (of the ferrite) was used. After a few minutes of mixing using a stirring bar, the clear nitrate solution changed into a viscous gel. The gel was carefully dried at $\approx$ 60°C and then crushed using a mortar and pestle. Synthesis was performed in a box furnace at the temperature of 450°C for 30 min with a heatng rate of 300°C/hr. Particle size of the synthesied powder was determined to be around 15 nm by TEM and XRD analyses.

27

Table I. Additive composition, linear shrinkage, bulk density and apparent porosity of the $Li_{0.3}Zn_{0.4}Fe_{2.3}O_4$ samples sintered at 1100°C for 3 hrs in air.

Additive Composition	Addition Method	Bulk Density (g/cm³)	Apparent Porosity(%)
no additives	-	3.83	18
1wt% CaO & 2wt% SiO_2	sol-gel	4.68	4
1wt% CaO & 2wt% SiO_2	batch-mixing	4.29	12
1wt% MnO_2 & 2wt% SiO_2	sol-gel	4.36	8
1wt% MnO_2 & 2wt% SiO_2	batch-mixing	4.22	10

An additional chemical procedure utilizing sol-gel reactions was adopted to modify the nanocrystalline ferrite composition with two additive systems, Ca/Si and Mn/Si. Appropriate amounts of TEOS(tetraethylorthosilicate), $Si(OC_2H_5)_4$, Ca isopropoxide, $Ca(OC_3H_7)_2$ and Mn acetate, $Mn(CH_3COO)_2\cdot 4H_2O$ corresponding to 2 wt% SiO_2, 1 wt% CaO and 1 wt% MnO_2 were dissolved completely in isopropanol as a solvent. Additive compositions are represented in Table I. After adding a small quantity of HCl as a catalyst for the gelation, the nanosized LiZn ferrites were added to the additive solution resulting in a slurry. Complete gelation of the additives occurred while the slurry was dried at room temperature. For comparison, the same amounts of silicon and calcium oxides (having microscale particle sizes) were batch-mixed with the synthesized nanocrystalline LiZn ferrites. The resultant powders processed by the two different addition methods were pressed uniaxially at ≈ 12,000 psi and sintered at 1100°C for 3 hrs in air. The heating rate was 300°C/hr.

Shrinkage behavior during sintering was investigated by a dilatometer. Bulk density and apparent porosity of the sintered samples were measured using the Archimedes' principle. Microstructures of the samples were observed using a scanning electron microscope (SEM, 1810, Amray Co.) after thermal etching. A sphere sample (≈ 2.5 mm in diameter) was placed in a uniform applied magnetic field for the measurement of saturation magnetization, $4\pi M_s$ by a perturbation method using a differential gaussmeter. The hysteresis loop properties such as the remanent magnetization, B_r and the coercive force, H_c were measured using toroidal samples at 400 Hz and a drive level of 5 times H_c. The equipment used included a function generator, a solid state amplifier and an oscilloscope.

RESULTS AND DISCUSSION

Densification Behavior

Sintering characteristics of the nanocrystalline $Li_{0.3}Zn_{0.4}Fe_{2.3}O_4$ samples fired at a low temperature of 1100°C are illustrated in Table I. It indicates that the additives, Ca/Si and Mn/Si, enhanced densification with a decreasing tendency of porosity regardless of addition method. In particular, Ca/Si was more promising in obtaining a high density value (4.68 g/cm³) at the low temperature. The sol-gel method seems to be more effective in improving sinterability than the conventional batch-addition. Our previous work[8] suggested that the sol-gel method could provide intimate mixing of additives in nanoscale, leading to a homogeneous distribution of the additives. It can be expected that the batch-mixing using microscale oxides of SiO_2 and CaO is less effective in distributing them uniformly in the nanocrystalline LiZn ferrites.

Fig. 1 shows shrinkage behavior of the pressed samples with the sol-gel derived additives (taken at a heating rate of 300°C/hr by a dilatometer). It is seen that the nanocrystalline LiZn ferrite containing no additives is very reactive during sintering. Shrinkage was found to begin around 400°C. Apparently, shrinkage of the samples with Ca/Si and Mn/Si occurred at much higher temperatures. The addition of Ca/Si was found to be much more effective (than that of Mn/Si) in suppressing shrinkage of the samples. Roles of the additives are not likely to be related to accelerating densification. In the case of Ca/Si, a temperature of 900°C can be regarded as an initial sintering temperature (700°C in the case of Mn/Si). The retardation of shrinkage by the

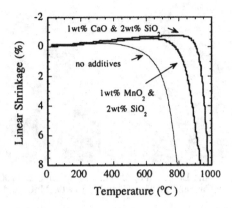

Fig. 1. Shrinkage behavior of nanocrystalline $Li_{0.3}Zn_{0.4}Fe_{2.3}O_4$ with sol-gel derived additives as a function of sintering.

additives may be associated with the final density values shown in Table I. The highest density was obtained in the case of addition of Ca/Si where retardation of shrinkage was remarkable at low temperatures.

Microstructural Characteristics

Microstructures of the nanocrystalline $Li_{0.3}Zn_{0.4}Fe_{2.3}O_4$ samples (without additives) sintered at 1100°C and 1200°C are shown in Fig. 2. This figure indicates abnormal grain growth of the LiZn ferrite, displaying enlarged grains with some large pores. These microstructural characteristics are typical of nanocrystalline ceramic samples sintered without any special treatment like hot-pressing or hot-forging. The large grains are believed to be induced from highly packed regions (i.e. agglomerates) of the nanocrystalline powder. The reactivity of the nanocrystalline powder in the highly packed region is likely to be very high at low temperatures, while particles in a loosely packed region are needed to be intimately contacted before coarsening. With increasing temperature, overall grain size tended to increase, but it exhibited the

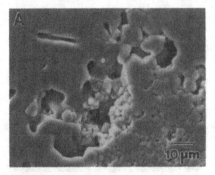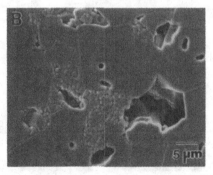

Fig. 2. Microstructures of the $Li_{0.3}Zn_{0.4}Fe_{2.3}O_4$ samples without additives, sintered at (A) 1100°C and (B) 1200°C for 3 hrs in air.

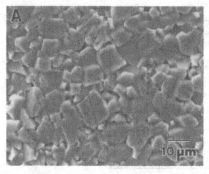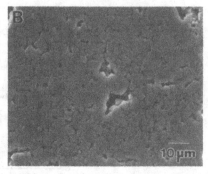

Fig. 3. Microstructures of the $Li_{0.3}Zn_{0.4}Fe_{2.3}O_4$ samples sintered at 1100°C for 3 hrs in air with sol-gel derived (A) 1wt% CaO and 2wt% SiO_2 and (B) 1wt% MnO_2 and 2wt% SiO_2.

same tendency of the abnormal grain growth (Fig. 2(B)). Large pores were found to remain after sintering at the high temperature of 1200°C. It is known that pores larger than grain size are not easy to eliminate even in the final stage of sintering[5,9]. Additionally, the pores do not retard or influence grain growth, primarily because such large pores are not mobile with grain boundaries[9].

A significant improvement in microstructure was achieved by adopting the chemical additives. Fig. 3 shows the microstructures of the $Li_{0.3}Zn_{0.4}Fe_{2.3}O_4$ samples with Ca/Si and Mn/Si, sintered at 1100°C for 3 hrs. No evidence of abnormal grain growth was found. It is presumed that the nanoscale gel of the additives penetrates into weak agglomerates of the nanocrystalline ferrites and helps in rearranging the nanocrystalline particles during the initial stage of sintering (since the gel may act as a viscous liquid at a sufficiently high temperature). The rearrangement of nanoparticles can contribute to eliminating heterogeneous regions (such as highly packed and loosely packed regions) which were observed in the samples without additives. In addition, calcium and silicon oxides are likely to act as grain growth inhibitors by exerting a drag force against grain boundary motion. Consequently, the chemical additives lead to homogeneous grain growth with retardation of densification as supported in the shrinkage behavior in Fig. 1. Fig. 4 shows the microstructures of the nanocrystalline LiZn ferrite samples containing the same additives, but processed by batch-mixing. As expected from the density values in Table I, batch-mixing was not effective in inducing homogeneous grain growth

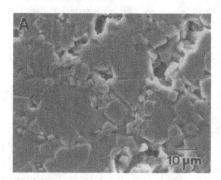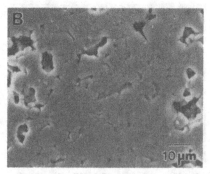

Fig. 4. Microstructures of the $Li_{0.3}Zn_{0.4}Fe_{2.3}O_4$ samples sintered at 1100°C for 3 hrs in air with batch-mixed (A) 1wt% CaO and 2wt% SiO_2 and (B) 1wt% MnO_2 and 2wt% SiO_2.

Table II. Magnetic properties of the $Li_{0.3}Zn_{0.4}Fe_{2.3}O_4$ samples sintered at 1100°C for 3 hrs

Additive Composition	$4\pi M_s$ (G)	B_r (G)	H_c (Oe)
no additives	4066	2061	1.54
1wt% CaO and 2wt % SiO_2 by sol-gel reaction	4210	1354	3.46
1wt% CaO and 2wt % SiO_2 by batch-mixing	4115	1294	2.65

compared to the sol-gel addition.

Correlation to Magnetic Properties

Table II shows an example of the influences of the Ca/Si addition and the processing methods on magnetic properties of the LiZn ferrite samples. The additives tended to cause saturation magnetization (M_s) and coercive force (H_c) to increase. Higher M_s and H_c values were obtained when the sol-gel addition method was used. Generally, saturation magnetization does not depend on grain size of a given sample, while the porosity can affect the value. The increased M_s in the case of Ca/Si addition may be related to the observed decrease in porosity although the nonmagnetic additives can affect the value adversely. A slightly higher M_s value of 4210 G was obtained in the case of sol-gel addition. Note that less porosity was observed by the sol-gel addition as compared in the microstructures shown in Fig. 3(A) and Fig. 4(A).

A change in coercive force can be analyzed in conjunction with grain size. As grain size decreases, coercive force tends to increase due to domain-wall pinning, resulting in a higher energy being required for switching[10]. Overall small average grain size obtained by the addition of Ca/Si may influence the H_c, resulting in a higher value of 3.46 Oe. It is noticeable that a highest value was obtained by the sol-gel addition where homogeneous grain growth was observed.

CONCLUSIONS

A chemical way to uniformly incorporate additives into a nanocrystalline ferrite powder was proven to be effective in improving microstructures and densification. Nanocrystalline $Li_{0.3}Zn_{0.4}Fe_{2.3}O_4$ ($\approx$ 15 nm particle size) was prepared at 450°C by a combustion synthesis using polyacrylic acid. Small amounts of several additives such as Ca/Si and Mn/Si were added to the synthesized nanocrystalline LiZn ferrite via sol-gel reactions. Homogeneous grain growth with less pores was obtained with the additives after firing at 1100°C for 3 hrs with a heating rate of 300°C/hr. The penetration of the gel additives into weak agglomerates was believed to be responsible for the promising microstructure and retardation of shrinkage at low temperatures. The microstructures were correlated to the obtained magnetic properties. The additives processed by the sol-gel method caused saturation magnetization and coercive force to increase.

ACKNOWLEDGMENTS

The support of Army Research Office (ARO), NYS Center for Advanced Ceramic Technology (CACT) at Alfred University and Electromagnetic Science Technologies Inc., Norcross, GA for the research project is acknowledged.

REFERENCES

1. P. L. Chen and I. W. Chen, J. Am. Ceram. Soc., **79**, p3129 (1996).

2. R. S. Mishra, C. E. Lesher and A. K. Mukherjee, J. Am. Ceram. Soc., **79**, p2989 (1996).

3. M. M. R. Boutz, L. Winnubst and A. J. Burggraaf, J. Am. Ceram. Soc., **78**, p121 (1995).

4. D. C. Hague and M. J. Mayo, J. Am. Ceram. Soc., **80**, p149 (1997).

5. D. J. Chen and M. J. Mayo, J. Am. Ceram. Soc., **79**, p906 (1996).

6. A. Morell and A. Hermosin, Am. Ceram. Soc. Bull., **59**. p626 (1980).

7. H. Yamamoto and T. Mitsuoka, IEEE Trans. on Magnetics, **30**, p5001 (1994).

8. Y. S. Cho and V. R. W. Amarakoon, J. Am. Ceram. Soc., **79**, p2755 (1996).

9. E. B. Slamovich and F. F. Lange, J. Am. Ceram. Soc., **75**, p2498 (1992).

10. A. C. Blenkenship and R. L. Huntt, J. Appl. Phys., **37**, p1066 (1966).

Part II

Characterization of Metallic Magnetic Oxides

Lattice deformation and magnetic properties in epitaxial thin films of $Sr_{1-x}Ba_xRuO_3$

Noburu Fukushima, Kenya Sano, Tatsuo Schimizu,
Kazuhide Abe and Shuichi Komatsu
Materials and Devices Research Labs., Research and Development Center,
Toshiba Corporation, Kawasaki 210, Japan

ABSTRACT

The crystal structure and magnetic properties in epitaxially grown $Sr_{1-x}Ba_xRuO_3$ on $SrTiO_3$ substrates were determined. Epitaxial $Sr_{1-x}Ba_xRuO_3$ exhibits a simple perovskite structure in the whole region of Ba/Sr ratio, in contrast to a complex hexagonal layered perovskite in the case of Ba-rich bulk $Sr_{1-x}Ba_xRuO_3$ which has plane-sharing oxygen octahedra. Tetragonal deformation was enhanced from the pseudo cubic structure of $SrRuO_3$ to a highly distorted tetragonal lattice in $BaRuO_3$. Electronic properties such as conductivity and magnetization were examined and compared to the results of band calculation in which a tetragonal distortion was taken into account. A metal-insulator transition was not observed in this system either in the experiment or in the simulation, and metallic conductivity was maintained in the whole region of Ba content. The ferromagnetic ordering which occurs at 160K in pseudo-cubic bulk $SrRuO_3$ was suppressed in $Sr_{1-x}Ba_xRuO_3$ films with increasing tetragonal deformation and Curie temperature decreased to 50K in $BaRuO_3$.

INTRODUCTION

Metallic perovskite oxides have been attracting attention because of their various electronic and magnetic properties such as the metal-insulator transition, magnetic ordering and superconductivity. This wide variation in their properties is caused by their relatively narrow bandwidth and strong electron correlation. The relation between electronic structure and lattice distortion, resulting in a competition between bandwidth and electron correlation have been extensively studied in various kind of compounds such as $RNiO_3$[1], $R_{1-x}A_xTiO_3$[2], AVO_3[2,3] and $R_{1-x}A_xMnO_3$[4] where R and A denote rare-earth ions and alkaline-earth ions of various ionic radii. So far the influences of the lattice distortion on the electrical and magnetic properties in these perovskite oxides though have been studied only in bulk samples and their lattice distortion is induced by means of application of external stress[5-7] or "chemical stress" introduced by substitution of ions of different ionic radii[1-4].

Epitaxial thin film growth on substrates with different lattice parameters can induce a lattice deformation due to the lattice mismatch between films and substrates. We have studied the relation between ferroelectric properties and the lattice distortion in epitaxially grown $Ba_{1-x}Sr_xTiO_3/SrRuO_3$ capacitors and found that the lattice mismatch-induced tetragonal distortion yields extremely high dielectric constants[8] or distortion-induced ferroelectricity[9] even in quite thin $Ba_{1-x}Sr_xTiO_3$ films.

35

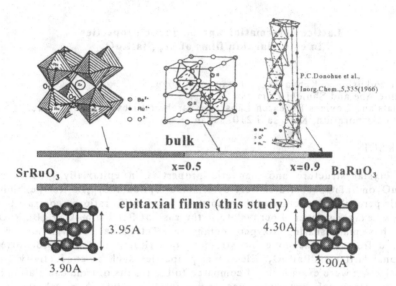

Fig.1. Crystal structures of bulk $Sr_{1-x}Ba_xRuO_3$
and $Sr_{1-x}Ba_xRuO_3$ epitaxial film.

This type of artificial deformation of perovskite oxides can be applied to other compounds such as electrically conducting and/or magnetic perovskite oxides. We have studied the crystal structure and magnetic properties of $Sr_{1-x}Ba_xRuO_3$ epitaxial films on $SrTiO_3$ substrates and found that $Sr_{1-x}Ba_xRuO_3$ epitaxial film maintained the simple perovskite structure in entire range of x as opposed to the complex hexagonal layered structure of Ba-rich bulk $Sr_{1-x}Ba_xRuO_3$ which has plane-sharing oxygen octahedra[10](Figure 1). Furthermore, the ferromagnetic ordering at 160K seen in bulk $SrRuO_3$[11] was found to be strongly suppressed in epitaxial $Sr_{1-x}Ba_xRuO_3$ films and that the Curie temperature decreased to 50K in tetragonal $BaRuO_3$. In addition to the experiments, first principle calculations were performed to elucidate the electronic structure in the distorted crystals and to predict the local atomic displacements caused by the tetragonal distortion, employing the total energy minimization. The calculation used here is based on the local density approximation (LDA), where ultrasoft pseudopotentials were used to describe the ions. The wave functions were expanded in a plane-wave basis set. 512 k points for integration over Brillouin zone were employed. Details of these calculations will be reported elsewhere.

EXPERIMENT and RESULTS

    $Sr_{1-x}Ba_xRuO_3$ films were deposited on $SrTiO_3$ (100) plane substrates by means of ordinary radio frequency sputtering using $SrRuO_3$ and hexagonal $BaRuO_3$ targets.   The crystal structure and lattice parameters were examined by means of X-ray diffraction on the 50nm-100nm thick $Sr_{1-x}Ba_xRuO_3$ films obtained. Figure 2a shows the result of a theta-2theta scan of $BaRuO_3$ on $SrTiO_3$ substrate, where peaks can be assigned to a tetragonal perovskite structure. Figure 2b is the result of a side inclination scan of $BaRuO_3$ and this and other side inclination scans implied $Sr_{1-x}Ba_xRuO_3$ films obtained in this study have the same a-axis length of 3.905A as the $SrTiO_3$ substrate. The phi scan of $BaRuO_3$ shown in Fig.2c revealed that (202) peaks precisely coincide with (101) peaks of $SrTiO_3$ substrate, which manifested the epitaxial growth of tetragonal $BaRuO_3$ film on $SrTiO_3$ substrate.

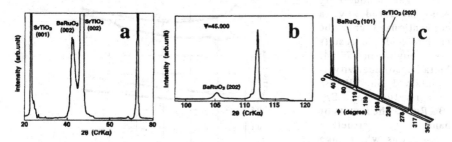

Fig.2. XRD pattern of $BaRuO_3$ epitaxial film; a)theta-2theta scan, b)side inclination scan on BaRuO3(202), c)phi scan on BaRuO3(101) and SrTiO3 substrate(202).

The c-axis length of $Sr_{1-x}Ba_xRuO_3$ film is shown in Figure 3. Epitaxially-grown films which have mismatch-induced lattice distortion tend to exhibit the relaxation of lattice parameters when the thickness of the films increased. The broad peaks of $BaRuO_3$ XRD pattern seen in Fig.2a and Fig.2b implies that $BaRuO_3$ film has the lattice relaxation to a certain extent. This relaxation is observed much more markedly in the films fabricated by the deposition method of laser evaporation[12,13] and molecular beam epitaxy[14,15]. In such films the crystal structure collapsed and lattice parameters relax to their original lengths of bulk crystals if their thickness is larger than the critical thickness[21]. On the other hand, sputtered films tend to maintain their distortion for larger thickness[16,17], as has been observed in the present study. This difference can be attributed to the higher kinetic energy of the incident particles in the sputtering.

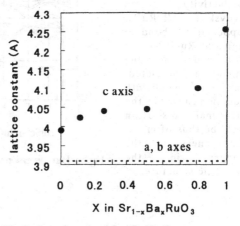

Fig.3. Axes length of $Sr_{1-x}Ba_xRuO_3$.

These results imply that epitaxial $Sr_{1-x}Ba_xRuO_3$ exhibits a simple perovskite structure in contrast to complex hexagonal layered perovskite of Ba rich bulk $Sr_{1-x}Ba_xRuO_3$; epitaxial growth can produce a new polytype oxide of this kind. No further information about precise crystal structure such as refined site positions can be obtained in such thin films on substrates of nearly identical structure. The total energy calculation based on band calculations predicted that the energy minimum was given when Ru-O-Ru bonding angle was 90 degree for both cubic and tetragonal structures.

The conductivity was measured by a conventional four probe method and the results are shown in Fig.4. Metallic conductivity is maintained in all Ba concentration of $Sr_{1-x}Ba_xRuO_3$. The results of band structure calculations which take into account the tetragonal distortion , did not show marked difference between the cubic $SrRuO_3$ and tetragonal $BaRuO_3$ in band dispersions (Fig.5) and thus predicted the metallic conductivity in perovskite $BaRuO_3$. Supposing the bonding angles of Ru-O-Ru are unchanged as the simulation predicted, the bandwidth reduction caused by the tetragonal distortion could be thus of minor effectiveness to the changes in the electronic structure.

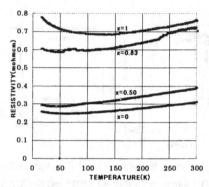

Fig.4. Temperature dependence of resistivity of $Sr_{1-x}Ba_xRuO_3$ epitaxial films.

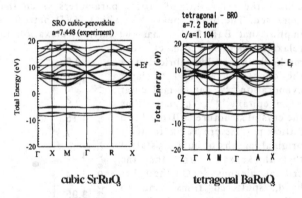

cubic $SrRuO_3$ tetragonal $BaRuO_3$

Fig.5. Band dispersion of cubic $SrRuO_3$ and tetragonal $BaRuO_3$.

The magnetization measurement was performed using the Quantum Design MPMS SQUID magnetometer. The influence of the substrates was carefully subtracted by blank measurements of $SrTiO_3$ substrates. The results are shown in Figure 6 where the in-plane magnetization of $Sr_{1-x}Ba_xRuO_3$ is plotted. Measurements were carried out for 100 nm thick, 10mmx10mm films. The deviations between the results of zero field cooling and field cooling measurements suggest the occurrence of the spin-freezing at lower temperatures for Ba rich samples. We now speculate that these deviations are attributed to a pinning of magnetic domains. On the other hand, the ferromagnetic

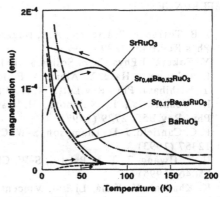

Fig.6. Temperature dependence of magnetization of epitaxial $Sr_{1-x}Ba_xRuO_3$ films. → arrows indicate the magnetization in zero field cooling while ← arrows indicate that in field cooling.

ordering seen in bulk $SrRuO_3$ is suppressed in these $Sr_{1-x}Ba_xRuO_3$ films with increasing Ba content and/or increasing tetragonal distortion. The Curie temperature decreased to 50K in tetragonal $BaRuO_3$. As band calculation predicted that the A site ion makes no contribution to the states near the Fermi level even in tetragonally distorted crystal as well as pseudo cubic crystal, these changes in magnetic properties observed in this study can be attributed to the lattice distortion. Although the change in electronic structure in the present system has not been clarified yet, the local environment change around Ru ions and the consequent change in electron population of Ru 4d levels could be speculated. As broad peaks are seen in XRD (Fig.2a,2b), the lattice relaxation should exist to a certain extent in $BaRuO_3$. Optimization of deposition process would provide more elongation of the c-axis and larger tetragonal distortion, and therefore the complete suppression of magnetic order might be expected. In this case, the ground state of this system, instead of ferromagnetism in pseudo cubic $SrRuO_3$, would be non-magnetic metal. The comparison would be of great interest between perovskite $BaRuO_3$ and non-magnetic layered perovskite Sr_2RuO_4 which exhibits superconductivity at low temperature[18].

Magnetic metallic perovskite oxides such as doped $LaMnO_3$ have been attracting attention as electrodes of colossal magnet-resistance heterostructure because of their high spin polarization[19]. As $SrRuO_3$ is predicted to have finite spin polarization by band calculations[20], the change in spin polarization which might be induced by the decrease of crystalline symmetry will be of interest from the viewpoint of such magnetic device applications. The epitaxial growth of magnetic metallic oxides reported here could provide a useful tool to optimize their electronic structure and magnetic properties, as well as a way to synthesize new polytypes.

REFERENCES

1. J. B. Torrance, P. Lacorre, A. I. Nazzal, E. J. Ansaldo and Ch. Niedermayer, Phys.Rev.B45, 8209 (1992)
2. Y. Tokura, J.Phys.Chem.Solids,53,1619(1992)
3. I. H. Inoue, I. Hase, Y. Aiura, A. Fujimori, Y. Haruyama, T. Maruyama and Y. Nishihara, Phys.Rev.Lett.,74,2539 (1995)
4. H. Yoshizawa,, R. Kajimoto, H. Kawano, Y. Tomioka and Y. Tokura, Phys.Rev.B55, 2729 (1997)
5. P. C. Canfield, J. D. Thompson, S-W. Cheong and L. W.Rupp, Phys.Rev.B47, 12357 (1993)
6. H. Y. Hwang, T. T. M.Palstra, S-W. Cheong and B. Batlogg, Phys.Rev.B52, 15046 (1995)
7. K. Khazeni, Y. X.Jia, Li Liu, Vincent H. Crespi, Marvin L. Cohen and A. Zettl, Phys.Rev.Lett., 76,295 (1996)
8. M. Izuha, K. Abe and N. Fukushima, Jpn.J.Appl.Phys., 36, 5866 (1997)
9. K. Abe and S. Komatsu, Jpn.J.Appl.Phys., part1, 33, 5297 (1994)
10. P. C. Donahue, L. Katz and R. Ward, Inorgic Chem., 4, 306 (1965); idem, ibid., 5, 335 (1966)
11. A. Callaghan, C. Moeller and R. Ward, Inorganic Chem.,5, 1572 (1996)
12. V. Srikant, E. J. Tarsa, D. R. Clarke, and J. S. Speck, J.Appl.Phys.,77,1517 (1995)
13. N. J. Wu, H. Lin, K. Xie, X.Y. Li and A. Ignatiev, Physica C232.,151 (1994)
14. Y. Yano, K. Iijima, Y. Daitoh, T. Terashima, Y.Bando, Y. Watanabe, H. Kasatani and H. Terasaki, J.Appl.Phys., 76, 7833 (1994)
15. Y. Yoneda, H. Kasatani, H. Terauchi, Y. Yano, T. Terashima and Y. Bando, J.Phys.Soc.Jpn., 62, 1840 (1994)
16. K. Abe and S. Komatsu, J.Appl.Phys., 77, 6461 (1995)
17. L. A. Wills and J. Amano, Material Research Society Proceedings: Ferroelectric Thin Films (1994)
18. Y. Maeno, H. Hashimoto, K. Yoshida, S. Nishizaki, T. Fujita, J. G. Bednorz and F. Lichtenberg, Nature 372, 532 (1994)
19. Yu Lu, X. W. Li, G. Q. Gong, G. Xiao, A. Gupta, P. Lecoeur, J. Z. Sun, Y.Y. Wang, J. Z. Sun, Y. Y. Wang and V. P. Dravid, Phys.Rev.B54, R8357 (1996)
20. P. B. Allen, H. Berger, O. Chauvet, L. Forro, T. Jarlborg, A. Junod, B. Revaz and G. Santi, Phys.Rev.B53, 4393 (1996)
21. H. Terauchi, Y. Watanabe, H. Kasatani, K. Kamigaki, Y. Yano, T .Terashima and Y. Bando, J.Phys. Soc. Jpn., 61, 2194 (1992)

MAGNETIC ANISOTROPY AND LATTICE DISTORTIONS IN THE DOPED PEROVSKITE MANGANITES

Y. SUZUKI[1], H.Y. HWANG[2], S-W. CHEONG[2], R.B. VAN DOVER[2], A. ASAMITSU[3], Y. TOKURA[3,4]

[1] Dept. Materials Science and Engineering, Cornell University, Ithaca, NY 14853
[2] Bell Labs, Lucent Technologies, 600 Mountain Ave., Murray Hill, NJ 07974
[3] Joint Research Center for Atom Technology, Tsukuba 305, Japan
[4] Dept. of Applied Physics, University of Tokyo, Tokyo, Japan

ABSTRACT

We have investigated the magnetic anisotropies of doped manganite materials in epitaxial thin film and single crystal form. Structural characterization, including x-ray diffraction, Rutherford backscattering spectroscopy and atomic force misocroscopy, indicate that our epitaxial films are single crystalline and have excellent crystallinity. Since lattice distortions greatly affect the magnetic and transport properties of this family of materials, it is not surprising to find the profound effect of strain in films due to the lattice mismatch between the substrate and film. Magnetic anisotropy results of single crystals, subject to no external stress, is compared to those of epitaxial films.

INTRODUCTION

The doped manganite perovskite materials $RE_{1-x}AE_xMnO_3$ (for trivalent rare earth ions RE and divalent alkaline earth ions AE) have been the focus of much renewed research efforts for potential magnetic recording applications [1-5]. As in many other perovskite systems with the ABO_3 structure, chemical or hydrostatic pressure gives rise to lattice distortions that change the B-O-B bond angle instead of changing the B-O bond distance, thus changing the matrix element b which describes electron hopping between B sites. In case of the doped manganites, electron hopping between manganese sites is correlated with the onset of ferromagnetic ordering so that lattice distortions gives rise to changes in Curie temperature, magnetoresistive properties as well as magnetic anisotropy. Lattice distortions can be in the form of chemical or hydrostatic pressure [6,7]. However in epitaxial thin samples the clamping effect of the substrate can give rise to additional lattice distortions, thus further affecting the magnetic and transport properties. The effects of strain on the magnetoresistive properties of epitaxial films has been studied by Jin et al.

and Kwon et al. [8,9]. LeCoeur et al. have studied magnetic have studied magnetic anistropy of $(La_{0.7}Sr_{0.3})MnO_3$ (LSMO) films [10].

In this paper we present a detailed study of the magnetic anisotropy of (001) and (110) LSMO films on (001) and (110) $SrTiO_3$(STO). Torque magnetometry and angular dependent magnetization measurements indicate that the magnetically easy axis is along the Mn-O-Mn or [100] and equivalent directions and the magnetically hard axis is along the [110] and equivalent directions. Careful study of LSMO films of varying thicknesses strongly indicate the important role of strain due to the lattice mismatch between the film and substrate. In order to determine the "intrinsic" magnetocrystalline anisotropy of LSMO, we carried out the same magnetic anisotropy measurements on single crystal samples of LSMO.

EXPERIMENTAL SETUP

We have studied both epitaxial films and single crystals of LSMO. In this paper, we will describe the orientation of the manganite films in terms of the pseudocubic lattices parameters $a'_{bulk}= b'_{bulk}= c'_{bulk}= 3.87$Å. The single crystals have a cubic perovskite structure with a rhombohedral distortion and a rhombohedral lattice constant of 7.78Å. The single crystal orientation will be described in terms of the rhombohedral lattice parameter. The epitaxial films were grown on (001) and (110) STO (a=b=c=3.91Å) substrates using pulsed laser deposition. LSMO films were grown at 700°C and 200mTorr O_2 with a KrF excimer laser running at 10Hz and 2J/cm². Films with thicknesses ranging from 500~3000Å were fabricated. However the data presented will be on ~2500Å thick films. The single crystals were grown by the float-zone method [11]. The resulting single crystal rod has the [100] direction at 17 degrees away from its rotational axis as measured by Laue diffraction. A disk sample (2.83mm radius and 0.57mm thick) with a (100) axis of rotation was then cut from the rod.

Structural characterization of the samples included x-ray diffraction, atomic force microscopy (AFM), Rutherford backscattering spectroscopy (RBS). X-ray diffraction was carried out on a four circle diffractometer with a Cu K_α radiation source (0.02° resolution). Two theta and phi scans were performed in order to ascertain epitaxy in the films. AFM of the film samples were carried out on a Digital Nanoscope III in order to determine surface morphology. RBS was use to determine the thickness, composition and crystallinity of our thin film samples.

Measurements of the magnetic anisotropy of the films and crystals were performed using vibrating sample and torque magnetometry with the magnetic field in the plane of the film and single crystal disk. A Lakeshore vibrating sample magnetometer 7307 was used to make the magnetization measurements and a custom-built torque magnetometer was used to measure the torque as a function of angle on both the epitaxial films and single crystals.

RESULTS

STRUCTURAL CHARACTERIZATION

X-ray diffraction measurements reveals that (001) LSMO grows on (001)STO while (110) LSMO grows on (110)STO (figure 1). Phi scans of the (011) and the (200) peaks for the (001) and (110) orientation films respectively indicate that there is in-plane alignment of the crystal axes. Figure 2 shows fourfold symmetry of a phi scan of the (011) peak in a (001) LSMO film indicating in-plane alignment. These oriented single crystalline films have excellent crystallinity with full width half maximum values of $\Delta\omega=0.11°$ as compared to the substrate which was measured to have $\Delta\omega \sim 0.02°$ which is resolution limited. RBS channeling of the LSMO films reveal excellent crystallinity with a figure of merit of 11%. While grain structure is difficult to determine from AFM, we observe that thinner films (≤ 1000Å) are on the average smoother with an rms roughness on the order of 20Å compared to thicker films with an rms roughness on the order of 40Å. Furthermore, AFM studies show structures typically 500Å on a side.

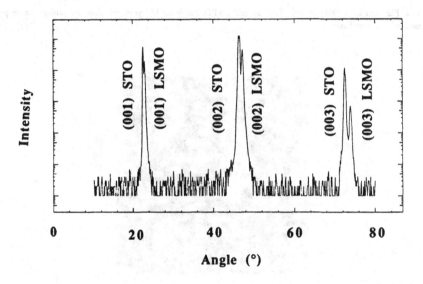

Figure 1. Two theta scan of a (001) LSMO film on (001) STO substrate indicates good epitaxy of the film on the substrate.

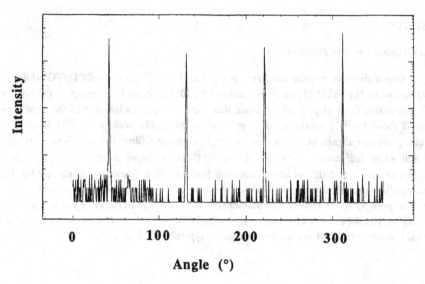

Figure 2. Phi scan of a (001) LSMO film on (001) STO has fourfold symmetry, thus indicating the in-plane alignment of the crystal axes.

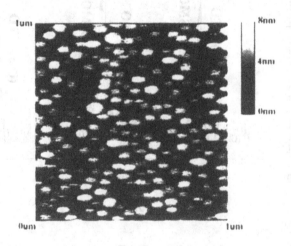

Figure 3. Atomic force microscopy shows rms roughness of 1.5nm for an LSMO film grown on STO.

MAGNETIC PROPERTIES

Since the magnetic properties are very sensitive to lattice distortions, it is not surprising that depending on microstructure and whether the film is under compression or tension due to its lattice mismatch with the substrate, epitaxial films of LSMO grown on different substrates by different groups exhibit different magnetic anisotropies. For example, Kwon et al. observe a perpendicular anisotropy in LSMO films on $LaAlO_3$ (grown under compression) but not on STO substrates (grown under tension) [9]. Lofland et al. observe a perpendicular anisotropy in $(La,Ba)MnO_3$ on $LaAlO_3$ substrates [11]. In our case the LSMO films are grown epitaxially on STO so that the films are grown under tension.

The magnetization loops of these films were measured in the plane of the film. Along the easy [100] and [010] directions in the plane of the (001) LSMO films, the loops are square with coercive fields (H_c)on the order of 5-10 Oe. Along the easy [001] direction in the plane of the (110) films, the H_c's are on the average larger (20-25 Oe) (Figure 4b). Magnetization loops with the field perpendicular and parallel to the film indicate that there is negligible perpendicular anisotropy, such as growth anisotropy, in these films [12].

To obtain a measure of the magnetic anisotropy in these films, we measure magnetization as a function of the angle that the applied field makes with respect to the crystal axes in the plane of the film. At a given angle, the magnetization is measured at different field values after the sample has been positively saturated. This technique identifies the easy axis as the direction with large remnant magnetization whereas the hard axis is the direction with negligible remnant magnetization as shown in Figure 5. Correlation of the structural data and the angular dependence of the magnetization confirms the easy axis to be along the [001] direction, while the hard axis is along the [1$\bar{1}$0] direction. In (110) films, the [001], [1$\bar{1}$0] and [1$\bar{1}$1] directions are all in the

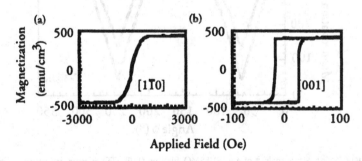

Figure 4. Magnetization loop of a (110) LSMO film on a (110) STO substrate along the (a) [1$\bar{1}$0] and (b) [001] directions.

plane of the film. We observe uniaxial anisotropy in the plane of these films with the easy direction along the [001] direction and the hard direction along the [1$\bar{1}$0] (Figure 4a). The field at which the angular dependence of the magnetization disappears is the "anisotropy field" H_K-1 kOe. This method specifies the "anisotropy field" as the field when M reaches M_s. We will henceforth refer to this field as the saturated anisotropy field H_{sat}. Therefore if there is a distribution of fields at which the spins in the film are saturated along the hard direction, the above method chooses the maximum field in the distribution. Given that we observe uniaxial anisotropy in (110) films, we expect that along the hard axis, there is zero remnant magnetization and a linear dependence of M on H. Thus we can also deduce an "anisotropy field" as the field corresponding to $Ms / \left(\dfrac{dM}{dH} \right)_{H=0}$. We will henceforth call this field the extrapolated anisotropy field H_{ex}. This H_{ex} value should coincide with H_{sat} when the distribution of saturation fields for spins along the hard direction is not broad but narrow. Therefore H_{ex} is a more appropriate value for the anisotropy field than H_{sat} to estimate an anisotropy constant consistent with torque measurements (figure 6). For example for the above (110) LSMO film, H_{ex} =410 Oe obtained from the magnetization data is consistent with the anisotropy constant K=8.4x10^{4}ergs/cm^3 deduced from the torque curve.

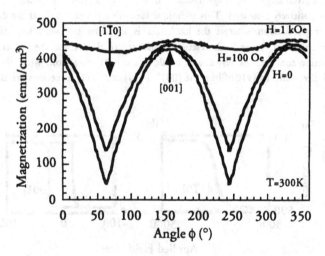

Figure 5. Magnetization versus angle φ of a (110) LSMO film on (110) STO at room temperature. The field is applied in the plane of the film as a function of angle φ as shown in the inset. The sample was initially poled positive and then the magnetization measurements were made at 1 kOe and zero field.

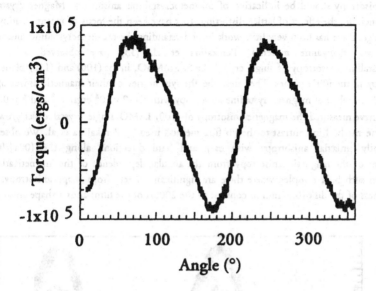

Figure 6. Torque curve of a (110)LSMO film on (110)STO reveals twofold symmetry that can only be explained by stress effects.

There are several possible explanations for the anisotropy that we observe in these (110) films: cubic or rhombohedral magnetocrystalline anisotropy or stress anisotropy. The torque curve cannot be fit to expressions of torque solely due to cubic magnetocrystalline anisotropy [12]. Furthermore rhombohedral magnetocrystalline anisotropy would predict a uniaxial anisotropy in both the (110) and (100) planes and none in the (001) plane. Since our (001) films show fourfold symmetry in the magnetic anisotropy[12], these predictions are inconsistent with our film results. If this magnetic anisotropy is mostly due to stress, the anisotropy constant has the form: $K_{stress}=3\lambda\sigma/2$ where λ is the magnetostriction constant and σ is the stress. The tensile stress in these LSMO films is estimated from the product of the Young's modulus and strain ($\sigma \cong 10^9-10^{10}$ dyne-cm^2); the strain in turn is estimated from the difference between bulk and film lattice constants of LSMO. The values of the magnetostriction constant deduced from the observed anisotropy constant and stress are consistent with values found in the literature ($\lambda \sim 10^{-4}$)[7].

Therefore magnetic anisotropy data from epitaxial films on STO indicate that lattice mismatch between the film and substrate induces a strain anisotropy of the order of $K=8.4 \times 10^4$ergs/cm^3. Further studies of the magnetic anisotropy as a function of film thickness have confirmed the effect of strain especially on thinner films [12]. However in single crystals the magnetic anisotropy should be indicative of magnetocrystalline anisotropy. Magnetocrystalline anisotropy includes the effects of lattice distortions that have been the focus of recent scrutiny. To our knowledge, there has been very little work in understanding the magnetocrystalline anisotropy of the doped manganite material. Perekalina et al. [13] have observed a uniaxial magnetocrystalline anisotropy of single crystal $(La,Sr,Pb)MnO_3$ in the (100) and (110) planes and no anisotropy in the (001) planes. They describe the symmetries of their magnetic anisotropy in terms of a rhombohedral magnetocrystalline anisotropy with $K_1 \sim 2 \times 10^4$ ergs/cm^3 and $K_2 = 0$.

We have measured the magnetic anisotropy of (100) LSMO single crystal disks grown by the float-zone method in contrast to the Pb flux method used by Perekalina et al. We observe a predominantly uniaxial anisotropy with easy and hard directions along the [001]/[010]. Measurement of the magnetic anisotropy from the angular dependence of the magnetization is precluded in such bulk samples where there are significant effects from shape anisotropy. The finite thickness of the disk (0.57mm in contrast to the 2000Å of the film) adds a shape anisotropy

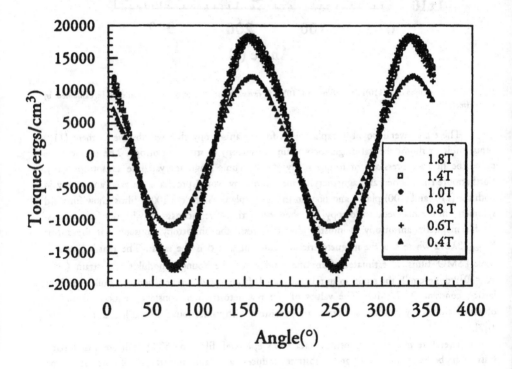

Figure 7. Torque curves of a (100) LSMO single crystal disk, at room temperature and at a variety of fields in the plane of the disk, shows that 600 Oe is required to saturate the sample.

contribution to such measurements. By contrast, torque measurements of a saturated sample preclude these shape effects. Figure 7 shows the torque curves for a single crystal sample in fields from 0.4 ~ 1.8 kOe. As soon as we reach a field of 600 Oe, we obtain a saturated torque curve. The torque curve is predominantly twofold with an anisotropy constant of 1.8×10^4 ergs/cm³ and a fourfold component that is 4% of the twofold contribution. Torque curves measured with clockwise and counterclockwise field rotation are identical, thus showing no evidence of rotational hysteresis(figure 8).

In our crystal samples, we obtain an anisotropy constant of $K = 1.8 \times 10^4$ ergs/cm³ in the (100) plane. The discrepancy in anisotropy values compared to those of Perekalina ($K_1 = 2 \times 10^4$ ergs/cm³) may be attributed to the presence of Pb in their samples which may result in structural distortions that affect the magnetic anisotropy.

For a cubic crystal, one would expect fourfold symmetry in the torque curve of the (100) plane. For a rhombohedral crystal, one would expect twofold magnetocrystalline anisotropy for

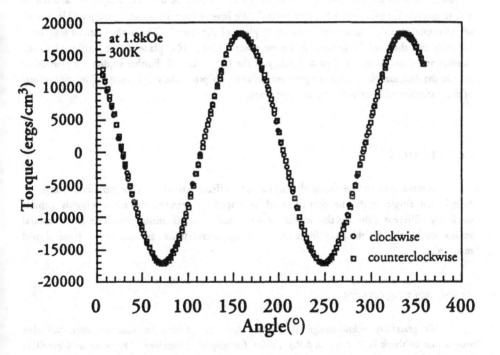

Figure 8. Torque curves of a (100) LSMO single crystal disk measured with the field rotated clockwise and counterclockwise indicate no effects due to rotational hysteresis.

an untwinned crystal. X-ray diffraction of our single crystal samples indicate that there is significant twinning in the single crystal. While a rhombohedral crystal with all four different kinds of twin domains should exhibit fourfold symmetry in the (100) plane, twinning may not necessarily change the underlying twofold symmetry of the crystal structure. Therefore the observed twofold symmetry is not inconsistent with rhombohedral magnetocrystalline anisotropy.

We mention some possible origins of the observed twofold symmetry here. Site ordering has been observed in the garnet and TbFe systems to give rise to an anisotropy [14-16]. Our uniaxial anisotropy cannot be explained by growth induced anisotropy due to site ordering since growth induced anisotropy would result in a perpendicular anisotropy in our single crystal disk samples. Our uniaxial anisotropy cannot be explained by misalignment of the (100) axis with the axis of rotation due to cleaving uncertainties or misalignment of the sample holder. The misalignment is to within 3 degrees. Given the dimensions of our disk sample, the calculated shape anisotropy is 4.5×10^5 ergs/cm^3. Even if our misalignment were 3 degrees, the shape anisotropy term in the above energy density expression would be only 1224 ergs/cm^3 and not be able to account for the entire 1.8×10^4 ergs/cm^3. We have eliminated some of the possible origins of the strong twofold anisotropy in the (100) plane of the crystal. These observations lead us to conclude that the twofold symmetry that we observe in the (100) plane can be attributed to the rhombohedral magnetocrystalline anisotropy of the LSMO crystal. Further studies are planned to detwin the disk sample at high temperatures in order to probe the magnetocrystalline anisotropy of these samples without twinning complications.

CONCLUSIONS

In conclusion we have shown that while strain effects dominate in the magnetic anisotropy of epitaxial single crystalline thin films of the doped manganites, the single crystals exhibit distinctly different behavior that cannot be attributed to stress anisotropy. The single crystal results shed light on the magnitude of the magnetocrystalline anisotropy of these doped manganites.

ACKNOWLEDGMENTS

We gratefully acknowledge T. Siegrist for x-ray diffraction measurements. We also would like to thank B. Batlogg and R.J. Felder for helpful discussions. The work at Cornell is supported by the NSF MRSEC and CAREER Award.

REFERENCES

1. R. V. Helmolt, J. Wecker, B. Holzapfel, L. Schultz and K. Samwer, Phys. Rev. Lett. 71 2331 (1993).
2. S. Jin, T.H. Tiefel, M. McCormack, R.A. Fastnacht, R. Ramesh and L.H. Chen Science 264 413 (1994).
3. H.L. Ju, C. Kwon, Qi Li, R.L. Greene and T. Venkatesan, Appl. Phys. Lett. 65 2108 (1994).
4. J.Z. Liu, I.C. Chang, S. Irons, P. Klavins, R.N. Shelton, K. Song and S.R. Wasserman, Appl. Phys. Lett. 66 32188 (1995).
5. M.E. Hawley, X.D. Wu, P.N. Arendt, C.D. Adams, M.F. Hundley and R.H. Heffner, Proc. Symp. Mater. Res. Soc. 401 531 (1995).
6. H.Y. Hwang, T.T.M. Palstra, S-W. Cheong and B. Batlogg, Phys. Rev. B 52 15046 (1995).
7. M.R. Ibarra, P.A. Algarabel, C. Marquina, J. Basco and J. Garcia, Phys. Rev. Lett. 75 3541 (1995).
8. S. Jin, T.H. Tiefel, M. McCormack, H.M. O'Bryan, L.H. Chen, R. Ramesh and D. Schurig, Appl. Phys. Lett. 67 557 (1995).
9. C. Kwon, K.-C. Kim, M.C. Robson, S.E. Lofland, S.M. Bhagat, T. Venkatesan, R. Ramesh and R.D. Gomez, submitted to J. Mag. Magn. Mater.
10. P. Lecoeur, P.L. Trouillard, Gang Xiao, A. Gupta, G.Q. Ging, X.W. Li, to be published in J. Appl. Phys.
11. S.E. Lofland, S.M. Bhagat, H.L. Lu, G.C. Xiong, T. Venkatesan, R. L. Greene, J. Appl. Phys. 79 5166 (1996).
12. Y. Suzuki, H.Y. Hwang, S-W. Cheong, R.B. van Dover, Appl. Phys. Lett. 71 142 (1997).
13. T.M. Perekalina, I.E. Lipinski, V.A. Timofeva and S.A. Cherkezyan, Sov. Phys. Solid State 32 1827 (1990).
14. E.M. Gyorgy, M.D. Sturge, L.G. Van Uitert, E.J. Heilner, W.H. Grodkiewicz, J. App. Phys.44 438 (1973).
15. A. Rosencwaig, W.J. Tabor,R.D. Pierce, Phys. Rev. Lett. 26 779 (1971).
16. F. Hellman, E.M. Gyorgy, R.C. Dynes, Phys. Rev. Lett. 68 1391 (1992).

Evidence for a Jahn-Teller Distortion in the CMR Layered Manganite $La_{1.4}Sr_{1.6}Mn_2O_7$

Despina Louca*, G. H. Kwei*, J. F. Mitchell**
*Los Alamos National Laboratory, Los Alamos, NM 87545
**Argonne National Laboratory, Argonne, IL 60439

ABSTRACT

The changes with temperature in the crystallographic structure of the two-layered $La_{1.4}Sr_{1.6}Mn_2O_7$, are *on average* quite small but the atomic pair density function analysis of pulsed neutron diffraction data shows that the lattice is locally distorted in accordance with the change in the transport properties. In particular, while no Jahn-Teller (JT) distortion is expected in the layered compounds because the octahedral bilayers are almost cubic, lattice distortions attributed to a large JT effect are present and are of comparable magnitude as in the cubic perovskite system. This could in turn explain the similarity in their properties. The number of the JT distorted sites is reduced with temperature concomitantly with the decrease in resistivity of the *ab*-plane.

INTRODUCTION

The discovery of colossal magnetoresistance (CMR) renewed the interest in manganites with the general composition of $(La/A)_{n+1}Mn_nO_{3n+1}$ [1-3] with A = Ba, Ca, Pb or Sr. The CMR effect results from a drastic drop in the resistivity with the application of a magnetic field. However, the mechanism that associates the magnetism to transport that gives rise to the CMR effect is not well understood. The double exchange (DE) interaction was originally proposed coupling the charge to the spin degrees of freedom in the ferromagnetic (FM) metallic state for the $n = \infty$ member ($La_{1-x}A_xMnO_3$) [4-7], but experimental and theoretical evidence suggested a mechanism requiring a contribution from the lattice [8-10]. Studies have shown that significant structural changes associated with the Jahn-Teller (JT) effect occur in Mn^{3+} systems and coincide with the transition in the properties [11-21] in accordance with the presence of lattice polarons. Similar properties have recently been observed in the layered compounds as well but the average crystallographic structure does not show evidence for a full JT distortion in this system.

In the layered compounds of the $n = 2$ series, $La_{2-2x}Sr_{1+2x}Mn_2O_7$, the average structure is tetragonal, made of a bilayer perovskite unit that spans the *ab*-plane of the crystal, separated by a single non-magnetic La/Sr-O layer along the *c*-axis [22] giving it a 2 dimensional character. The separation of the octahedra in the *c*-axis inhibits both the inter-layer magnetic coupling and the electrical conductivity. The magnetic coupling of the *ab*-plane is an order of magnitude bigger than that along the *c*-axis [23]. This is understood in terms of the presence of the non-magnetic (La/Sr)-O layer. Similarly, the transport measurements in single crystals show that the resistivity along the *c*-axis is about 2 orders of magnitude greater than across the *ab*-plane [23].

Our pulsed neutron powder diffraction study on $La_{1.4}Sr_{1.6}Mn_2O_7$ with x = 0.3 indicates that the lattice is actively involved in the properties of this system similarly to the perovskites. In this study, we show that local lattice distortions stem from a large Jahn-Teller (JT) effect which is reduced in the FM state. A magnetic neutron scattering measurement on a

La$_{1.2}$Sr$_{1.8}$Mn$_2$O$_7$ sample with x = 0.4 provided evidence for the existence of short range antiferromagnetic (AFM) correlations in the paramagnetic (PM) state [24]. It was suggested that the presence of such domains might induce deviations from the average crystallographic structure. The existence of local deviations from the average structure in the PM phase may be connected with the domains of AFM correlations. Also, the transition in the resistivity of the *ab*-plane for compositions with 30 % of doping occurs at a temperature closer to the one observed in the cubic perovskites with an equivalent amount of doping. This temperature is higher than that of the 3D FM transition [25, 26]. The structural changes observed in the present study are consistent with the transition in the transport.

EXPERIMENT

The sample was prepared by a standard ceramic method. No significant (less than 1-2 %) secondary phases were identified in the diffraction data. The sample was characterized by resistance and magnetization measurements and the FM transition occurs at 116 K which coincides with the insulator-metal (IM) transition temperature. The neutron diffraction data were collected using the Special Environment Powder Diffractometer (SEPD) of the Intense Pulsed Neutron Source (IPNS) of the Argonne National Laboratory. Data were collected from 300 K down to 20 K. The structure function, S(Q), was corrected for background, absorption, incoherent scattering, multiple scattering as well as for inelastic scattering (Placzek correction [27]). The multiple scattering correction procedure was carefully calibrated so that it gives a correct PDF for crystalline Ni powder. S(Q) was Fourier transformed to obtain the pair density function (PDF), $\rho(r)$,

$$\rho(r) = \rho_0 + \frac{1}{2\pi^2 r} \int_0^\infty Q[S(Q) - 1]\sin(Qr)dQ \qquad (1)$$

where ρ_0 is the average number density of the material and Q is the momentum transfer. The PDF is a real space representation of atomic density correlations and has been quite effective in determining the local atomic structure of amorphous as well as crystalline materials [28]. The PDF analysis provides direct information with regard to the local structure without a requirement of long range structural periodicity.

RESULTS

The PDF determined from the data collected at 300 K (symbols) is compared to a model PDF calculated from the crystallographic structure with the I4/mmm symmetry (solid line) (Fig. 1). The model PDF is constructed of δ-functions centered at positions corresponding to the probability of finding a particular atomic pair. The parameters for the model are obtained from the Rietveld refinement of the same data used for the PDF analysis. The sum of the partial PDF's for each atom provides the total PDF of the crystal convoluted with a gaussian function to simulate thermal and quantum zero point vibrations. A single, wide Mn-O peak is observed in the model PDF centered at ~ 1.95 Å with a FWHM of 0.154 Å. This peak is negative because the neutron scattering length of manganese is negative. The experimental PDF agrees well with the PDF determined from the model at longer range (beyond 3.5 Å), but a clear difference is observed between the model and experiment at shorter range. Only one

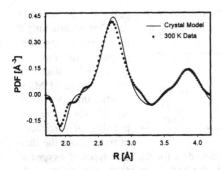

Figure 1: The PDF of $La_{1.4}Sr_{1.6}Mn_2O_7$ determined at 300 K (symbols) compared to a model PDF (solid line) for the crystallographic structure. Note that the first peak which corresponds to the Mn-O bonds within the octahedron is clearly split at 1.92 and at 2.12 Å corresponding to short and long bonds, respectively. This split is absent in the model.

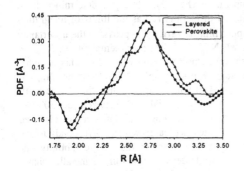

Figure 2: The local structure of the layered compound is compared to the one obtained from the pure $LaMnO_3$ perovskite at 300 K. Note that the split in the Mn-O peak at 1.95 and at 2.12 Å is as strong as the one observed in the perovskite. This serves as a strong indication that the Jahn-Teller distortion is present in this system as in the perovskites.

Mn-O peak is seen in the crystal structure model which represents the average structure whereas two types of Mn-O bonds are clearly identified in the local structure at this temperature, one at 1.92 Å and another at 2.12 Å. No long bonds are expected in the layered system because the octahedra are almost perfectly symmetric in the average structure.

In $La_{1.4}Sr_{1.6}Mn_2O_7$, the nominal valence of Mn is 3.3+ but the actual electronic state of Mn can vary from 3+ to 4+. If carriers are localized on Mn sites, some crystal regions will become hole-rich consisting of Mn^{4+} whereas others will be hole-poor containing Mn^{3+} ions. Thus the local octahedral environment is particularly sensitive to the charge distribution and, in turn, provides direct evidence for the response of the lattice to the change in the transport. In a system with Mn^{3+} and Mn^{4+} ions, it is possible to identify each ion type structurally from the distinct environment associated with the Mn^{3+} and Mn^{4+} octahedra. A Mn^{3+} octahedron is elongated because of the JT effect induced by the presence of an electron in the e_g orbital. Of the six Mn-O bonds around manganese, some of them will be longer than average because of Coulomb repulsion. This is clearly the case in the pure perovskite (with Mn^{3+} ions only) where a distribution of 4 short and 2 long bonds at distances ranging from 1.87 - 2.00 and from 2.11 - 2.22 Å, respectively, is observed [29]. On the other hand, the octahedron associated with the Mn^{4+} is undistorted because it is JT-free and all Mn-O bonds are of the same average length. An intermediate electronic state between these two extremes can also exist with an average charge on Mn of 3.3 where the carriers in this case are uniformly distributed over all sites and all Mn-O bonds are of about the same length.

The PDF of the layered manganite $La_{1.4}Sr_{1.6}Mn_2O_7$ determined at 300 K is compared to that of pure $LaMnO_3$ [14-16] in Fig. 2. The first two negative peaks at 1.92 and 2.12 Å correspond to the two types of Mn-O bond distances within the octahedron. It is quite unexpected to see that the Mn-O peak in the layered material is split as in the pure perovskite with the

Table 1: Mean amplitude of oscillations determined from the thermal factor $B = 8\pi^2 \langle u_{ii}^2 \rangle$ at room temperature for the three types of oxygen atoms along the a, b and c-axes.

	$\langle u_{11}^2 \rangle^{1/2}$ (Å)	$\langle u_{22}^2 \rangle^{1/2}$ (Å)	$\langle u_{33}^2 \rangle^{1/2}$ (Å)
O1 (axial, octahedron)	0.130	0.130	0.100
O2 (planar)	0.095	0.084	0.138
O3 (axial, rocksalt)	0.122	0.122	0.148

peaks at 1.95 and 2.15 Å. This is surprising if the average crystallographic structure is considered since it gives a range of Mn-O bonds from 1.93 - 2.03 Å. However, the amplitudes of oscillation obtained from the thermal factors of the Rietveld refinement of the diffraction data for the three types of oxygen atoms in the structure are quite large as seen in Table 1 which increases the uncertainty in the position of the atoms. The long Mn-O bond in the layered system is comparable in size to that found in the perovskites and suggests that Mn^{3+} ions are present in the PM state. As the magnitude of the distortion is large as in the cubic system, it probably originates from a JT effect, sometimes referred to as a pseudo JT effect because of the absence of a cubic environment in the layered material but with similar consequences [30].

As the crystal goes through the IM transition one can use the integrated intensity of the first Mn-O peak to quantify the changes in the octahedral environment. This peak corresponds to the number of short bonds in the octahedron, N_{Mn-O}, and its integrated intensity demonstrates how the short and long bonds vary with temperature (Fig. 3). N_{Mn-O} is determined from $4\pi r^2 \rho(r)$ and normalized by the scattering lengths, with r_{min} and r_{max} as the limits of integration taken from 1.75 to 2.10 Å. The cutoff of the upper limit is varied and provides the uncertainty of 0.1 in this calculation. In the crystalline model, N_{Mn-O} is 6. At room temperature, $N_{Mn-O} = 5.0$ and corresponds to about 50 % of the Mn sites with no long bonds (no JT). This is higher than expected if the charge were localized at a single Mn site (with 30 % of doping, $N_{Mn-O} = 4.6$) which suggests that the charge is already partially delocalized at 300 K. With cooling, the number of distorted sites decreases which corresponds to the enhancement in the charge mobility due to its delocalization in the metallic phase. But the fact that $N_{Mn-O} = 5.6$ rather than 6 at 20 K suggests that about 20 % of the Mn sites maintain a local JT distortion.

The change in the MnO_6 octahedral environment is qualitatively reflected in the temperature dependence of the oxygen-oxygen correlations. The peak at 2.75 Å which corresponds, among other pairs, to the short oxygen-oxygen bond distance within the octahedron changes with temperature in the way shown in Fig. 4. The peak height increases from 300 to 20 K which corresponds to the decrease in the number of JT distorted sites with temperature. The octahedra become

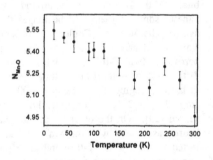

Figure 3: The number of short Mn-O bonds, N_{Mn-O}, as a function of temperature. The uncertainty is determined by varying the upper limit of integration of the peak. As the temperature is lowered from 300 K, N_{Mn-O} increases corresponding to an increase in the number of short bonds, or a decrease in the number of Mn sites with the JT distortion. Between 200-300 K the structural changes observed coincide with changes in the resistivity along the easy axis.

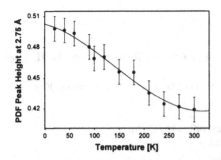

Figure 4: The PDF peak height at 2.75 Å as a function of temperature. This peak includes the short O-O bonds of the octahedron in addition to La/Sr-O bonds. On cooling, the peak height exhibits an increase of 16 %. The increase in the peak height reflects the decrease in the number of JT distorted sites.

more symmetric as the charge becomes delocalized. The coherence in the O-O correlations contribute to producing an increase in the peak height with decreasing temperature.

DISCUSSION

The PDF analysis has shown that a large JT distortion is present locally in the layered manganites as in the perovskites. The distortion changes with temperature in accordance with the transport properties. As observed in Fig, 3, N_{Mn-O} increase between 200 and 300 K. This coincides with the decrease in the transport along the ab-plane shown in refs. 25 and 26 where the resistivity drops along the easy axis with cooling. The temperature at which this occurs is close to the T_c of the perovskite with an equivalent amount of doping. Due to the 2D nature of the layered structure, the DE interaction between the bilayers is weak because interlayer magnetic coupling is much weaker than intralayer coupling. But the DE interaction may occur within the octahedral bilayer and resemble that in perovskites. This takes place at a temperature that precedes the 3D FM ordering in the layered system. At the 3D FM ordering, ρ_c transition occurs but with a very small change in the total transport since ρ_c is 3000 times larger than ρ_{ab} [25].

In the perovskites, the cooperative JT phenomena in the ab-plane gives rise to the orthorhombic splitting. In the layered manganites, however, the octahedral bilayers are almost cubic on average which suggests the absence of the JT distortion in this system. But the local structure indicates that the JT distortion is present even in the layered system and is of comparable magnitude to the one found in the perovskites. Although the average crystal bondlengths show no evidence for the distortion, the large amplitude of thermal vibrations provides an agreement between the local and average structures.

The JT distorted sites might be associated with regions of AFM domains observed by Perring et al. [24] in the PM phase. If almost half of the sites are distorted by the JT effect in the PM state, it is likely that charge hopping is only allowed in regions where no distortion is present. Within the regions of the distortion the local magnetic order can be antiferromagnetic. In conclusion, this work has shown that a local JT effect is present in the layered manganites that is equal in magnitude to that in the perovskites.

ACKNOWLEDGMENTS

The authors would like to acknowledge valuable discussions particularly with T. Egami and M. F. Hundley. They also thank A. R. Bishop, J. L. Sarrao, S. Trugman, R. H. Heffner and H. Röder for helpful conversations and S. Short for helping out with the data collection on SEPD. Work at the Los Alamos National Laboratory is performed under the auspices of the U.S. Department of Energy under contract W-7405-Eng-36. The IPNS is supported by the U.S. Department of Energy, Division of Materials Sciences under contract W-31-109-Eng-38.

REFERENCES

1. J. Volger, Physica **20**, 49 (1954).
2. S. Jin, T. H. Tiefel, M. McCormack, R. A. Fastnacht, R. Ramesh and L. H. Chen, Science **264**, 413 (1994).
3. Y. Moritomo, Y. Tomioka, A. Asamitsu, Y. Tokura and Y. Matsui, Phys. Rev. B **51**, 3297 (1995).
4. C. Zener, Phys. Rev. **82**, 403 (1951).
5. E. O. Wollan and W. C. Koehler, Phys. Rev **100**, 545 (1955).
6. J. B. Goodenough, Phys. Rev. **100**, 564 (1955).
7. P. G. de Gennes, Phys. Rev. **118**, 141 (1960).
8. Y. Tokura, A. Urushibara, Y. Moritomo, T. Arima, A. Asamitsu, G. Kido and N. Furukawa, J. Phys. Soc. Japan B **63**, 3931 (1994).
9. A. J. Millis, P. B. Littlewood, and B. I. Shraiman, Phys. Rev. Lett. **74**, 5144 (1995).
10. H. Röder, J. Zang and A. R. Bishop, Phys. Rev. Lett. **76**, 1356 (1996).
11. P. Dai et al., Phys. Rev. B **54**, R3694 (1996).
12. S. J. L. Billinge, R. G. DiFrancesco, G. H. Kwei, J. J. Neumeier and J. D. Thompson, Phys. Rev. Lett. **77**, 715 (1996).
13. M. F. Hundley and J. J. Neumeier, Phys. Rev. B **55**, 11511 (1997).
14. D. Louca and T. Egami, J. Appl. Phys. **81**, 5484 (1997).
15. D. Louca, T. Egami, E. Brocha, H. Röder and A. R. Bishop, Phys. Rev. B **56**, R8475 (1997).
16. D. Louca and T. Egami, submitted Phys. Rev. B (1997).
17. H. Y. Hwang, S.-W. Cheong, P. G. Radaelli, M. Marezio and B. Batlogg, Phys. Rev. Lett. **75**, 914 (1995).
18. T. A. Tyson, J. Mustre de Leon, S. D. Conradson, A. R. Bishop, J. J. Neumeier and J. Zang, Phys. Rev. B **53**, 13985 (1996).
19. W. Archibald, J.-S. Zhou and J. B. Goodenough, Phys. Rev. B **53**, 14445 (1996).
20. G.-M. Zhao, K. Conder, H. Keller and K. A. Müller, Nature (London) **381**, 676 (1996).
21. P. G. Radaelli, M. Marezio, H. Y. Hwang, S.-W. Cheong and B. Batlogg, Phys. Rev. B **54**, 8992 (1996).
22. S. N. Ruddlesden and P. Popper, Acta Crystall. **11**, 54 (1958).
23. Y. Moritomo, A. Asamitsu, H. Kuwahara and Y. Tokura, Nature **380**, 141 (1996).
24. T. G. Perring, G. Aeppli, Y. Moritomo and Y. Tokura, Phys. Rev. Lett. **78**, 3197 (1997).
25. T. Kimura, Y. Tomioka, H. Kuwahara, A. Asamitsu, M. Tamura and Y. Tokura, Science **274**, 1698 (1996).
26. T. Kimura, A. Asamitsu, Y. Tomioka and Y. Tokura, Phys. Rev. Lett. **79**, 3720 (1997).
27. G. Placzek, Phys. Rev. **86**, 377 (1952).
28. B. H. Toby and T. Egami, Acta Crystallogr. A **48**, 336 (1992).
29. J. F. Mitchell, D. N. Argyriou, C. D. Potter, D. G. Hinks, J. D. Jorgensen and S. D. Bader, Phys. Rev. B **54**, 6172 (1996).
30. I. B. Bersuker, *Electronic Structure and Properties of Transition Metal Compounds* (John Wiley & Sons, Inc., New York, 1996), p. 297.

Mn K-EDGE X-RAY ABSORPTION SPECTROSCOPY (XAS) STUDIES OF La$_{1-x}$Sr$_x$MnO$_3$

S. M. Mini*[†], J. F. Mitchell[†], D. G. Hinks[†], Ahmet Alatas$^{\lozenge}$[†], D. Rosenmann[†],
C. W. Kimball*[†], and P.A. Montano[#][†]

*Northern Illinois University, Department of Physics, DeKalb, IL 60115
† Materials Science Division, Argonne National Laboratory, Argonne, IL 60439
$^{\lozenge}$Illinois Institute of Technology, Chicago, IL
Dept. of Physics, University of Illinois, Chicago, IL 60680

ABSTRACT

Systematic Mn K-edge x-ray absorption spectroscopy (XAS) measurements on samples of La$_{1-x}$Sr$_x$MnO$_3$, which are precursors to colossal magnetoresistive (CMR) materials, are reported. Detailed results on the edge or chemical shift as a function of Sr concentration (hole doping) and sample preparation (air vs oxygen annealed), are discussed. For comparison, a systematic XANES study of the Mn K-edge energy shift, denoting valence change in Mn, has been made in standard manganese oxide systems. Contrary to expectations, the variation in near-edge energies for Mn in La$_{0.725}$Sr$_{0.275}$MnO$_3$ were small when compared to the difference between that for manganese oxide standards of nominal valence of +3 and +4 (Mn$_2$O$_3$ and MnO$_2$).

INTRODUCTION

While Ln$_{1-x}$A$_x$MnO$_3$ systems are noted for their giant magneto resistance (GMR) behavior [1], the system La$_{1-x}$Sr$_x$MnO$_3$ is prototypical of colossal magneto resistive (CMR) materials as introduced by Asamitsu et al. [2]. This perovskite system exhibits unusual coupling which is presumed to depend on the mixed valent Mn^{+3}/Mn^{+4} lattice.

LaMnO$_3$, without an divalent ion or hole substitution for La^{+3}, exhibits ferromagnetism, insulator-metal transitions, and GMR, provided a sufficient proportion of Mn^{+4} is introduced by La or Mn deficiency [3]. Chemical substitution of La^{+3} by Sr^{+2} introduces holes into the e$_g$ orbitals, that are mobile and mediate an interatomic ferromagnetic interaction between the Mn atoms and the material becomes ferromagnetic and metallic below T$_c$ [4, 5].

Occurrence of ferromagnetism and metallic behavior of La$_{1-x}$A$_x$MnO$_3$ has been explained by Zener's double exchange mechanism [6] for the hopping of an electron from Mn^{+3} (t$_{2g}^3$e$_{2g}^1$) to Mn^{+4} (t$_{2g}^3$) via the oxygen. Double exchange and superexchange interactions [7] control the ferromagnetism in these systems. The Mn^{+3}-O-Mn^{+4} superexchange interaction is ferromagnetic, while that of Mn^{+3}-O-Mn^{+3} or Mn^{+4}-O-Mn^{+4} are not. There is also evidence, in the case of layered perovskite La$_{1.2}$Sr$_{1.8}$Mn$_2$O$_7$, under hydrostatic pressure, of exchange striction, which reflects the competition between super and double exchange [8].

The change in structural properties of $La_{1-x}Sr_xMnO_{3\pm\delta}$ as a function of oxygen partial pressure [9] indicate that the properties are strongly affected not only by the strontium content, x, but also by the oxygen nonstoichiometry, δ. It has been found that decreasing $P(O_2)$ yields smaller cation vacancy concentrations and smaller Mn^{+3}/Mn^{+4} ratio [10]. Undoped and Sr-doped $LaMnO_3$ exhibited reversible oxidation-reduction behavior. where the perovskites can be excess, stoichiometric or deficient in oxygen content depending on he specific condition. Metal vacancies are assumed for the oxygen excess condition and oxygen vacancies are assumed for the oxygen deficient condition. In high $P(O_2)$ region, metal vacancies on both La and Mn sites are the predominant defects [11].

The energy required, in X-ray Absorption Near Edge Spectroscopy (XANES), to excite a 1s electron to a delocalized p level at the Mn K-edge varies as the number of electron holes associated with the Mn atom changes. The calculated value of the K-edge for Mn mesh is 6539 eV. However, it requires more energetic photons to excite an electron away from the core potential as the positive charge on the Mn atom that absorbs the photon is increased. This effect manifests itself as a shift in the overall energy position of the absorption edge (called an edge shift) to higher energies with increasing net charge. For example, the Mn K-edge of MnO_2 (nominal valence Mn^{+4}) will have a shift to several eV higher energy when compared to that of Mn_2O_3 (nominal valence Mn^{+3}).

In order to quantify the energy shift of Mn in different oxide environments, we use a method involving the determination of an energy moment of the normalized absorption cross-section μ_0 and treating the absorption edge as a step function [12]. This method is more representative of overall charge density at the manganese atom in each compound than other approaches which attribute detailed meaning to isolated or individual features of the spectrum [13]. By taking an systematic overall approach, which is more akin to the concept of valence, the details of spatial charge distribution and electronic hybridization as well as complicated pre-edge structure may be avoided. In this manner, the edge shift observed for the Mn_3O_4, Mn_2O_3, $La_{1-\delta}Mn_{1-\delta}O_3$ and MnO_2 are real and reflect a difference in the Mn valence.

Formal valence is only one factor contributing to the charge environment of the Mn atoms. Other factors such as coordination numbers and the relative electronegativities strongly influence the actual charge environment in solids. Qualitatively, the shape of the XANES signal can be correlated with the coordination number of the element of interest [14].

SAMPLE PREPARATION AND EXPERIMENT

$La_{1-x}Sr_xMnO_{3+\delta}$ samples were synthesized by co-precipitation of carbonates from a solution of the corresponding nitrates. Starting materials consisted of the high purity oxides La_2O_3 (Johnson-Matthey REacton, 99.99%) and MnO_2 (Johnson-Matthey Puratronic, 99.999 %); $SrCO_3$ (Johnson-Matthey Puratronic, 99.999%) was used as a Sr source. La_2O_3 was prefired in flowing oxygen at 1000 °C for several hours to decompose residual carbonates, and the MnO_2 was treated in flowing oxygen at 425 °C and slowly cooled (1 °C/min) to room temperature. $SrCO_3$ was used as-received. Prefired MnO_2 was dissolved in concentrated HCl and the solution was slowly evaporated to dryness. The residue was then redissolved in a minimum of H_2O. To remove chlorides, concentrated HNO_3 was added to the solution, which was again heated to dryness. The resulting pink solid was dissolved in H_2O. Concentrated nitric acid was slowly added to a stoichiometric mixture of $SrCO_3$ and the prefired La_2O_3. This solution was then added to the manganese solution and the cations coprecipitated by slow addition of a saturated $(NH_4)_2CO_3$ solution. The supernatant was then decanted and the precipitate dried and pulverized to yield a fine, light pink precursor powder. The precursor powder was slowly heated (0.5 °C/min) in flowing oxygen to 250 °C to remove water. The temperature was then ramped at 10 °C/min to 1000 °C and held at this temperature for 10 hours before rapidly cooling to room temperature. The resulting material is an extremely fine black powder.

The XANES measurements were made in step scan mode in transmission geometry at beamline X6B at the National Synchrotron Light Source (NSLS, BNL). Si(111) crystals are used to monochromatize the synchrotron radiation. The $\Delta E/E \approx 10^{-4}$. Manganese mesh spectra was recorded simultaneously (in the backchannel) so that the edge position of the manganese could be calibrated. The inflection point in the first resolved peak of the manganese mesh was chosen to be the zero of the energy scale. This is relatively easy to locate and reproduce, since the derivative of the spectrum has a sharp peak at that point. In this manner, the edge shifts (the change in valence of the Mn) of the Mn K-edges for all manganese containing systems could be compared.

RESULTS AND DISCUSSION

The XANES spectra shown in figure 1 were taken at room temperature. A standard manganese mesh spectrum was taken simultaneously so that the backchannels could be lined up on top of each other and the front channel (sample) shifted by the same amount. The inflection point in the first resolved peak of the standard Mn mesh of the near edge was chosen to be the zero of the energy scale. This is relatively easy to locate and reproduce, since the derivative of the spectrum has a sharp peak at that point. Therefore, the edge shift observed for the Mn_2O_3 $La_{0.725}Sr_{0.275}MnO_3$, and MnO_2 in figure 1, are real and reflect a difference in the Mn valence, which are, nominally +3, +3.275 and +4.

All of the spectra from the Mn oxide standards, exhibit different shapes which are indicative of their respective Mn-oxygen coordination number or the position of the Mn in the different crystal structures. It has been shown [15] for transition metal oxides that the various shapes of the transition metal K-edge reflect the coordination number of the transition metal with oxygen. For example the shape of the Mn K-edge of Mn_2O_3 is like that of the other 4-fold transition metal elements with +3 octahedral coordination with oxygen.

From previous work [16] we have found that the shape of the Mn K-edge of $La_{1-x}Sr_xMnO_3$ is exactly the same for x = 0.000 and 0.275, so that one may infer that with the addition of Sr, the electronic configuration of Mn remains intact. The slight shift (< 0.2 eV) of the Mn edge to higher energy for $La_{0.725}Sr_{0.275}MnO_3$ indicates an increase of the valence on the Mn atom. For the standard Mn oxide compounds, as shown in figure 1, the different shapes of the Mn K-edges are more distinguishable and the edge shifts are much larger (> 2 eV) indicating very different coordination numbers and Mn electronic configurations.

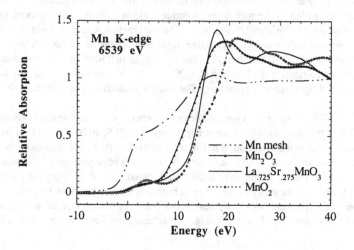

Figure 1. Mn K-edge of room temperature XANES comparing La$_{0.725}$Sr$_{0.275}$MnO$_3$ with a nominal Mn valence of +3.275 to standard Mn mesh, Mn$_2$O$_3$ with a nominal Mn valence of +3, and MnO$_2$ with a nominal valence of +4.

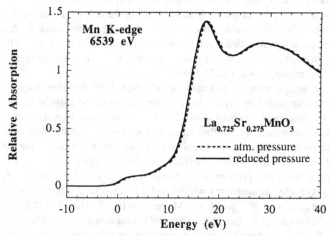

Figure 2. Comparison of Mn K-edge spectra of La$_{0.725}$Sr$_{.275}$MnO$_3$ prepared at atmospheric and reduced pressure.

Figure 2, shows an edge shift to higher energy and therefore higher Mn valence for Mn K-edge spectra of La$_{0.725}$Sr$_{0.275}$MnO$_3$ samples produced at atmospheric pressure and reduced pressure, respectively. Nominally, $La_{0.725}Sr_{0.275}MnO_3$ has a Mn valence of +3.275. However, figure 2 indicates that the sample prepared at atmospheric pressure has an overall Mn valence slightly higher than that of the sample produced at reduced pressure. This suggests that the sample produced at atmospheric pressure has less oxygen in its system and a higher Mn^{+4} content than that produced at reduced pressure.

It has been shown by Mahendiran, et al. [17] that reduced pressure $La_{1-\delta}Mn_{1-\delta}O_3$ and hole doped $LaMnO_3$ exhibit cation vacancies as opposed to anion vacancies. In Mahendiran's experiments the Mn^{+4} content of the samples was determined independently by redox titrations methods. In the high $P(O_2)$ region, metal vacancies on both La and Mn sites are the predominant defects [11].

It has been suggested that in the perovskite system $LaMnO_3$, the origin of Mn^{+4} cannot be due to oxygen excess since the perovskite structure cannot accommodate excess oxygen and that high percentages of Mn^{+4} in $La_{1-\delta}Mn_{1-\delta}O_3$ materials, are created by the random presence of cation vacancies in both La and Mn sites [18]. In fact, a large suppression of magnetoresistance due to cation disorder is evidenced [19] in the CMR precursor systems, indicating that the cation disorder and size effects in magnetoresistive manganese oxide perovskites is an important consideration.

To summarize, we have shown that the Mn K-edge XANES, gives information regarding the valence changes on the Mn and therefore the electronic configuration in manganese oxides as a function of hole doping as well as whether the sample is produced in atmospheric or reduced pressure. The Mn K-edge XANES is very sensitive to hole doping, indicating that the Mn valence in $La_{1-x}Sr_xMnO_3$ increases as a function of Sr doping and also to the increase or decrease in the oxygen content, as shown in the difference in the XANES at atmospheric and reduced pressure. The XANES edge shifts maybe be correlated to the Mn^{+4} content through independent means (redox titration or density measurements for the oxygen content).

ACKNOWLEDGMENTS

This work was supported by ARPA / Office of Naval Research (ONR), the State of Illinois under IBHE / HECA Grant, and by US DOE-BES Materials Sciences under contract #W-31-109-ENG-38.

REFERENCES

1. C.N.R. Rao, Chem. Commun. 2217 (1996).
2. A. Asamitsu, Y. Moritomo, Y. Tomioka, T. Arima, and Y. Tokura, Nature **373**, 407 (1995).
3. A. Arulraj, R. Mahesh, G.N. Subbanna, R. Mahendiran, A.K. Raychaudhuri, C.N. R. Rao, J. Solid State Chem, **127** 87 (1996).
4. C.N. R. Rao, A.K. Cheetham, R. Mahesh, Chem. Mater. **8** 2421 (1996)
5. J.H. van Santen, G.H. Jonker, Physica **16** 599 (1950).
6. C. Zener, Phys. Rev. **82** 403 (1951).

7. J.B. Goodenough, Prog, Solid State Chem. **5** 149 (1971).

8. D.N. Argyriou, J.F. Mitchell, J.B. Goodenough, O. Chmaissem, S. Short, and J.D. Jorgensen, Phys. Rev. Lett. **78** 1568 (1997).

9. H.U. Anderson, J. H. Kuo and D.M. Sparlin, in: Proc. First Intern. Symp. Solid Oxide Fuel Cell (Electrochem. Soc., 1089) pp. 111-128.

10. John F. Mitchell, D.N. Argyriou, C.D. Potter, D.G. Hinks, J.D. Jorgensen, and S.D. Bader, Phys Rev. B **54**, 6172 (1996).

11. J.H. Kuo, H. U. Anderson, and D.M. Sparlin, J. Solid State Chem. **83**, 52 (1989) and J.H. Kuo, H. U. Anderson, and D.M. Sparlin, J. Solid State Chem. **87** 55 (1990).

12. E.E. Alp, G.L. Goodman, L. Soderholm, S. M. Mini, M. Ramanathan, G.K. Shenoy and A.S. Bommannavar, J. Phys.: Condens. Matter **1** 6463 (1989).

13. M. Croft, D. Sills, M. Greenblatt, C. Lee, S.-W. Cheong, K.V. Ramanujachary and D. Tran, Phys. Rev. B **55**, 8726 (1997).

14. M. Lenglet, J. Delepine, J. Lopitaux, J. Durr, J. Kasperek and R. Bequignat, J. Solid State Chem. **58**, 194 (1985).

15. G.S. Knapp, B.W. Veal, H.K. Pan and T. Klippert, Solid State Comm. **44** 1343, 1982.

16. S.M. Mini, John Mitchell, D.G. Hinks, C.W. Kimball, P.A. Montano, P.Lee, D. Rosenmann, A.V. Tkachuk, and A. Udani, Mat. Res. Soc. Symp. Proc. **437**, 85 (1996).

17. R. Mahendiran, S.K. Tiwary, A.K. Raychaudhuri, T.V. Ramakrishnan, R. Mahesh, N. Rangavittal, and C. N. R. Rao, Phys. Rev. B 53, 3348, (1996).

18. J.A.M. Van Roosmalen, E.H.P. Cordfunke, R.B. Helmholdt, and H.W. Zandbergen, J.Solid State Chem. **110**, 100 (1994) and J.A.M. Van Roosmalen and E.H.P. Cordfunke J.Solid State Chem. **110**, 109 (1994)

19. Lide M. Rodriguez-Martinex and J. Paul Attfield, Phys. Rev. B **54**, 15622 (1996).

X-RAY INDUCED INSULATOR-METAL TRANSITIONS IN CMR MANGANITES

V. KIRYUKHIN[1,*], D. CASA[1], B. KEIMER[1], J.P. HILL[2], A. VIGLIANTE[2],
Y. TOMIOKA[3], Y. TOKURA[3,4]
1) Dept. of Physics, Princeton University, Princeton, NJ 08544
2) Dept. of Physics, Brookhaven National Laboratory, Upton, NY 11973
3) Joint Research Center for Atom Technology (JRCAT), Tsukuba, Ibaraki 305, Japan
4) Dept. of Applyed Physics, University of Tokyo, Tokyo 113, Japan
*) Present Address: Dept. of Physics, 13-2154, MIT, Cambridge, MA 02139

ABSTRACT

In this work we report a study of the photoinduced insulator-to-metal transition in manganese oxide perovskites of the formula $Pr_{1-x}Ca_xMnO_3$. The transition is closely related to the magnetic field induced insulator-to-metal transition (CMR effect) observed in these materials. It is accompanied by a dramatic change in the magnetic properties and lattice structure: the material changes from an insulating charge-ordered canted antiferromagnet to a ferromagnetic metal. We present an investigation of the transport and structural properties of these materials over the course of the transition (which usually takes about an hour to complete). The current-voltage characteristics exhibited by the material during the transition are highly nonlinear, indicating a large inhomogeneity of the transitional state. Possible practical applications of this novel type of transition are briefly discussed. We also report a high-resolution x-ray diffraction study of the charge-ordering in these materials. The temperature dependent charge ordering structure observed in these compounds is more complex than previously reported.

INTRODUCTION

Perovskite manganites of the general formula $A_{1-x}B_xMnO_3$ (where A and B are trivalent and divalent metals, respectively) have recently attracted considerable attention by virtue of their unusual magnetic and electronic properties [1]. These properties result from an intricate interrelationship between charge, spin, orbital and lattice degrees of freedom that are strongly coupled to each other. The parameters of the system can be substantially varied by choosing different A and B in the structural formula and by changing the doping level x. This provides the opportunity to control the value and relative strength of important interactions in these materials and synthesize materials exhibiting a vast variety of experimental phenomena.

The most extensively investigated property of the perovskite manganites is the so-called phenomenon of Colossal Magnetoresistance (CMR) [1]. It can be viewed as a magnetic field induced transition from either paramagnetic semiconducting phase (eg. in $La_{0.7}Ca_{0.3}MnO_3$) or from the insulating charge-ordered phase (eg. in $Pr_{0.7}Ca_{0.3}MnO_3$) to the ferromagnetic metallic phase. The large interest to the CMR phenomenon was generated in part due to the possibility of practical applications in magnetic recording. We have recently reported [2] that in $Pr_{0.7}Ca_{0.3}MnO_3$ the transition from the insulating antiferromagnetic (AFM) to metallic ferromagnetic (FM) state can be driven by illumination with x-rays at low temperature

(T<40K). This transition is accompanied by significant changes in the lattice structure, and can be reversed by thermal cycling. In this work we present the study of x-ray illumination effects on the structural and transport properties of $Pr_{1-x}Ca_xMnO_3$, x=0.3-0.5, and also in $Pr_{0.65}Ca_{0.245}Sr_{0.105}MnO_3$. All these materials undergo the photoinduced insulator-metal transition in a certain region of their magnetic phase diagram. Like the closely related CMR phenomenon, this effect also has potential for practical applications.

The x=0 and x=1 members of the $Pr_{1-x}Ca_xMnO_3$ family are insulating antiferromagnets with the manganese ion in the Mn^{3+} and Mn^{4+} valence states respectively [3]. For intermediate x, the average Mn valence is non-integer and the material is a paramagnetic semiconductor at high temperatures. At low temperatures, a variety of charge and magnetically ordered structures is observed (Fig. 1), Ref. [3, 4]. The ground state of $Pr_{1-x}Ca_xMnO_3$ with x=0.3-0.5 is a charge-ordered antiferromagnetic insulator. The 1:1 charge ordering (CO) of Mn^{3+} and Mn^{4+} ions occurs below T_{CO}=200-230K. It is associated with the corresponding lattice distortion leading to the doubling of the crystallographic unit cell. The CO transition is followed by the Néel transition at T_N=150-170K. Compounds with x=0.3-0.4 undergo an additional low temperature transition at which the system acquires a spontaneous magnetic moment (spin-canting transition). Thus, the low temperature state of $Pr_{1-x}Ca_xMnO_3$ with x=0.3-0.5 is the insulating antiferromagnetic charge ordered state.

Application of a magnetic field to this charge ordered insulating state was found to induce an insulator-metal transition, at which the resistivity changes by more than ten orders of magnitude [4]. The metallic phase is ferromagnetic due to the double-exchange interaction [6] between localized 3d t_{2g} spins (S=3/2) of Mn ions mediated by conduction 3d e_g carriers. Due to the carrier delocalization, charge ordering is destroyed in the metallic phase [7]. The field induced transition is associated with large hysteresis [4, 7]; the x=0.3 compound in fact remains metallic after the field is reduced to zero at low temperature, but reverts to charge-ordered insulating state on subsequent heating above 60K. Inside the hysteresis region, either the insulating or the metallic state is metastable, and external perturbations could, in principle, induce a transition to the ground state phase. We find that x-ray illumination constitutes an example of such an external perturbation that induces a transition to the metallic phase in the hysteretic region of the magnetic phase diagram. This effect, a remarkable example of simultaneous structural, magnetic, and insulator-metal transition, is a clear manifestation of strong coupling between charge, lattice and magnetic degrees of freedom in this system. Interesting in itself, the phenomenon of the photoinduced insulator-metal transition in manganite perovskites also provides a new tool for investigation of the microscopic properties of these intriguing compounds.

EXPERIMENT

Single crystals of $Pr_{1-x}Ca_xMnO_3$ were grown by the floating zone technique described elsewhere [4]. Samples with x=0.3, 0.4, 0.5, and also $Pr_{0.65}Ca_{0.245}Sr_{0.105}MnO_3$ were chosen for the experiment. We have performed simultaneous x-ray diffraction and electrical resistance measurements, so that the structural and transport properties of the sample can be correlated. X-rays were scattered from a polished surface of the sample on which two silver epoxy contacts for resistance measurements were placed. The sample was loaded into a superconducting magnet designed for x-ray diffraction measurements. The experiment was performed on beamline X22B at the National Synchrotron Light Source at Brookhaven National Laboratory. The x-ray energy was 8 keV, and the photon flux was $5 \times 10^{10} s^{-1} mm^{-2}$.

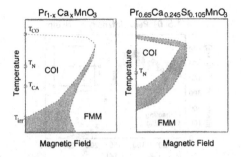

Figure 1: Sketches of the magnetic phase diagram of $Pr_{1-x}Ca_xMnO_3$, $0.3 < x < 0.4$ (left), and of $Pr_{0.65}Ca_{0.245}Sr_{0.105}MnO_3$ (right). The hysteretic regions are shaded. COI - charge ordered insulating phase, FMM - ferromagnetic metallic phase. The values of the relevant temperatures are given in the text. (Adapted from Refs. [4] and [5].)

X-ray measurements were performed in the (H K 0) reciprocal space zone, so that the superlattice reflections due to the lattice distortion accompanying the charge ordering (H K/2 L), K odd, with L=0 could be accessed. (The reflections are indexed on an orthorhombic, though nearly cubic, lattice with room temperature lattice constants a=5.426Å, b=5.478Å, $c/\sqrt{2}$=5.430Å, Ref. [3].) Momentum resolutions of $\sim 0.005 - 0.01$Å^{-1} were achieved. Some of the x-ray measurements, not involving electrical resistivity studies, were performed in a closed cycle refrigerator on beamline X22C. The DC resistance between the contacts deposited on the sample surface was measured by either 2 or 4 contact technique. Supplementary neutron scattering measurements were also conducted. They were performed on the H7 spectrometer at the High Flux Beam Reactor at Brookhaven National Laboratory with 14.7-meV neutrons.

RESULTS

As a $Pr_{1-x}Ca_xMnO_3$, $0.3 < x < 0.5$, sample is cooled through the charge ordering transition, superlattice diffraction peaks appear in the reciprocal space positions (H K/2 L), K odd [3]. The intensity of one such reflection, (4, 1.5, 0), was monitored using neutron diffraction for x=0.3 sample and is shown in Fig. 2 as a function of temperature. The system is in the charge ordered state below $T_{CO} \sim$200K. (The CO peak intensity is slightly depressed below T$\sim$100K which is likely the result of spin-canting transition at $T_{CA} \sim$110K, but this effect is small.) As was mentioned in the Introduction, application of the magnetic field drives the system to the metallic state and destroys the charge ordering. This transition is irreversible below $T_{irr} \sim$60K. We find that below T_{irr} x-ray irradiation destroys the charge ordering and induces an insulator-metal transition similar to that driven by the magnetic field.

The effect of x-ray irradiation at T=4K for $Pr_{0.7}Ca_{0.3}MnO_3$ is shown in Fig. 3. Note, that in this and in subsequent figures zero time corresponds to the moment when the resistance becomes measurable (approximately 1 to 3 minutes after the x-rays are switched on). When the sample is being illuminated with x-rays, the charge ordering peak intensity is diminishing, while the conductance is rising. The properties of the sample do not change when the

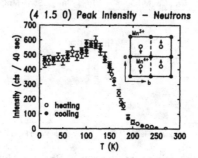

Figure 2: Intensity of the (4, 1.5, 0) superlattice reflection characteristic of the low-temperature charge ordered state of $Pr_{0.7}Ca_{0.3}MnO_3$, measured by neutron diffraction on a single crystal sample. The filled (open) symbols represent data taken on cooling (heating). The inset shows a two-dimensional section of the charge ordering pattern with the primary lattice distortion.

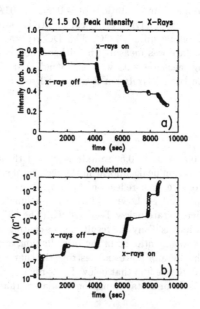

Figure 3: X-ray exposure dependence of the peak intensity of the (2, 1.5, 0) charge ordering superlattice reflection (a), and of the electrical conductance (b) at T=4K in $Pr_{0.7}Ca_{0.3}MnO_3$.

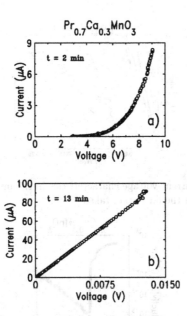

Figure 4: Current-voltage characteristics measured after different x-ray exposures at T=4K.

x-rays are turned off. The transition to the metallic phase is essentially complete after approximately 20 minutes of x-ray irradiation. After that, the sample remains in the metallic phase whether or not x-ray radiation is present.

While the metallic state generated after prolonged x-ray exposure exhibits conventional ohmic conductivity (Fig. 4b), Fig. 4a shows that the current-voltage characteristics after short exposure are remarkably nonlinear (intermediate regime). (We therefore quoted conductance rather than conductivity in Fig. 3.) In this regime, the current-voltage characteristics are often non-reproducible, showing irregular jumps, as shown in Fig. 5. This behavior reflects the inhomogeneity of the intermediate state and is due to formation and destruction of various current paths in the sample. The data of Fig. 5 also shows that application of voltage across the sample (or, equivalently, injection of current carriers into it) drives the system towards the metallic state. A possible explanation of this effect is that when the charge carriers are forcedly moved (delocalized) they revive the ferromagnetic interaction and convert a portion of the sample into the metallic state, producing a new current path. Recently, it was reported that application of the sufficiently large voltage to the $Pr_{0.7}Ca_{0.3}MnO_3$ sample at low temperature converts it into the metallic state [8], providing an argument in favor of this explanation. However, in this case the material stays metallic only as long as the voltage is applied, and reverts to the insulating state when the voltage is reduced to zero.

Due to the finite x-ray penetration depth in this compound, only a narrow surface region should undergo the transition to the metallic phase. We obtain $\rho \sim 5 \times 10^{-4}\Omega cm$ for the ohmic resistivity after prolonged x-ray exposure, using the calculated for our scattering geometry x-ray penetration depth of about $2\mu m$ as the depth of the conducting channel between the contacts. This is (to within the errors) identical to the resistivity measured

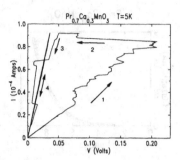

Figure 5: An example of current-voltage characteristics in the intermediate regime. Arrows show the sequence at which the data were taken.

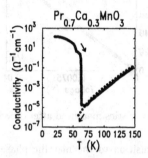

Figure 6: Conductivity measured on cooling before x-ray illumination (dotted curve) and on heating after illumination with x-rays for a moderate amount of time (solid curve). The conductivity on cooling was measured directly. On heating, the conductance data were converted to conductivity by considering a conducting channel produced by the x-ray irradiation described in the text.

without x-rays above the critical magnetic field [4], suggesting that the photoinduced and field-induced metallic phases are identical. Further strong evidence for this assertion comes from a comparison of the metastability boundaries of these two metallic states. Figure 6 shows that the x-ray induced conductivity is annealed out on heating above 60 K, which coincides with the annealing temperature T_{irr} of the magnetic-field-induced phase [4]. Although the magnetization was not measured directly, the close analogy to the magnetic-field-induced transition implies that the magnetic properties of the sample change dramatically with illumination, from canted antiferromagnetic to ferromagnetic. To our knowledge, this is the first example of a photoinduced antiferromagnet-to-ferromagnet transition.

We have studied the effects of x-ray illumination at various temperatures and magnetic fields inside and outside of the hysteretic region of the magnetic phase diagram. Our general conclusion is that in all samples (except for the x=0.5 sample, see below) the x-ray illumination induces the transition to the metallic phase *inside* the hysteretic region; outside of this region we do not observe any measurable changes in the structure or resistivity of the samples. The closer we move to the boundaries of the hysteretic (metastability) region, the harder is to induce the transition. This is reflected in the slower transition rates under the

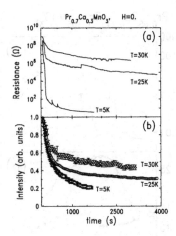

Figure 7: X-ray exposure dependence of the electrical resistance (a), and (2, 1.5, 0) charge ordering peak intensity at various temperatures in zero field in $Pr_{0.7}Ca_{0.3}MnO_3$.

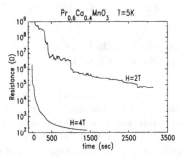

Figure 8: X-ray exposure dependence of the electrical resistance in the $Pr_{0.6}Ca_{0.4}MnO_3$ sample at T=5K in various magnetic fields.

same x-ray flux. This tendency is illustrated by the data of Figs. 7 and 8: the transition proceeds faster at lower temperatures and higher magnetic fields.

In the x=0.5 sample it was possible to induce the insulator-metal transition at low temperatures outside of the hysteretic region of the phase diagram reported in [4]. Namely, the transition is induced at T=5 K, H=8 Tesla. Thus, the boundaries of the metastable region for this sample should be reconsidered. When the magnetic field is turned off, the sample returns to the insulating state at H=0. This transition is not sharp and occurs between 2 and 4 Tesla. The discrepancy between our measurements and those of [4] is probably due to the large times required for the insulator-metal conversion (high energy barrier between the states) at low magnetic fields.

While x-ray irradiation outside of the metastability region does not result in any observable changes in the sample resistance or the CO peak intensity, it nevertheless significantly affects the sample properties. For example, the x=0.4 sample at T=5K undergoes the insulator-metal transition at H~7 Tesla when not subject to x-ray irradiation. But if this

71

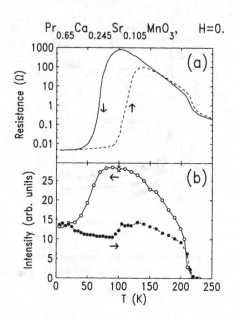

Figure 9: The temperature dependence of the sample resistance (a), and the (2, 1.5, 0) CO peak intensity (b) in $Pr_{0.65}Ca_{0.245}Sr_{0.105}MnO_3$ on heating and on cooling.

sample is irradiated at H=0, T=5K for approximately 10 minutes (and then the x-rays are switched off), it undergoes the transition between 5.0 and 5.5 Tesla. We also find that x-ray irradiation at low temperature and zero field results in faster transition rates when the sample is subsequently irradiated in the field (in the hysteretic region of the phase diagram). Similar effects were observed in the x=0.5 sample. Therefore, x-ray radiation induces changes in the sample also outside of the metastability region of the phase diagram, even though these changes do not reveal themselves in the resistivity or structural measurements.

The last sample we discuss here, $Pr_{0.65}Ca_{0.245}Sr_{0.105}MnO_3$, has its CO phase shifted to higher temperatures (Fig. 1), and the metallic phase is reentrant at low T, Ref. [5]. The temperature dependencies of the electrical resistance and the CO peak intensity for this sample are shown in Fig. 9. Note, that the charge ordering does not disappear in the low temperature metallic phase. Also, contrary to the resilts of Ref. [5], the CO peak intensity does not recover in the insulating phase on heating. This may be the result of the x-ray illumination effects, since the x-ray intensity in our experiment was much higher than that of the Ref. [5]. The nature of the coexisting metallic conductivity and charge ordering is intriguing and will be the subject of the further investigation. We have investigated the x-ray illumination effects in the hysteretic region at H=0, T=70K and T=100K. In both cases, we were able to induce the transition to the conducting phase. The x-ray irradiation effect on the charge ordering was in this case substantially smaller than in $Pr_{0.7}Ca_{0.3}MnO_3$: at T=70K the reduction of the CO peak intensity was about 20% after 5 hours of the x-ray exposure, and at T=100K the reduction was not observed (less than 5%) after 2 hours of irradiation (when the transition to the conducting phase is essentially complete).

In our previous paper [2] we have concluded that the photoinduced transition is caused by x-ray photoelectrons and secondary electrons generated in collision (such mechanisms as oxygen diffusion, beam heating, and other experimental artifacts were carefully eliminated). The scenario that we proposed was the following. The insulating state in these materials is believed to be associated with strong polaronic self-trapping of the carriers mediated by the Jahn-Teller distortion of the Mn^{3+} ion [9]. In the CO phase, these distortions order, giving rise to the CO diffraction peaks. In the conducting ferromagnetic phase the electrons are delocalized, and the lattice distortion is absent. When an electron gets photoexcited out of the polaronic self-trap, the lattice relaxes, which is unambiguously demonstrated by the data of Fig. 3. The electron does not get captured again since the distortionless conducting phase is (meta)stable. Any means of removal of the electron from the polaronic trap should induce the transition. It was recently shown, that forced carrier injection (application of high voltages) [8], and visible light together with carrier injection [10] also drive the system into the metallic state. However, in these cases the induced conductivity is not persistent. This was attributed to the collapse of the conducting paths due to the elastic strain exerted to them by the insulating matrix [10].

Inhomogeneity and lattice strain seem to play an important role in these materials. It was recently shown that two separate phases associated with the metallic and the insulating phases coexist in the x-ray irradiated material [11]. These phases possess different lattice constants, which results in the lattice strain. These observations, along with the results of [8, 10] and our current-voltage characteristics measurements, demonstrate the inhomogeneity of the photoinduced state in the investigated compounds. Therefore, x-ray irradiation of the sample most likely results in the creation of the FM metallic clusters, and when the percolation threshold is reached, the metallic conductivity is attained. The FM clusters might also be created outside of the hysteretic region (not reaching the concentration necessary for the metallic conductivity), facilitating the transition upon subsequent application of higher magnetic field. This could explain why the field-induced transition occurs at lower fields if the sample is irradiated outside of the metastability region. The authors of [11] have also reported that lattice strain develops below T_{CO} even above the irreversibility temperature T_{irr}. Based on this observation, they proposed that the FM clusters are present in the system at any $T < T_{CO}$, inducing the lattice strain and also being responsible for the ferromagnetic moment below T_{CA}. (Such clusters were observed in different perovskite materials well above Curie temperature [12].) However, recent scanning Hall probe measurements of the surface magnetization contradict this hypothesis [13]. In any case, a complex strained state is realized in these materials at low temperatures. X-ray irradiation creates clusters of the FM phase in the insulating AFM matrix, resulting in a unique example of microscopic phase separation in this system.

In addition to the lattice strain and possible phase separation, the CO state in these materials is more complex than previously reported [3, 11]. Our latest x-ray measurements show that the CO diffraction peaks are split into two components (the peaks are shifted from each other along the (H, H, 0) direction in the reciprocal space). Thus, two different ordering wavevectors are present in the system. One of the components is broader in the q-space than the other, and therefore we denote these peaks as "broad" and "sharp". Their intensities show strikingly different, hysteretic temperature dependencies, as shown in Fig. 10. The behavior of the broad component above T_{irr} is consistent with the neutron diffraction data of Fig. 2, and therefore it may reflect bulk behavior of the material. The sharp component, on the other hand, is almost certainly due to surface effects, unless the properties of the sample are strongly affected by the x-ray irradiation even at these high temperatures. These

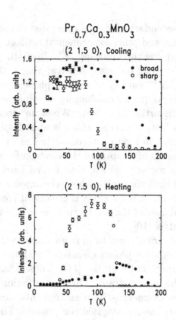

Figure 10: Temperature dependencies of the "broad" and "sharp" components observed in the vicinity of the (2, 1.5, 0) position on cooling and on heating in $Pr_{0.7}Ca_{0.3}MnO_3$.

two peaks can come from different orthorhombic twin domains present in the sample. The observed difference in the scattering angle δq (0.2° at the x-ray energy 8 keV) is roughly consistent with that calculated from the lattice parameters given in [11]. In this case, doubling of the unit cell in both a and b crystallographic directions is present near the surface of the sample. Neither of the two components can come from the suggested FM clusters: δq calculated from the lattice constants of the insulating and the metallic phases [11] at this wavevector (about 0.06° at T=20K) is much smaller than experimentally observed. The data of Fig. 10 is puzzling and demonstrate the complex nature of the lattice distortion near the surface in the CO phase. Surface effects are known to play an important role in manganites (see, for example, the discussion of low-field magnetoresistance in Ref. [1]); more experimental work is needed to understand them in detail. The fact that the surface properties of these materials are different from the bulk ones should be taken into account when investigating low-field tunneling magnetoresistance and other grain boundary related effects. Also, the x-ray data of Ref. [11], including the observation of the lattice strain below T_{CO}, may be affected by the surface effects. We should note that the difference between the bulk and surface properties (including different correlation lengths) was observed before in other materials [14], but remains largely unexplained. To separate the surface and bulk behavior unambiguously, high resolution neutron diffraction measurements are needed.

The observation of the photoinduced insulator-metal transition in CMR manganites opens the way for a variety of further experimental studies. Detailed structural and transport measurements should reveal the microscopic nature of the intriguing low-temperature state of these materials. The understanding of this phase will help to elucidate the physics underlying the complex properties of the manganites. From the point of view of applications, the unique properties of the manganites may prove useful in x-ray detection and lithography. Using x-ray lithography, it may be possible to pattern very small ferromagnetic structures into these materials, which would open up new possibilities for both fundamental and applied research on magnetism.

CONCLUSIONS

The main result reported in this work is the observation of the x-ray induced transition from the antiferromagnetic charge-ordered insulating into ferromagnetic metallic state. The transition can be induced in the hysteretic (metastable) regions of the magnetic phase diagram of the investigated compounds. Substantial changes in the lattice structure ("melting" of the charge lattice) are observed during the transition. The material can be annealed back into the insulating phase by heating above the irreversibility temperature, T_{irr}. The charge-ordered phase at low temperatures possesses a complex structure, which remains to be understood.

The study of this novel type of transition helps to elucidate the physics responsible for the complex phenomena exhibited by the manganites. First, it illustrates the importance of the electron-lattice coupling thought to be responsible for the transport properties of these materials. Second, it provides clues for understanding of the complex nature of their low-temperature state. Therefore, the discovered effect also provides a useful tool for investigation of the properties of the manganites. Many of these properties are still not understood completely, and new phenomena will certainly be discovered. We believe, that manganite materials will continue to provide a rich ground for both fundamental and applied research in the future.

ACKNOWLEDGEMENTS

The work at Princeton University was supported by NSF grants DMR-9303837 and DMR-9701991, and by the Packard and Sloan Foundations; the work at Brookhaven National Laboratory was supported by the US Department of Energy under contract DE-AC0276CH00016; this work was also supported in part by NEDO and the Ministry of Education, Japan.

References

[1] For a review, see A. P. Ramirez, J. Phys.: Condens. Matter **9**, p. 8171 (1997)

[2] V. Kiryukhin, D. Casa, J. P. Hill, B. Keimer, A. Vigliante, Y. Tomioka, and Y. Tokura, Nature **386**, p. 813 (1997)

[3] Z. Jirak, S. Krupicka, Z. Simsa, M. Dlouha, and S. Vratislav, J. Magn. Magn. Mat. **53**, p. 153 (1985)

[4] Y. Tomioka, A. Asamitsu, H. Kuwahara, Y. Moritomo, and Y. Tokura, Phys, Rev. B **53**, p. R1689 (1996)

[5] Y. Tomioka, A. Asamitsu, H. Kuwahara, and Y. Tokura, J. Phys. Soc. Japan **66**, p. 302 (1997)

[6] C. Zener, Phys. Rev. **82**, p. 403 (1951); P. W. Anderson and H. Hasegava, Phys. Rev. **100**, p. 675 (1955)

[7] H. Yoshizawa, H. Kawano, Y. Tomioka, and Y. Tokura, Phys. Rev. B **52**, p. R13145

[8] A. Asamitsu, Y. Tomioka, H. Kuwahara, and Y. Tokura, Nature **388**, p. 50 (1997)

[9] A. J. Millis, P. M. Littlewood, and B. I. Shraiman, Phys. Rev. Lett. **74**, p. 5144 (1995)

[10] K. Miyano, T. Tanaka, Y. Tomioka, and Y. Tokura, Phys. Rev. Lett. **78**, p. 4257 (1997)

[11] D. E. Cox, P. G. Radaelli, M. Marezio, S-W. Cheong, Phys. Rev. B **57**, to be published

[12] J. W. Lynn, R. W. Erwin, J. A. Borchers, Q. Huang, A. Santoro, J-L. Peng, and Z. Y. Li, Phys. Rev. Lett. **76**, p. 4046 (1996); J. M. De Teresa, M. R. Ibarra, P. A. Algarabel, C. Ritter, C. Marquina, J. Blasco, J. García, A. del Moral, Z. Arnold, Nature **386**, p. 256 (1997)

[13] D. Casa, K. Moler, B. Keimer, Y. Tomioka, and Y. Tokura, unpublished

[14] S. R. Andrews, J. Phys. C **11**, 3721 (1986); T. R. Thurston, *et al.*, Phys. Rev. Lett. **70**, 3151 (1993); Q. J. Harris, *et al.*, Phys. Rev. B **52**, 15420 (1995); G. M. Watson, *et al.*, Phys. Rev. B **53**, 686 (1996)

RESONANT X-RAY FLUORESCENCE SPECTROSCOPY AT THE V L-EDGES OF VANADIUM OXIDES

L.-C. DUDA, C. B. STAGARESCU†, J. E. DOWNES, K. E. SMITH, and G. DRÄGER[2]
[1]Department of Physics, Boston University, Boston, MA 02215
[2]Fachbereich Physik der Martin-Luther-Universität Halle-Wittenberg, D-06108 Halle, Germany

ABSTRACT

We have studied resonant V L_α-fluorescence spectra of vanadium oxides with V in several different oxidation states. The spectra are dominated by the O $2p$-contribution centered at about 6 eV below the top of the valence band (VB-top). The V $3d$-contribution, found close to the VB-top, increases with decreasing valency of the vanadium atoms. Resonant inelastic (Raman) x-ray scattering is fairly weak in these compounds and overlaps with the ordinary fluorescence spectrum. Large spectral changes of V L_α-fluorescence in the metal-insulator transition of V_2O_3 have been observed.

INTRODUCTION

Vanadium oxides exist in several different valency states and display an interesting variety of magnetic and metal-insulator-transition (MIT) phenomena. Although many of the V-oxides have partially filled V $3d$-bands they all exhibit insulating phases at some temperature. Therefore electron correlations or, as suspected in the case of VO_2, crystal-distorting electron-phonon interactions must be present.

Numerous experimental studies of the valence and conduction band of vanadium oxides exist [1-7] in literature and band structure calculations [8] have had various degrees of success trying to explain the experimental results. X-Ray and ultraviolet photoemission spectroscopies have been used [1,2,4] to map out the valence band contributions of the O $2p$- and V $3d$-states close to the top of the valence band whereas x-ray absorption spectroscopy [3] reflects the conduction band.

Soft x-ray emission spectroscopy

Soft x-ray emission spectroscopy (SXES) is a powerful probe of the valence band but so far there exist only few reports concerning V-oxides [4,5,6,7]. Recently, SXES has gained renewed interest since high-brightness synchrotron radiation has become available for energy-selective excitation. Moreover, inelastic or Raman x-ray scattering - present in threshold excited, i.e *resonant*, soft x-ray emission spectra has been shown [9] to yield complementary information. While ordinary x-ray fluorescence reflects the partial density of states (DOS) (which in our case is related to the V $3d$-band) resonant inelastic x-ray scattering (RIXS) can be viewed as an energy-loss that occurs due to low-energy dd-excitations (Fig. 1). In contrast to optical absorption spectroscopy these excitations are fully allowed in RIXS. Significantly, RIXS as a spectroscopy is not core-hole lifetime resolution limited, as are ordinary x-ray fluorescence or x-ray absorption spectroscopy. Therefore RIXS is, in principal, capable of resolving subtle changes in the electronic structure, that occur in certain metal-insulator transitions.

Figure 1 shows a schematic picture of the underlying processes. Most importantly, while in ordinary fluorescence (Fig. 1A) there is a relaxation step (Fig. 1A-3) that allows the electronic

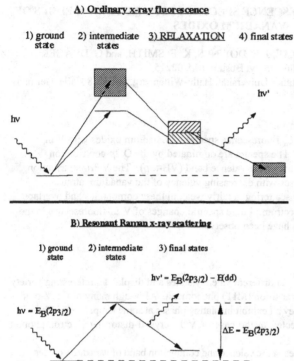

A) Ordinary x-ray fluorescence

1) ground state 2) intermediate states 3) RELAXATION 4) final states

hv

hv'

B) Resonant Raman x-ray scattering

1) ground state 2) intermediate states 3) final states

$hv = E_B(2p_{3/2})$

$hv' = E_B(2p_{3/2}) - E(dd)$

$\Delta E = E_B(2p_{3/2})$

Figure 1 Schematic of the ordinary x-ray fluorescence process (A) and resonant Raman x-ray scattering (B).

system to dissipate energy there is no such step in the scattering process. Therefore the inelastic contributions will be found to follow the elastic peak in the spectrum. Moreover, the core hole exists only in a virtual state in the scattering and hence does not contribute to life time broadening of the spectrum.

The present study was undertaken, partly, in order to assess the feasibility of studying RIXS in V-oxide compounds. We studied resonantly excited L-fluorescence spectra of V_2O_5, V_6O_{13}, VO_2, V_2O_3, VO, and the superconductor V_3Si (in order of decreasing valency). We find that the dominating part of the spectra of all oxides lie about 6 eV below the top of the valence band (VB-top). This is due to strong hybridization of the V $3d$- and O $2p$-states. Pure V $3d$-states are found at the VB-top as a pronounced shoulder. This part of the spectrum increases with decreasing valency of the V-atoms. RIXS overlaps with ordinary fluorescence close to the VB-top and is fairly weak, making it difficult to distinguish both contributions. Large spectral changes have been found in the MIT of V_2O_3.

EXPERIMENT

The vanadium oxide samples were small (approximately 2 x 2 mm²) single crystals, except for VO which was a powder. The samples were inserted into the vacuum as grown. Soft x-ray absorption (SXA) and soft x-ray emission (SXE) measurements were performed at the undulator beam line X1B at the National Synchrotron Light Source, Brookhaven National Laboratories. Absorption spectra were recorded in the total electron yield mode by measuring the sample drain current. Emission spectra were recorded using a Rowland-mount grazing-incidence grating spectrometer [10] using a 5m, 1200 lines/mm grating in first order of diffraction at a resolution of about 0.8 eV. The acquisition time for individual SXE spectra was several hours. The base pressure in the experimental system was 1.0 x 10⁻⁹ Torr. This vacuum is quite adequate since both SXA and SXE are primarily bulk probes, and surface phenomena were not under investigation.

RESULTS

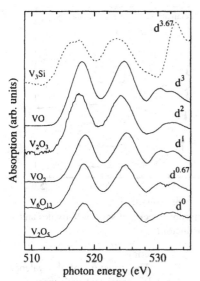

Figure 2 V L-absorption spectra of several V-oxides (solid lines) and a metallic superconductor (dashed line). The nominal number of d-electrons is shown on the right side.

Figure 2 displays, in order of increasing valency (top to bottom), the V L-absorption spectra of all the compounds under investigation. Note that the monochromator energy band pass used for V_2O_3 and VO was somewhat larger (1 eV) than for the other compounds (0.8 eV). Nevertheless, the V L-spectra of the oxides are all similar in shape and energy position. The spin-orbit doublet has an apparent separation of approximately 7 eV in all compounds. The band gap of the compounds increases from top to bottom but we observe no significant systematic changes. However, note that V_3Si is metallic and is significantly different in shape. We ascribe the lack of pre-edge structure for the V L-absorption of the oxides, found in other publications, to the low spectral resolution used in our experiment.

For the O K-edge, on the other hand, we observe much stronger changes with oxidization state. The O K-edge is found at energies just above the V L_2-edge. The double-peak structure at 530 eV - 533 eV has been shown to come from the (unoccupied) π^*- and σ^*-bands derived from a molecular orbital picture. As the d-band is successively filled (bottom to top) we observe a decrease in intensity of the σ^*-band. This is consistent with the fact that the σ^*-orbitals have a stronger effect on the bonding properties than the non-bonding π^*-orbitals.

The absorption spectra are essential to enable us to choose a proper excitation energy for obtaining resonant x-ray fluorescence spectra. High-energy excitation or unselective excitation, such as high-energy electron bombardment, is known to give a fluorescence spectrum in which L_α- and L_β-emission bands overlap. Threshold excitation with monochromatic synchrotron radiation, however, has a twofold effect. Firstly, we may stay below the excitation threshold for L_β-emission and eliminate overlapping contributions, and secondly, we limit the number of intermediate states thus ``purifying'' the spectrum.

Figure 3 shows a series of V L_α-spectra for all the compounds excited at an energy of 515 eV with a band pass of about 2.5 eV. The spectra are normalized to have equal maxima. The excitation energy was chosen to lie below the threshold of exciting V $3p_{1/2}$- and O 1s-electrons. However, we find intensity also at energies around 523 eV which we attribute to O K-emission excited by higher-order radiation from the monochromator.

The dominating feature in the spectra is found at about 506eV - 507 eV (arrows in Fig. 3). It is due to V 3d-states that are heavily hybridized with O 2p-states at about 6 eV below the VB-top. The position of the peak maxima and thus the O 2p-states shift to lower energies as the band is gradually filled.

The VB-top is situated below the left dashed line that defines the elastic peak region in Fig. 3. The intensity between the high-energy flank of the dominating peak and the VB-top arises from pure, unhybridized V 3d-states and from inelastic scattering contributions. This is emphasized in the top panel of Fig. 4 as shaded area. The bottom panel of Figure 4 re-displays this portion as the difference between the dots representing the raw spectrum and the solid line. This illustrates how the intensity of this portion of the spectrum increases with increased filling of the V 3d-band. The high-energy peaks are due to elastic scattering while the low-energy peaks reflects the position of the maximum in the V 3d density of states (DOS) overlapped with inelastic

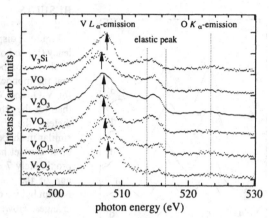

Figure 3 V L-emission spectra of several V-oxides and a metallic superconductor (top).

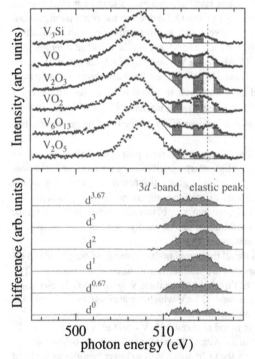

Figure 4 Top panel: shaded area shows the VB-top of each x-ray emission spectrum. Bottom panel: extracted VB-top. Dashed lines indicate approximate position of elastic peak and unhybridized 3d-band.

scattering.

The structure of the valence band is consistent with theoretical DOS calculations and x-ray photoemission results [2]. A detailed comparison is impaired, however, by low statistical accuracy and low resolution of our present data. We emphasize that the cross section for dd-excitations is comparable or lower than ordinary fluorescence from these compounds. A way to increase resolution and to quench the ordinary fluorescence channel might lie in excitation at the V $3p$-edge, explained in the following. The binding energy of $3p$-electrons is some ten times lower which in turn allows roughly a tenfold increase of absolute energy resolution. Moreover, ordinary M-fluorescence is considerably weaker than L-fluorescence for $3d$-transition metals which would allow one to study dd-excitations at the M-edge without being obscured by ordinary fluorescence.

We also applied resonant V L-fluorescence spectroscopy to study the metal-insulator transition of V_2O_3. Two spectra were taken in immediate succession: one spectrum with the sample at room temperature, and another spectrum with the sample cooled by liquid-nitrogen (Fig. 5). The sample was glued with conductive silver epoxy to a Ta-foil which in turn was tightened to the Cu cooling-rod. A thin sheet of sapphire was placed between the Ta-foil and the Cu-rod in order to avoid electrical grounding for sample-current absorption measurements. The temperature was monitored with a thermocouple device attached to the Ta-foil sample holder, a few centimeters away from the sample. The MIT of V_2O_3 occurs at T=168 K, and low-temperature SXE measurements commenced as the temperature stabilized at about 155 K which ensured that the sample temperature was well below T_C.

The spectra in Fig. 5 were normalized to equal height of the elastic peak (arrow). Clearly, we observe that intensity in the main peak is shifted to higher binding energies upon cooling below T_C. This is emphasized by the differently hatched areas. Vertically hatched areas reflect intensity gains, while horizontally hatched areas reflect intensity losses over the room temperature spectrum. Moreover, the VB-top exhibits an intensity gain for the insulating phase. Partly, this could be due to a relative increase of inelastic scattering compared to ordinary fluorescence, induced by the opening of a band gap and hence allowing less relaxation. The trends in the spectral changes are similar to what has been observed in UPS on (Cr-doped) V_2O_3 [2]. Note that there appears to be differences in the O K-emission but this is subject to an ongoing study to be

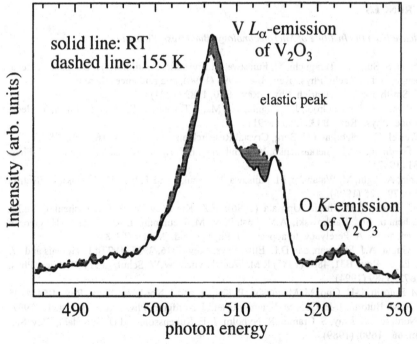

Figure 5 Resonant V L_α-emission of V_2O_3 at room temperature (solid line) and below its metal-insulator transition (dashed line). The peak centered around 523 eV is O K-emission excited by higher order monochromator-radiation.

presented elsewhere. The band gap itself is too small to be evident at the resolution used here but, as pointed out above, inelastic scattering at the $3p$-edges might be able to reveal more subtle changes.

CONCLUSIONS

Resonant fluorescence spectroscopy at the V L-edges of vanadium oxides has been shown to give important information about the valence band structure. It was possible for the first time to monitor the metal-insulator transition of V_2O_3 by resonant soft x-ray fluorescence spectroscopy. Future improvements concern increased resolution of the secondary monochromator and by further limiting the band pass of the excitation radiation. Resonant inelastic scattering at the $3p$-edges is proposed to reach this goal.

ACKNOWLEDGMENTS

This work was supported in part by the National Science Foundation under DMR-9504948 and DMR-9501174. L.-C.D. gratefully acknowledges scholarship support from the Swedish Natural Research Council (NFR). On of us (G.D.) would like to thank Prof. S. Horn (Institut für Physik der Univ. Augsburg) for kindly lending us the samples.

REFERENCES

†: On leave from the Institute of Microtechnology, Bucharest, Romania.

1. S. Shin, S. Suga, M. Taniguchi, M. Fujisawa, H. Kanzaki, A. Fujimori, H. Daiman, Y. Ueda, K. Kosuge, and S. Kachi, Phys. Rev. **B41**, 4993 (1990), and references therein
2. K.E. Smith and V.E. Henrich, Phys. Rev. B50, 1382 (1994)
3. M. Abbate, F.M.F. de Groot, J.C. Fuggle, Y.J. Ma, C.T. Chen, F. Sette, A. Fujimori, Y. Ueda, K. Kosuge, Phys. Rev. **B43**, 7263 (1991)
4. F. Werfel, G. Dräger, and U. Berg, Crystal Research and Technology **16**, 119 (1981)
5. S.P. Friedman, V.M. Cherkashenko, V.A. Gubanov, E.Z. Kurmaev, and V.L. Volkov, Z. Phys. **B46**, 31 (1982)
6. S. Shin, A. Agui, M. Watanabe, M. Fujisawa, Y. Tezuka, and T. Ishii H. Kanzaki, J. Electron Spectros. **79**, 125 (1996)
7. L.-C. Duda, PhD Thesis, Uppsala (1996); E.Z. Kurmaev, V.M. Cherkashenko, Yu.M. Yarmashenko, St. Bartkowski, A.V. Postnikov, M. Neumann, L.-C. Duda, J.-H. Guo, J. Nordgren, and V.A. Perelyaev, to appear in J. Phys.: Cond. Matter (1998)
8. . M. Gupta, A.J. Freeman, and D.E. Ellis, Phys. Rev. **B16**, 3338 (1977); F. Gervais and W. Kress, Phys. Rev. **B31**, 4809 (1985) R.M. Wentzkovitch, W.W. Schulz, and P.B. Allen, Phys. Rev. **B72**, 3389 (1994)
9. S. M. Butorin, J.-H. Guo, M. Magnuson, P. Kuiper, and J. Nordgren, Phys. Rev. **B54**, 4405 (1996);S. M. Butorin, J.-H. Guo, M. Magnuson, and J. Nordgren, Phys. Rev. **B55**, 4242 (1997)
10. J. Nordgren, G. Bray, S. Cramm, R. Nyholm, J.-E. Rubensson, and D. Wassdahl, Rev. Sci. Instrum. **66**, 1690, (1989)

PHASE DIAGRAM AND ANISOTROPIC TRANSPORT PROPERTIES OF $Nd_{1-x}Sr_xMnO_3$ CRYSTALS

H. KUWAHARA* (kuwahara@tsukuba.rd.sanyo.co.jp), T. OKUDA*, Y. TOMIOKA*, T. KIMURA*, A. ASAMITSU*, and Y. TOKURA*,**
*Joint Research Center for Atom Technology (JRCAT), 1-1-4 Higashi, Tsukuba, Ibaraki 305, Japan
**University of Tokyo, 7-3-1 Hongo, Bunkyo-ku, Tokyo 113, Japan

ABSTRACT

We have investigated electronic transport and magnetic properties of perovskite-type $Nd_{1-x}Sr_xMnO_3$ crystals with change of controlled hole-doping level ($0.30 \leq x \leq 0.80$). The electronic phase diagram of $Nd_{1-x}Sr_xMnO_3$ was obtained by systematic measurements of magnetization (magnetic structure), resistivity, and lattice parameter. We have also studied the anisotropic transport properties of $x=0.50$ and 0.55 crystals with different magnetic structures: CE-type antiferromagnetic (AF) structure for $x=0.50$ and A-type layered AF one for $x=0.55$. In the case of the $x=0.55$ crystal, the metallic behavior was observed within the ferromagnetic (F) layers, while along the AF-coupling direction the crystal remains insulating over the whole temperature region. The observed large anisotropy is due to the magnetic as well as orbital-ordering induced confinement of the spin-polarized carriers within the F sheets. The nearly isotropic transport behavior has been confirmed for the CE-type AF charge-ordered state in the $x=0.50$ crystal.

INTRODUCTION

The system investigated here, $Nd_{1-x}Sr_xMnO_3$, is derived by decreasing the one-electron bandwidth (W) from the canonical double-exchange system, $La_{1-x}Sr_xMnO_3$ with the maximal W. With decrease of W, instabilities competing with the ferromagnetic (F) double-exchange (DE) interaction become to be pronounced. Such instabilities, i.e., the antiferromagnetic (AF) superexchange, Jahn-Teller, orbital-ordering, and charge-ordering (CO) interactions, complicate the phase diagram as compared with the $La_{1-x}Sr_xMnO_3$ system. The present $Nd_{1-x}Sr_xMnO_3$ system, for which high quality crystals with wide hole-doping levels can be easily obtained, is suitable to investigate such instabilities. In particular, only few studies have so far been made at the transport properties of the overdoped manganites ($x>0.50$). In this paper, we have investigated the transport and magnetic properties of Nd-based crystals over a wide hole-doping region ($0.30 \leq x \leq 0.80$). We report here the electronic and magnetic as well as crystal-structural phase diagram for $Nd_{1-x}Sr_xMnO_3$ system as a function of hole-doping level x.

In this study, we have found an anomalous metallic state in the overdoped region ($x>0.52$) for $Nd_{1-x}Sr_xMnO_3$ crystals. The neutron scattering study on the $Nd_{1-x}Sr_xMnO_3$ system has revealed that its magnetic structure state is the layered (A-type) AF [1]. The simple DE mechanism does not account for the AF metallic states. Kawano et al. suggested a possible anisotropy of transport reflecting the anisotropic magnetic structure in the layered AF state [1], i.e., ferromagnetically coupled layers are expected to be metallic, while the AF direction to be insulating. From this viewpoint, we have investigated the anisotropic transport properties of the A-type AF phase for $Nd_{1-x}Sr_xMnO_3$ ($x=0.55$) crystal at various temperatures down to ~30mK, as compared with the CE-type AF one for $x=0.50$ crystal.

EXPERIMENT

Crystals of $Nd_{1-x}Sr_xMnO_3$ ($0.30 \leq x \leq 0.80$) were grown by the floating zone method with use of a lamp-image furnace, details are published in Ref. [2]. Inductively coupled plasma spectrometry (ICP) on the respective grown crystals indicated that the stoichiometry is nearly identical to the prescribed ratio with accuracy of x within ±0.01. Powder X-ray diffraction (XRD) apparatus equipped with temperature-controllable cryostat was used to check the crystal quality and to determine the lattice parameters. Rietveld refinement of XRD pattern for the pulverized crystal indicated that all peaks are indexed without impurity phase. In order to investigate the anisotropic transport properties, we have carefully cut out the samples along the pseudo-cubic principal directions from the melt-grown ingot, which is based on the results of the X-ray back Laue measurements: The cutting accuracy is about 0.5 degrees. Resistivity measurements were performed using the conventional four-probe method. Resistivity in the dilution-temperature region was measured by using the AC resistance bridge. Magnetization was measured at a field of 0.5 T after cooling down to 5 K in zero field using a SQUID magnetometer.

RESULTS AND DISCUSSION

Phase Diagram

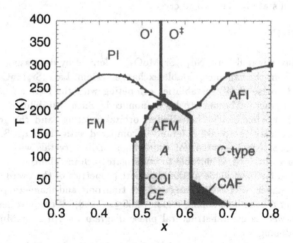

FIG 1: Electronic phase diagram of $Nd_{1-x}Sr_xMnO_3$ crystal. The abbreviations mean paramagnetic insulator (PI), ferromagnetic metal (FM), CE-type antiferromagnetic charge-ordered insulator (COI), A-type antiferromagnetic metal (AFM), C-type antiferromagnetic insulator (AFI), and canted antiferromagnetic insulator (CAF).

First of all, we show in Fig. 1 the electronic/magnetic and lattice phase diagram of $Nd_{1-x}Sr_xMnO_3$ crystal. As one can immediately notice from the figure that the ferromagnetic (F) metallic state due to the DE interaction dominates the low-doped region. With increase of hole-doping level x (decrease of e_g electron density), ground states are changed to the antiferromagnetic (AF) one due to the superexchange interaction of t_{2g} electrons. In particular, the electronic and magnetic properties critically vary depending on x near the commensurate value of x=0.50, in which the charge- and orbital-ordering transition accompanies the concomitant CE-type AF ordering occurs below T_{CO}=155 K.

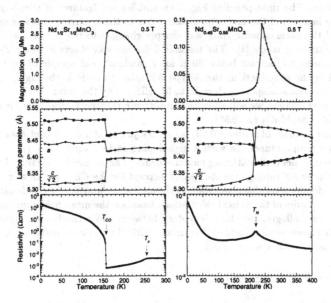

FIG 2: Temperature dependence of magnetization (top), lattice parameter (middle), and resistivity (bottom) in the crystals of $Nd_{1-x}Sr_xMnO_3$: x=0.50 (left, cited from Ref. 3) and x=0.55 (right) samples, which are randomly cut (the current direction is not specified). Lattice parameters are represented in the *Pbnm* settings. The abbreviations mean charge-ordering (T_{CO}), Curie (T_c), and Néel (T_N) temperatures.

We show in the left panels of Fig. 2 the CO transition of the x=0.50 crystal [3]. Upon the CO phase transition, the resistivity jumps by more than two orders of magnitude from a typically metallic value and the F magnetization disappears, indicating the simultaneous F-to-AF transition. In accord with these changes in electric and magnetic properties, the lattice parameters of the orthorhombically distorted perovskite show a distinct change as shown in the middle panel of Fig. 2. The CO transition is of the first order in nature and accompanies hystereses. The neutron scattering study [1] confirmed this F metal to AF insulator transition to be the phase change to the charge/orbital-ordered state, in which the nominally Mn^{3+} and Mn^{4+} species show a real space ordering (checkerboard-like) on the (001) plane of orthorhombic lattice (*Pbnm* notation). The $3x^2-r^2/3y^2-r^2$ type orbitals show superlattice along b-axis on the same plane, as evidenced by neutron [1], X-ray [4], and electron diffraction measurements [5]. Reflecting such an orbital ordering, the spin ordering shows a complicated sublattice structure as illustrated in the left panel of Fig. 3: The spin structure in charge-ordered state below T_{CO} is the $4 \times 4 \times 2$ unit cell in the pseudo-cubic perovskite setting and the stripes of the homovalent Mn ions are along the c-axis with collinear AF coupling of spins.

Such a charge-ordered state completely disappears and the F metallic state is stabilized when x is decreased below ~0.48. As x is increased beyond $x = 0.52$, on the other hand, the A-type layered AF structure (like LaMnO$_3$, shown in the right panel of Fig. 3) shows up. In the overdoped region ($0.52 < x < 0.63$), the metallic conduction is observed for the A-type layered AF state. The right panel of Fig. 2 exemplifies the features of the A-type layered AF state. The increase of the magnetic moment due to the F correlations was quenched at $T_N=220$K and the paramagnetic state was sharply changed to the AF one, as revealed by the neutron scattering study [1]. The metallic behavior was observed below T_N although the resistivity turns to increase below 80 K for a randomly cut sample of $x=0.55$ crystal. As mentioned in the introduction, the A-type AF spin structure is the highly anisotropic, therefore, the metallic transport is thought to be reflected by the nature of the conducting F plane. In the next section, we will show the anisotropic transport properties for the A-type AF state of Nd$_{1-x}$Sr$_x$MnO$_3$ ($x=0.55$).

With further increase of x above 0.63, resistivity jumps at T_N and continues to increase with decreasing temperatures. In such a high-doping region, the C-type (stripe-type) AF order is observed by neutron scattering measurements. The magnetic structure changes from the F to the C-type AF through the A-type AF (except for the CE-type AF at $x=0.50$) with increase of the doping level. Such variation of the electronic/magnetic ground state is consistent with recent results of theoretical calculations based on the mean-field approximation [6]. There exists a crystallographic phase boundary between two orthorhombic structures: the one is the well known orthorhombic O′ structure with the lattice constants $c/\sqrt{2} < b < a$, and the other is O‡ with $a \sim b < c/\sqrt{2}$.

Anisotropic Transport Properties

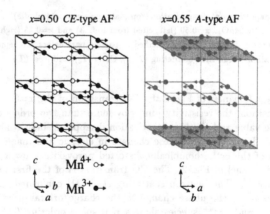

FIG 3: Schematic diagrams of the charge-ordered CE-type antiferromagnetic structure for Nd$_{1-x}$Sr$_x$MnO$_3$ $x=0.50$ crystal (left, Ref. 7) and the layered (A-type) antiferromagnetic one for $x=0.55$ (right, Ref. 1).

We show in Fig. 4 anisotropy of transport for $x=0.50$ (left) and $x=0.55$ (right) crystals. In the case of $x=0.55$ crystal showing the layered AF state, the highly anisotropic electrical transport was observed, corresponding to the anisotropic magnetic structure while much less for the $x=0.50$ case. The metallic behavior was observed within the F layers (ab plane), while along the AF-coupling direction (c) the crystal remains insulating over the whole temperature region. The anisotropy ratio of resistivity of the AF to F direction, ρ_c/ρ_{ab}, is $\sim 10^4$ at low temperatures. This value is merely a lower bound for the anisotopy, since the multi-domain structures cannot thoroughly be eliminated in crystal. The observed large anisotropy in spite of the nearly cubic lattice structure is due to the confinement of the spin-polarized carriers within the F sheets. No spin-canting along the AF direction indicates that carriers are confined within the F plane and DE mechanism along the c-axis is quenched.

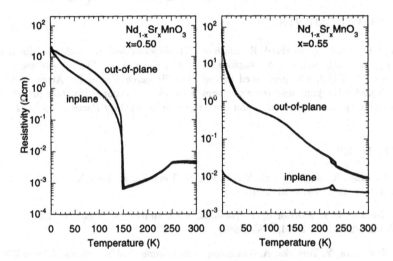

FIG 4: Temperature dependence of resistivity for $Nd_{1-x}Sr_xMnO_3$ crystals: $x=0.50$ (left) and $x=0.55$ (right). Anisotropic transport properties were measured along the c direction (labeled as out-of-plane) and in the ab plane (inplane). See also Fig. 3.

In accord with these magnetic and electronic transitions, a crystallographic structural change occurs: The ab plane (F plane) expands and the c-axis (AF direction) shrinks below T_N (right middle panel of Fig. 2). These changes in lattice parameters imply that the $x^2 - y^2$ type orbitals lie in the F plane and the transfer interaction along the AF direction almost vanishes. Namely, the $x^2 - y^2$ orbitals are stabilized to maximize the transfer interaction or the kinetic exchange of carriers within the ab (F) plane. In addition, no sign of super-lattice is observed below T_N within the F plane by electron diffraction measurement [5], which is also consistent with the above mentioned orbital-ordering accompanying no CO. A similar anisotropic behavior is observed in the A-type AF phase of the analogous compound $Pr_{0.45}Sr_{0.55}MnO_3$ ($T_N=220$ K) crystal: the AF propagation vector is $[\frac{1}{2} \frac{1}{2} 0]$ direction, which is different from the case of $Nd_{0.45}Sr_{0.55}MnO_3$.

CONCLUSIONS

We have obtained the electronic phase diagram for $Nd_{1-z}Sr_zMnO_3$ crystal by systematic measurements. In particular, we have revealed anisotropic transport properties in pseudo-cubic perovskite manganites with the layered AF spin structure. The ground state of this system is found to be a diffuse metal with an extremely large anisotropy. The observed huge anisotropy implies the magnetic confinement of the spin-polarized carriers within the F plane. The change in lattice parameters at T_N suggests that the $x^2 - y^2$ type orbitals order in the F plane and the transfer interaction along the AF direction is quenched. By analogy of giant magnetoresistance (GMR) effects for the AF coupled metallic superlattices, this layered AF state of perovskite-type manganite might be an another candidate for a spin-valve-type GMR material because of its analogy to the AF-coupling spin structure.

ACKNOWLEDGMENTS

The authors would like to thank R. Kajimoto, H. Kawano, and H. Yoshizawa for neutron scattering measurements, and N. Nagaosa for helpful discussions. The present work, partly supported by NEDO, was performed in the Joint Research Center for Atom Technology (JRCAT) under the joint research agreement between the National Institute for Advanced Interdisciplinary Research (NAIR) and the Angstrom Technology Partnership (ATP).

REFERENCES

1. H. Kawano, R. Kajimoto, H. Yoshizawa, Y. Tomioka, H. Kuwahara, and Y. Tokura, Phys. Rev. Lett. **78**, 4253 (1997).

2. H. Kuwahara, Y. Moritomo, Y. Tomioka, A. Asamitsu, M. Kasai, R. Kumai, and Y. Tokura, Phys. Rev. B **56**, 9386 (1997).

3. H. Kuwahara, Y. Tomioka, A. Asamitsu, Y. Moritomo, and Y. Tokura, Science **270**, 961 (1995).

4. S. Shimomura et al., unpublished.

5. J. Q. Li, Y. Matsui, H. Kuwahara, and Y. Tokura, unpublished.

6. T. Mizokawa and A. Fujimori, Phys. Rev. B **56**, R493 (1997); W. Koshibae Y. Kawamura, S. Ishihara, S. Okamoto, J. Inoue, and S. Maekawa, J. Phys. Soc. Jpn. **66**, 957 (1997); R. Maezono, S. Ishihara, and N. Nagaosa, preprint.

7. E. O. Wollan and W. C. Koehler, Phys. Rev. **100**, 545 (1955).

STOICHIOMETRY AND MAGNETIC PROPERTIES OF IRON OXIDE FILMS

D.V.DIMITROV*, G.C.HADJIPANAYIS *,V.PAPAEFTHYMIOU ** ,
A.SIMOPOULOS ***

*Department of Physics and Astronomy, University of Delaware, Newark, DE 19716

**Department of Physics, University of Ioannina, Ioannina, Greece

***NCSR Demokritos, 15310 Aghia Paraskevi, Attikis, Athens, Greece

ABSTRACT

The stoichiometry, structural and magnetic properties, and Mössbauer spectra of reactively sputtered Fe-O films were studied as a function of the O_2 partial pressure during the deposition. By increasing the amount of O_2 films with the following crystallographic structures and stoichiometry were fabricated; amorphous Fe-O, mixture of Fe and FeO, offstoichiometric single-phase Fe_xO, mixture of FeO and Fe_3O_4, single-phase Fe_3O_4, mixture of Fe_3O_4 and γ-Fe_2O_3, mixture of Fe_3O_4 and α-Fe_2O_3, and single-phase α-Fe_2O_3. The Verwey transition in Fe_3O_4 films was observed in the coercivity versus temperature curve and in the thermomagnetic data. FeO and α-Fe_2O_3 films showed anomalous ferromagnetic-like behavior, in contrast to their antiferromagnetic nature in bulk. The unusual magnetic properties were attributed to the formation of clusters of tetrahedraly coordinated Fe^{3+} ions in FeO and to uncompensated surface spins. in α - Fe_2O_3 films.

INTRODUCTION

Extensive studies on the magnetic [1], microstructural [2] and Mössbauer properties [3-4] of iron oxide small particles and thin films have been done for the last decades, because of their vast use in magnetic recording. Other technological applications of iron oxides include heterogeneous catalysis [5[, corrosion [6] and redox reactions in environmental science and waste remediation [7]. Recently it was found, that the surface chemistry (in particular thin oxide layer) changes substantially the magnetic properties of small Fe particles [8]. In contrast to the large interest in small Fe particles, there were few studies on exchange coupled Fe/Fe-O bilayers [9-10]. The second system has the advantage to be geometrically simpler, which makes it easier to be theoretically modeled and compared to experimental studies. A necessary prerequisite for a systematic study of Fe/Fe-O bilayer system is the ability to deposit thin Fe-O films with controlled thickness and stoichiometry as well as a good understanding of their magnetic properties. The passivation process used to create a surface oxide layer on small Fe particles creates a mixture of iron oxides. Another approach to fabricate iron oxides is to use reactive sputtering. In this process, during the deposition some amount of O_2 or water vapor is added to the inert gas, usually Ar, leading to the oxidation of the film during its growth.

The goal of this work was to study the dependence of the stoichiometry and crystallographic structure, magnetic properties and Mössbauer spectra on the O_2 flow during the deposition process.

Mat. Res. Soc. Symp. Proc. Vol. 494 © 1998 Materials Research Society

EXPERIMENT

Fe-O films were prepared by reactive dc magnetron sputtering on water cooled substrates in a mixture of Ar and O_2 gases. The base pressure of the system was $2x10^{-7}$ Torr. The total pressure during the deposition was 5 mTorr. The O_2/Ar ratio was varied in the range between 0.4 and 4 %, by changing the flow of O_2 and keeping the flow of Ar constant. The sputtering power was 21 watts which led to deposition rates between 1.6 and 1.9 Å/s. The thickness of the films was kept about 3000 Å. Kapton and microscopic glass slides were used as substrates. The stoichiometry and crystallographic structure were determined using X-ray diffraction (XRD) and Mössbauer spectroscopy. Rutherford backscattering spectroscopy (RBS) was used to measure the thickness of the films. The magnetic properties in the temperature range between 10 and 300 K were measured using SQUID magnetometer with applied fields up to 55 kOe.

RESULTS AND DISCUSSION

By varying the O_2/Ar ratio iron oxides with different stoichiometry and their mixtures were obtained. Table I summarizes the dependence of the crystallographic structure and stoichiometry as a function of the O_2 flow.

Table I: Stoichiometry and crystallographic structure of Fe-O films as a function of the O_2 flow during deposition

O_2 flow (cc/min)	Stoichiometry
0.4-0.5	Amourphous
0.6	α-Fe and FeO
0.7	Single phase Fe_xO
0.8	FeO and Fe_3O_4
0.9	Single phase Fe_3O_4
1.0 – 1.6	Fe_3O_4 and γ-Fe_2O_3
1.6-3.0	Fe_3O_4 and α-Fe_2O_3
3.0 – 4.0	Single phase α-Fe_2O_3

The possibility of depositing single-phase γ-Fe_2O_3 films is still under investigation. Fig.1 shows the X-ray of some representative samples. The intensity of the X-ray peaks was substantially smaller and the peaks were wider in all films with mixed stoichiometry as compared to the ones in single-phase oxides. This indicates that the grain growth is obstructed in films with mixed stoichiometry. Our main objective was to study the conditions leading to single phase iron oxides and their magnetic and Mössbauer properties.

The magnetic properties of Fe_3O_4 films, in particular the saturation magnetization ($M_s = 380$ emu/cc at 10 K) and coercivity ($H_c = 680$ Oe at 10 K and 220 Oe at 300 K), were consistent with those of other studies [11]. The Verwey transition caused a minimum in the coercivity and a large increase in the magnetization around 120 K (Fig.2). Above the Verwey temperature ($T_v = 119$ K in bulk) the Fe^{2+} ions are randomly distributed on the octahedral (B) positions in the oxygen face centered cubic sublattice, but below it they become ordered. As a consequence of the ordering, there is a second order phase transition accompanied with a change of the crystallographic system from cubic to orthorombic. In addition, there is a large change, by an order of magnitude, of the crystal anisotropy, which leads to the observed minimum in the H_c and the steep increase in the magnetization at T_v.

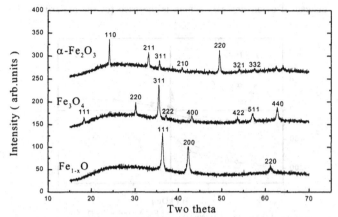

Figure 1: X-ray spectra of $Fe_{1-x}O$ (bottom), Fe_3O_4 (middle) and
α-Fe_2O_3 (top) films

FeO and α-Fe_2O_3 films showed ferromagnetic-like behavior (Fig.3) with very large coercivities at low temperature (Fig.4). The saturation magnetization of these samples were 220 emu/cc (FeO) and 55 emu/cc (α-Fe_2O_3) at low temperatures, and at room temperature decreased to 200 emu/cc and 45 emu/cc respectively. These results are very interesting, considering that both FeO and α-Fe_2O_3 are antiferromagnets in bulk form. The ferromagnetic behavior in these samples can not be explained on the basis of the presence of a second phase (α-Fe in FeO) and (γ-Fe_2O_3 in α - Fe_2O_3) because of the following reason. If this was the case, the presence of α-Fe and that of γ-Fe_2O_3 should be about 15 % of the sample volume in order to account for the experimental values of the saturation magnetization. This possibility was ruled out using Mössbauer spectroscopy, which found no traces of a second Fe-O phase in both sets of samples.

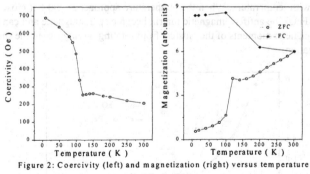

Figure 2: Coercivity (left) and magnetization (right) versus temperature
of Fe_3O_4 film

Mössbauer spectra (Fig.5) indicated that FeO films were offstoichiometric ($Fe_{1-x}O$), with x about 0.1. As a result of the offstoichiometry a fraction of the Fe ions are in triple ionized states (Fe^{3+}). Fe^{3+} ions were found to occupy positions with tetrahedral (A) oxygen coordination, instead of octahedral (B), which is the case for the majority of Fe^{2+} in FeO.

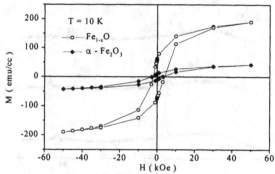

Figure 3: Hysteresis loops of $Fe_{1-x}O$ and α-Fe_2O_3 films

Experimental [12-13] and theoretical studies [14] on bulk $Fe_{1-x}O$, have shown that the Fe^{3+} ions occupy tetrahedral positions and are surrounded by cation vacancies. It was also shown [14] that it is energetically favorable for those defects to form clusters by sharing a corner or an edge cation vacancies. The Fe^{3+} ions are in an environment close to that of Fe_3O_4 and it is to be expected that the cluster possess a net magnetic moment. It is important to note that these clusters can not be considered as a Fe_3O_4 phase precipitated in FeO matrix, because if this was the case the amount of Fe_3O_4 phase should have been about 40 % of the total volume in order to account for the saturation magnetization of the sample. X-ray and Mössbauer studies clearly ruled out the presence of any noticeable amount of Fe_3O_4 in these samples. The Fe^{3+} clusters are coherently embedded and as a result there exists a strong exchange coupling between them and the antiferromagnetic matrix of FeO. This explains the sharp increase of the magnetization at 200 K (Fig.6) in zero field cooled (ZFC) and the large shift in the hysteresis loop of a field cooled (FC) samples (Fig.7). At low temperatures the exchange interactions between the clusters and the antiferromagnetic matrix does not allow the magnetic moments to rotate along the applied field, resulting in a small change of the magnetization with temperature. When the temperature approaches 200 K (close to the Neel temperature of FeO) the antiferromagnetic matrix becomes paramagnetic and can not prevent any more the magnetic moments of the clusters from rotating, giving rise to the sharp increase in the magnetization.

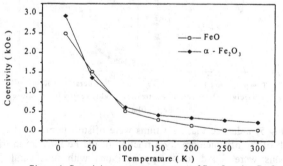

Figure 4: Coercivity versus temperature of $Fe_{1-x}O$ and α-Fe_2O_3 fil

The Mössbauer spectra of α-Fe$_2$O$_3$ films (Fig.5) indicated the presence of two magnetic components. Both components are typical for α-Fe$_2$O$_3$, the first with bulk properties and the second with substantially lower hyperfine field characteristic for the surface Fe ions of the α-

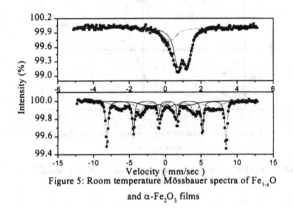

Figure 5: Room temperature Mössbauer spectra of Fe$_{1-x}$O
and α-Fe$_2$O$_3$ films

Fe$_2$O$_3$ grains. About 60 % of the Fe atoms are in surface-like environment, similar to the results of previous studies on ultrafine α-Fe$_2$O$_3$ particles [15]. For the Fe ions at the surface some of the exchange interactions are broken, which leads to randomization of the surface spins [16].

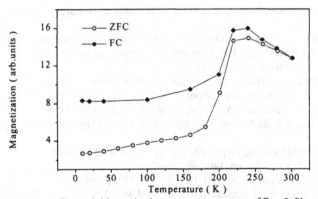

Figure 6: Magnetization versus temperature of Fe$_{1-x}$O film

As a result of the randomization of the surface spins a complete compensation of the magnetic moments in α-Fe$_2$O$_3$ grains is not possible. A net uncompensated magnetic moment is created on the α-Fe$_2$O$_3$ grains. The hysteresis properties of α-Fe$_2$O$_3$ films are result of the existence of uncompensated magnetic moments. As in the case of FeO films there is a strong exchange coupling between the ferromagnetic and antiferromagnetic constituents, indicated by the large shift in the hysteresis loop of field cooled samples.

In summary, an interesting hysteretic behavior was observed in thin Fe$_{1-x}$O and α-Fe$_2$O$_3$ films. This unusual properties are attributed to the net magnetic moment of clusters of Fe^{3+} ions in Fe$_{1-x}$O and uncompensated magnetic moment of the surface component in α-Fe$_2$O$_3$ films.

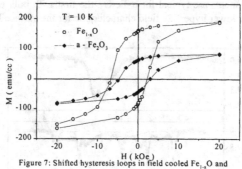

Figure 7: Shifted hysteresis loops in field cooled $Fe_{1-x}O$ and
α-Fe_2O_3 films

ACKNOWLEDGMENTS: This work was supported by NSF grant DMR 9307676

REFERENCES

1.Z.J.Zhou, and J.J.Yan, J.Magn.Magn.Mater 115,pp. 87-98, . (1992)

2.C.Hwang, M.M.Chen, and G.Castillo, J.Appl.Phys. 63 (8), pp.3272-3274, (1988)

3. S.Morup, F.Bodker, P.V.Hendriksen, and S.Linderoth, Phys. Rev.B 52, (1995)

4.A.F.Lehlooh, S.Mahmood, and I. Abu-Aljarayesh, J.Magn.Magn.Mater. 136, pp.143-148, (1994)

5.H.H.Kung, Transition Metal Oxides: Surface Chemistry and Catalysis, Elsevier, New York, (1989)

6.R.K.Wild, Surface Analysis: Techniques and Applications, Special Publication 84, edited by D.R.Randel and W.Neagle, Royal Society of Chemistry, London, (1990)

7.T.D.Waite, Rev.Mineral. 23, p. 559 (1990)

8.S.Gangopadhyay, G.C.Hadjipanayis, S.I.Shah, C.M.Sorensen, K.J.Klabunde, V.Papaefthymiou, and A.Kostikas, J.Appl.Phys. 70 (10), pp. 5888-5890, (1991)

9.R.R.Ruf, and R.J.Gambino, J.Appl.Phys. 55 (6), pp. 2628-2630, (1984)

10.J.W.Schneider, A.M.Stoffel, and G.Trippel, IEEE Trans. Magn., MAG-9 (3), pp. 183-185, (1973)

11.Y.K.Kim, and M.Oliveria, J.Appl. Phys. 75 (1), pp. 431-437, (1994)

12.W.L.Roth, Acta Cryst. 13, pp. 140-149, (1960)

13.F.Koch, and J.B.Cohen, Acta Cryst. B25, pp. 275-287, (1969)

14.C.R.A.Catlow and B.E.F.Fender, J.Phys. C: Solid State Physics, 8, pp. 3267-3279, (1975)

15.15.A.M.Van der Kraan, Phys. Stat. Sol. A 18, pp. 215-226, (1973)

16.R.H.Kodama and A.E.Berkowitz, E.J.McNiff. Jr. and S.Foner, Phys.Rev.Lett , 77 (2), pp. 394-397, (1996)

PARAMAGNETIC SUSCEPTIBILITY OF THE CMR COMPOUND $La_{1-x}Ca_xMnO_3$

D. H. Goodwin[1], J. J. Neumeier[1], A. H. Lacerda[2], and M. S. Torikachvili[3]
[1]Department of Physics, Florida Atlantic University, Boca Raton, FL 33431
[2]National High Magnetic Field Laboratory, Pulse Facility, Los Alamos National Laboratory
Los Alamos, NM 87545
[3]Department of Physics, San Diego State University, San Diego, CA 92182

ABSTRACT

Measurements and analysis of the paramagnetic susceptibility of $La_{1-x}Ca_xMnO_{3+y}$ ($0 \leq x \leq 1$) are reported. The magnetization was also measured in large fields ($H \leq 18$ tesla) for a specimen of $La_{0.79}Ca_{0.21}MnO_{3+y}$. The paramagnetic susceptibility is observed to be strongly enhanced for most compositions studied, including a number of antiferromagnetic compositions. This behavior is attributed to the existence of magnetic polarons.

INTRODUCTION

Transition metal oxides exemplified by the chemical formula $La_{1-x}Ca_xMnO_{3+y}$ have been investigated for over 40 years [1-4]. In the range $0 < x \leq 0.5$ ferromagnetism is observed to occur with T_c values as high as 280 K [1,2,5]. The electrical resistivity displays a metal-semiconductor transition at T_c for $0.20 \leq x \leq 0.5$ [5], metallicity and ferromagnetism are observed simultaneously in this range. An exceptionally large magnetoresistivity occurs in these materials and is often referred to as colossal magnetoresistance or CMR. Currently, it is believed that a strong interplay between magnetic, electrical, and lattice degrees of freedom [6-8] leads to the CMR effect [4-7,9,10].

Electrical transport properties of these compounds have been studied extensively but their magnetic properties have been the focus of less attention. Although the magnetic phase diagram is known [5], many aspects of the magnetic properties as a function of doping in the entire range $0 \leq x \leq 1$ have not yet been investigated. It has been noted that a strong enhancement of the paramagnetic susceptibility occurs [5,11] for ferromagnetic specimens. This enhancement exists to temperatures as high as $4T_c$ and is not a result of ferromagnetic fluctuations [11]. A model proposed by Tanaka et al. [11] provided good agreement with data for a $x = 0.2$ specimen. This model assumes that the double-exchange interaction [3] provides a magnetic coupling between the magnetic moments of neighboring Mn ions, thus resulting in a paramagnetic system composed of groups of magnetically interacting ions, mixed with non-interacting ions. The groups of interacting Mn ions correspond to the magnetic polarons which are thought to play a significant role in determining the electrical properties of these oxides [6-8,12-14].

In this work we present some preliminary results on our measurements of the magnetic susceptibility of $La_{1-x}Ca_xMnO_{3+y}$ in the composition range $0 \leq x \leq 1$. In particular we investigate the enhancement of the paramagnetic susceptibility.

EXPERIMENTAL

Samples were prepared using standard solid state reaction. High purity (> 99.99%) La_2O_3, $CaCO_3$ and MnO_2 powders were weighed in stoichiometric amounts and mixed with a mortar and pestle for 10 minutes. The La_2O_3 powder was dried at 600 °C overnight just prior to use. The resulting mixture was placed in an alumina crucible and reacted at 1100 °C for twenty hours. The specimen was then reground and reacted at 1300 °C for 20 hours. This step was repeated twice. The sample was then reground, pressed into pellets and fired at 1325 °C for 20 hours. The pellets were then reground, pressed and reacted at 1325 °C for 17 hours.

Specimens with x > 0.5 were slow cooled to 30 °C at a rate of 5 °C/min. The remaining specimens were removed from the furnace at 1000 °C. The magnetic susceptibility was measured using a commercially available SQUID magnetometer in an applied field of H = 4000 Oe over the temperature range 5 K ≤ T ≤ 400 K. High field measurements of the magnetic susceptibility were conducted at the National High Magnetic Field Laboratory (NHMFL) in Los Alamos. These experiments utilized the 20 tesla superconducting solenoid into which a vibrating sample magnetometer was adapted.

RESULTS

An example of the measurements we have conducted is illustrated in Fig. 1 where the magnetic susceptibility χ is plotted as a function of absolute temperature for the sample $La_{0.79}Ca_{0.21}MnO_{3+y}$ which becomes metallic below T_c. At 5 K χ saturates to a value which corresponds to 3.71 μ_B/Mn-ion. This value corresponds to full ferromagnetic alignment of the magnetic moments. A value of $T_C = 192.6$ K is consistent with the inflection point in the curve of Fig. 1, Arrott plots (i.e. M^2 versus H/M), and the peak in the electrical resistivity.

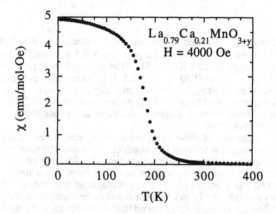

Fig. 1. Magnetic susceptibility versus temperature for a specimen of $La_{0.79}Ca_{0.21}MnO_{3+y}$ at 4000 Oe.

In Fig. 2 a plot of $1/\chi$ versus T from 150 K to 400 K is presented. Ideally, $1/\chi$ should be linear in this region and follow the relation $1/\chi = (1/C)(T - \Theta)$ where C is the Curie constant and Θ is the Curie-Weiss temperature. It is clear in Fig. 2 that $1/\chi$ deviates from linearity, this deviation is particularly severe near T_c. Above 300 K, the data are fairly linear. If the data are fitted in the region 350 K ≤ T ≤ 400 K, we obtain values of C = 4.20 emu-K/mole-Oe and Θ = 238 K. Using C, we calculate an effective paramagnetic moment of p_{eff} = 5.78 μ_B/Mn-ion. This is significantly enhanced over the value of 4.68 μ_B/Mn-ion expected for a mixture of 79% Mn^{3+} and 21% Mn^{4+}. We have conducted similar fits specimens at a number of x values. In our analysis, we neglect contributions due to Pauli paramagnetism and diamagnetism which account for less than 1% of the overal magnitude of χ. For consistency, all fits were done in the region 350 K ≤ T ≤ 400 K. The temperature limit of our instrument is 400 K.

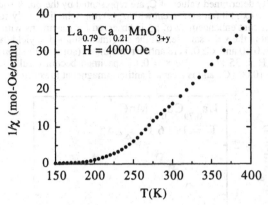

Fig. 2. Inverse of χ plotted versus temperature. T_c for this specimen is 192.6 K.

In order to analyze the obtained effective paramagnetic moments, we need a relation for the value of $p_{eff}(x)$ expected for a conventional paramagnetic system. The x = 0 specimen has four unpaired electrons in the Mn d shell with parallel spin. Due to quenching of the orbital angular momentum, J takes on the value of S. Thus, p_{eff} can be calculated from the equation $p_{eff} = g[S(S + 1)]^{1/2}$, using g = 2 and S = 2 this yields p_{eff} = 4.89 μ_B/Mn-ion. Similarly, the x = 1 specimen has S = 3/2 yielding a value of p_{eff} = 3.87 μ_B/Mn-ion. The value of p_{eff} as a function of x is then given by the relation $p_{eff}(x) = 4.87(1-x) + 3.87x$, in units of μ_B/Mn-ion; this is

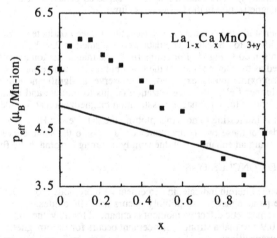

Fig. 3. The effective paramagnetic moment p_{eff} is plotted versus x for the specimens investigated in this study. The solid line is the value of p_{eff} expected for the appropriate value of x assuming the orbital angular momentum is quenched.

represented by the solid line in Fig. 3. The experimental values of p_{eff}, which are calculated from the experimentally determined values of C, are represented by the black squares in Fig. 3. The experimental values exhibit fair agreement with $p_{eff}(x)$ (the solid line) only for specimens of $0.7 \leq x \leq 0.9$. A significant enhancement of p_{eff} is observed for specimens with $x < 0.7$. There is also an enhancement in the $x = 1$ sample. Of the specimens studied to yield the data in Fig. 3, those with $x = 0, 0.05, 0.10$ and $x \geq 0.5$ are antiferromagnetic (some with strong ferromagnetic canting) in the range $H < 35$ kOe. The $x = 0.15$ specimen becomes fully ferromagnetic in magnetic fields above 10,000 Oe and is a canted antiferromagnet at lower fields.

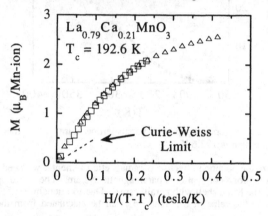

Fig. 4. Magnetization versus magnetic field for a specimen of $La_{0.79}Ca_{0.21}MnO_{3+y}$ with $T_c = 192.6$ K. The data points were obtained at 235 K (triangles) and 270 K (squares). The dashed line is the expected behavior of the magnetization in the Curie-Weiss limit.

In order to further study the paramagnetic susceptibility, we conducted measurements of the magnetization in fields up to 18 tesla. The data were obtained at 235 K (triangles) and 270 K (squares) and are presented in Fig. 4. For comparison, the magnetization expected in the Curie-Weiss limit is plotted as the dashed line. In the low field region, where the best agreement with the Curie-Weiss approximation is expected, we observe a significant enhancement of the magnetization. At higher fields, the magnetization begins to saturate and attains values which are a significant fraction of the ferromagnetic saturation magnetization at 5 K (M(5K, 35 kOe) = 3.71 μ_B/Mn-ion). It is interesting to note that plotting the data versus $H/(T-T_c)$ (as is done here) illustrates that the data at these two temperatures collapse onto the same curve. This suggests that we are not observing an enhancement due simply to strong ferromagnetic fluctuations.

DISCUSSION AND CONCLUSIONS

The results presented herein indicate, in agreement with previous work [5,11], that a strong enhancement of the paramagnetic susceptibility occurs in CMR oxides. The results of Fig. 3 establish that the paramagnetic effective moment is enhanced for all values of x studied except in the range $0.7 \leq x \leq 0.9$ and that a strong enhancement occurs for ferromagnetic as well as some antiferromagnetic compositions.

Magnetic polarons are thought to play an important role in the CMR oxides [5-8,12-19]. Polarons are generally thought of as a lattice distortion which has an associated distortion of charge, however, they can also possess a magnetic character [20]. A magnetic polaron consists of a conduction electron and a group of magnetic moments which propagates through the lattice by aligning the magnetic moments along its path. As the temperature is lowered below T_c, the magnetic polarons begin to delocalize thereby allow formation of the metallic state [7,13]. We suspect that the enhancement of the paramagnetic susceptibility is associated with the existence of magnetic polarons. If this is correct, the results above suggest that magnetic polarons exist in both antiferromagnetic as well as ferromagnetic $La_{1-x}Ca_xMnO_{3+y}$.

Previous work by Tanaka et al. [11] on a specimen of $La_{0.8}Ca_{0.2}MnO_3$ illustrated that the magnetic susceptibility at 11.16 kOe could be fit to a model which assumed that a complex of Mn magnetic moments were coupled via the double exchange interaction [3]. In their model, they assumed that two magnetic species were present, one is magnetically independent Mn ions and the other a complex consisting of four Mn^{3+} ions and one hole; the latter has a total spin of S = 15/2. A Boltzmann distribution was used to estimate the number of complexes which existed at a given temperature and the activation energy agreed well with that observed in the electrical resistivity. The complex of four Mn^{3+} ions and a hole corresponds to a magnetic polaron.

In the future, we will conduct detailed comparisons of our results with the model of Ref. 11. In particular, it will be interesting to see if this model is also capable of predicting the field dependence of the paramagnetic susceptibility and ultimately, if such modeling can provide quantitative information regarding the statistical distribution of the polarons.

ACKNOWLEDGMENTS

Work at Florida Atlantic University was supported by an FAU-sponsored Research Initiation Award. Work at the NHMFL-Los Alamos was performed under the auspices of the National Science Foundation, the State of Florida, and the US Department of Energy.

REFERENCES

1. G. H. Jonker and J. H. Van Santen, Physica 16, 337 (1950).
2. J. H. Van Santen and G. H. Jonker, Physica 16, 599 (1950).
3. C. Zener, Phys. Rev. 81, 440 (1951); 82, 403 (1951).
4. J. Volger, Physica 20, 49 (1954).
5. P. Schiffer, A. P. Ramirez, W. Bao, and S.-W. Cheong, Phys. Rev. Lett. 75, 3336 (1995).
6. R. M. Kusters, J. Singelton, D. A. Keen, R. McGreevy, and W. Hayes, Physica B 155, 362 (1989); K. N. Clausen, W. Hayes, D. A. Keen, R. M. Kusters, R. L. McGreevy, and J. Singleton, J. Phys. Condens. Matter 1, 2721 (1989).
7. M. F. Hundley, M. Hawley, R. H. Heffner, Q. X. Jia, J. J. Neumeier, J. Tesmer, J. D. Thompson, and X. D. Wu, Appl. Phys. Lett. 67, 860 (1995).
8. C. H. Booth, F. Bridges, G. H. Kwei, J. M. Lawrence, A. L. Cornelius, and J. J. Neumeier, to appear in Phys. Rev. Lett. 80 (1998).
9. R. von Helmholt, J. Wecker, B. Holzapfel, L. Schultz, and K. Samwer, Phys. Rev. Lett. 71, 2331 (1993).
10. S. Jin, T. H. Tiefel, M. McCormack, R. A. Fastnacht, R. Ramesh, and L. H. Chen, Science 264, 413 (1994).
11. J. Tanaka, H. Nozaki, S. Horiuchi, and M. Tsukioka, J. Physique Lett. 44, L-129 (1983).
12. M. Jaime, M. B. Salamon, M. Rubinstein, R. E. Treece, J. S. Horwitz, and D. B. Chrisey, Phys. Rev. B 54, 11914 (1996); M. Jaime, H. T. Hardner, M. B. Salamon, M. Rubinstein, P. Dorsey, and D. Emin, Phys. Rev. Lett. 78, 951 (1997).
13. M. F. Hundley and J. J. Neumeier, Phys. Rev. B 55, 11511 (1997).
14. S. J. L. Billinge, R. G. DiFrancesco, G. H. Kwei, J. J. Neumeier, and J. D. Thompson, Phys. Rev. Lett. 77, 719 (1996).
15. A. J. Millis, P. B. Littlewood, and B. I. Shraiman, Phys. Rev. Lett. 74, 5144 (1995).
16. A. J. Millis, Phys. Rev. B 53, 8434 (1996).
17. H. Röder, J. Zang, and A. R. Bishop, Phys. Rev. Lett. 76, 1356 (1996).
18. A. J. Millis, B. I. Shraiman, and R. Mueller, Phys. Rev. Lett. 77, 175 (1996).
19. G.-M. Zhao, K. Conder, H. Keller, and K. A. Müller, Nature 381, 676 (1996).
20. D. Emin, M. S. Hillary, and N.-L. H. Liu, Phys. Rev. B 35, 641 (1987).

EFFECT OF DOMAIN STRUCTURE ON THE MAGNETORESISTANCE OF EPITAXIAL THIN FILMS OF FERROMAGNETIC METALLIC OXIDE SrRuO3

R.A. RAO, D.B. KACEDON, C.B. EOM*
Department of Mechanical Eng. and Materials Science, Duke University, Durham, NC 27708
*email : eom@acpub.duke.edu

ABSTRACT

We have grown epitaxial ferromagnetic metallic oxide $SrRuO_3$ thin films with different domain structures on (001) $LaAlO_3$ and miscut (001) $SrTiO_3$ substrates. The effect of crystallographic domain structures on the magnetization and magnetoresistive behavior of epitaxial $SrRuO_3$ thin films has been studied. Magnetization measurements on the single domain film on 2° miscut (001) $SrTiO_3$ substrate showed that the in-plane [$\bar{1}$10] direction, which is aligned along the miscut direction, is the easier axis for magnetization compared to the [001] direction. This film also showed a strong anisotropic magnetoresistance (AMR) effect of ~ 8% in magnitude. In contrast, the $SrRuO_3$ thin film on (001) $LaAlO_3$ substrate shows identical magnetization and magnetoresistance behavior in two orthogonal directions on the film due to the presence of 90 domains in the plane. For both the films, large negative magnetoresistance effects (~10%) were observed when the current and the applied magnetic field are parallel. The magnetoresistance behavior is explained in terms of suppression of spin fluctuations near T_c and the AMR effect.

INTRODUCTION

The discovery of colossal magnetoresistance (CMR) in doped perovskite manganite thin films,[1] has generated a lot of interest in the magnetoresistance of perovskite magnetic oxide thin films. Among the many perovskite magnetic oxides, the $Sr_{1-x}Ca_xRuO_3$ system is especially attractive, because of its interesting magnetic and transport properties.[2,3] $SrRuO_3$ is an itinerant ferromagnet with a Curie temperature (T_c) of ~ 160K in the bulk material. It has $GdFeO_3$-type pseudocubic perovskite structure with a bulk lattice parameter of 3.93Å. Due to a slight orthorhombic distortion, single domain $SrRuO_3$ thin films can be distinguished from 90 misoriented two domain films by off-axis azimuthal scans using a four circle X-ray diffractometer. $SrRuO_3$ is also structurally and chemically similar to the $LaMnO_3$-based colossal magnetoresistive (CMR) materials. As in CMR materials, magnetic ordering decreases electrical resistivity in $SrRuO_3$ rendering it attractive for magnetotransport studies.[4] As the amount of Ca doping (x) is increased, ferromagnetism in the parent material is gradually suppressed and the T_c decreases, until a completely paramagnetic material is obtained above 70% doping. Furthermore, the orthorhombic distortion of the perovskite unit cell increases and the lattice parameter decreases as x is increased.

$SrRuO_3$ exhibits a strong magnetocrystalline anisotropy in single crystal bulk[5] and single domain thin film[6] samples. Therefore, the growth of epitaxial and single crystal thin films by various techniques including 90 off-axis sputtering,[7] molecular beam epitaxy,[8] and pulsed laser ablation[9,10] allows us to explore the intrinsic anisotropic properties of this material. We have recently controlled the growth mechanisms[11] and domain structure[12] of epitaxial $SrRuO_3$ thin films on miscut (001) $SrTiO_3$ and (001) $LaAlO_3$ substrates and studied the magnetotransport properties[13] of these films. In this report we elaborate on the domain structure dependence of the magnetization and magnetotransport properties of epitaxial $SrRuO_3$ thin films.

EXPERIMENT

The SrRuO₃ thin films were deposited from a 2" diameter stoichiometric composite target in a 90° off-axis sputtering technique.[14,15] The films were deposited on a (001) SrTiO₃ substrate miscut by 2° toward [010] and an exact (001) LaAlO₃ substrate. The deposition atmosphere consisted of 80 mTorr O_2 and 120 mTorr Ar and the substrate block temperature was typically held at 680°C. After the deposition the chamber was immediately vented in O_2 to a pressure about 300 Torr and the samples were then cooled down. The films were approximately 3000Å thick as determined from their Rutherford backscattering spectra using the simulation program RUMP.[16]

At these growth conditions, the SrRuO₃ (110) plane grows readily on the SrTiO₃ (001) plane, as the energetically favorable direction for fast growth. All the crystallographic planes and directions for SrRuO₃ referred to in this work are based on the orthorhombic unit cell. The growth mechanisms and surface morphology of the films were studied by an atomic force microscope (AFM) and a scanning tunneling microscope (STM). The epitaxial arrangement and crystalline structure of the films were analyzed with a four circle diffractometer. The magnetization of the films was studied using a SQUID magnetometer. Finally, the films were patterned into a cross geometry with ion-milling and the magnetotransport properties were measured along two orthogonal directions within the plane using an Oxford Instruments MAGLAB 2000™ materials characterization system.

RESULTS AND DISCUSSION

Growth Mechanisms and Domain Structure

On 2° miscut SrTiO₃ substrate, the terrace length is shorter than the critical surface diffusion length of the film adatoms at the growth temperature. Therefore, the film grows completely by step flow mechanism, as indicated by the straight steps observed in the STM images of the film surface.[11] X-ray diffraction showed that the film had a purely (110) texture normal to the substrate. The in-plane epitaxial alignment was determined from off-axis azimuthal (φ) scans of the (221) reflection. Two peaks were observed indicating a single domain texture with an in plane epitaxial arrangement of SrRuO₃[1̄10]//SrTiO₃[010] and SrRuO₃[001]//SrTiO₃[100].[12] Due to the coherent growth this film has a strained lattice with in-plane lattice parameters close to that of the substrate (3.90Å) and an out-of-plane lattice parameter of 3.96Å.

In contrast, the film on LaAlO₃ substrate has a three dimensional island growth mechanism due to the large lattice mismatch with the substrate.[11] These three dimensional islands are 3000 - 5000Å in base diameter and ~ 500Å in height, which leads to a rough surface with a root mean square roughness of 190Å. Furthermore, due to the incoherent growth, this film has a strain-free lattice with both in-plane and out-of-plane lattice parameters same as that of the bulk material (3.93Å). The out-of-plane film texture is a mixture of (110) and (001) normal to the substrate. Both the (110) and (001) grains have two 90° domains in the plane present in equal volume fraction.

Magnetization and Magnetotransport Properties

The influence of the domain structure on the magnetization of the films was studied with a SQUID magnetometer. The samples were cooled in a magnetic field of 0.05T applied parallel to the film surface. Magnetization measurements on the single domain film (on 2° miscut SrTiO₃) showed an anisotropic behavior with the in-plane [1̄10] direction being the easier axis for magnetization compared to the [001] direction, which is in agreement with previously reported results.[6] Accordingly, this film showed a fully saturated magnetization at low temperatures (~5K), when the field is applied along the [1̄10] direction, as shown in Fig. 1(a). However, an unsaturated and lower magnetization (~65% of the value in the [$\bar{1}$10] direction) is obtained when the field is in the [001] direction. The Curie temperature of this film is ~ 120K, which is smaller than the bulk value. This T_c suppression is attributed partly to the lattice strain of the film.[17] Furthermore, it is believed that this film might be a little deficient in Ru, which could also contribute to the T_c suppression. In contrast, the film on LaAlO$_3$ shows identical unsaturated magnetization, as shown in Fig. 1(b), when the field is applied along any of the two orthogonal directions on the film surface due the presence of 90° domains in the plane of the film. As this film is strain-free, the Curie temperature is close to the bulk value ~ 160K.

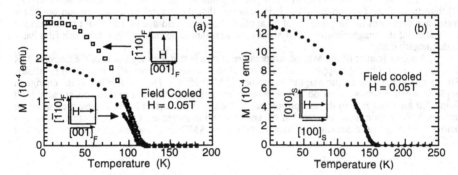

Fig. 1. Magnetization versus temperature for a SrRuO$_3$ thin film on (a) 2° miscut (001) SrTiO$_3$ substrate along two orthogonal directions and (b) exact (001) LaAlO$_3$ substrate. The inset shows the applied field direction with respect to the crystallographic directions of the film and substrate denoted by "F" and "S" respectively.

In order to investigate the correlation between the magnetization and resistivity, magnetoresistance (MR) measurements were performed on these films. For the single domain SrRuO$_3$ thin film on 2° miscut (001) SrTiO$_3$ substrate, the resistivity was measured in four possible combinations of **J** (current density) and **H** (applied field) directions with respect to the in-plane [$\bar{1}$10] and [001] directions of the film (i.e., **J** ∥ **H** ∥ [$\bar{1}$10]; **J** ∥ **H** ∥ [001]; **J** ⊥ **H**, **J** ∥ [001]; and **J** ⊥ **H**, **J** ∥ [$\bar{1}$10]). In contrast, for the film on exact (001) LaAlO$_3$ substrate, the resistivity was measured along the [010] and [100] directions of the substrate in two orientations (**J** ∥ **H** and **J** ⊥ **H**) because of the presence of 90 domains in the plane. The temperature dependence of resistivity was measured by zero-field cooling and field warming.

The temperature dependence of resistivity for the films in zero applied field shows the characteristic change of slope at the Curie temperature (T_c) signifying a ferromagnetic phase transition. The T_c of the film on SrTiO$_3$ is ~ 120 K which is significantly lower than that of the film on LaAlO$_3$ substrate (T_c ~ 160K), as confirmed by the magnetization measurements. Above T_c the resistivity increases linearly with temperature and is expected to continue without saturation.[18] The magnetoresistance is calculated as : MR = [ρ(H) - ρ(0)]/ρ(0), where ρ(H) and ρ(0) are the resistivity at a field H and at zero field respectively.

All the films show a negative magnetoresistance that varies with temperature and applied magnetic field. Figure 2 compares the MR as a function of temperature for the SrRuO$_3$ films on the two different substrates. The field and current directions are shown by arrows in the inset and are with respect to the crystallographic directions of the film and substrate denoted by the subscript "F" and "S" respectively. Several interesting features are observed from the figure. All the films display a large negative MR at temperatures just below T_c in all orientations of the current and field which has already been attributed to the suppression of spin fluctuations.[19] For the single domain film, the largest MR among all 4 combinations of **J** and **H** directions is observed at low temperatures, when the current and field are parallel to the [$\bar{1}$10] direction, shown in Fig. 2(b). This is due to the fact that the [$\bar{1}$10] direction is the easier axis of magnetization compared to the

[001] direction. The spins are more easily aligned in the [$\bar{1}$10] direction which accounts for the larger change in resistivity due to the magnetic field.

The film on LaAlO$_3$ shows the largest MR of ~ 10% at low temperatures and high fields. For this film, the negative MR begins to decrease after having reached a maxima at low temperatures of about 30 - 40K. This could be related to the temperature dependence of the magnetocrystalline anisotropy. At low temperatures a higher field would be required to change the orientation of spins in the material and thus, the negative MR decreases. A similar maxima in the MR at low temperatures has been observed by Klein *et al.* in their films on SrTiO$_3$ which have a higher T$_c$.[6] Such behavior is not observed in our films on SrTiO$_3$ probably due to its low T$_c$. Furthermore, the MR effect of this film is higher (maximum MR ~ 10% at low temperature with J ∥ H) than that of the film on SrTiO$_3$. It is believed that the crystallographic domain boundary resistance decreases in the presence of an applied magnetic field due to a decrease in the relative orientation of the magnetization of adjacent crystallographic domains. Therefore, this film has a higher negative MR.

Another feature of the MR vs. temperature curves is that the magnitude of the negative MR is always larger when the applied field is parallel to the current, especially at low temperatures. This is due to the anisotropic magnetoresistance (AMR) effect. The AMR effect is the change in angle between the magnetization and current because of magnetization rotation observed at high fields that are smaller than the anisotropy field. Such an AMR effect has been observed in single domain SrRuO$_3$ thin films at a field of 6T and is found to increase as the temperature is lowered below T$_c$.[20] The single domain film shows a maximum AMR of -8% at low temperatures in a field of 8T.

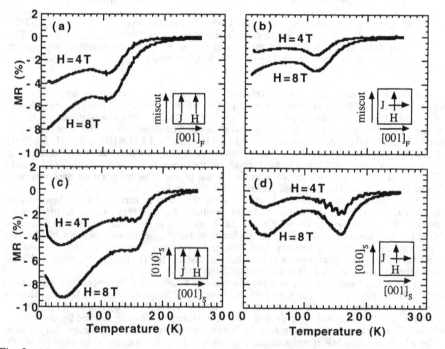

Fig. 2. Magnetoresistance vs. temperature for a SrRuO$_3$ thin film on (a) and (b) 2 miscut (001) SrTiO$_3$ substrate along two orthogonal directions and (c) and (d) exact (001) LaAlO$_3$ substrate. The inset shows the applied field direction with respect to the crystallographic directions of the film and substrate denoted by "F" and "S" respectively.

The MR was also measured as a function of applied field in the four orientations of field and current. Figure 3(a) shows this data at 5K for the single domain SrRuO$_3$ thin film with the current along the [1̄10] direction. A strong hysteresis is observed in the MR behavior at low fields. The hysteresis effect itself is related to the magnetization hysteresis. The peak in the MR hysteresis corresponds to the coercive field and the point of overlap between the forward and backward sweeps of the field corresponds to the saturation field. As the applied field is increased beyond saturation, the magnetization of the sample does not change significantly, but magnetization rotation occurs due to the AMR effect contributing to a larger negative MR when J ∥ H. As the film is single domain, the anisotropy in MR reflects the inherent magnetocrystalline anisotropy of SrRuO$_3$, which is attributed to the large spin-orbit coupling of Ru.

Figure 3(b) shows magnetoresistance as a function of field for the multi domain SrRuO$_3$ film on LaAlO$_3$ substrate at 5K. The MR is shown in two orthogonal directions, with the field parallel to current and field perpendicular to current. The MR is significantly larger when the current is parallel to the field, as observed in the single domain film. The hysteresis in MR is observed in both orientations of the field and the current. However, compared to the single domain film on SrTiO$_3$, the hysteresis on this film is smaller. It is believed that the film on LaAlO$_3$ does not have any magnetic domain walls, because the grain size of these films (as observed from the STM and AFM images)[11] is of the same order as the spacing between magnetic domain walls (~ 2000Å) observed by Lorentz microscopy imaging.[21] Therefore, the crystallographic domain boundaries in this film act as magnetic boundaries as well and the change in relative orientation of the magnetization of adjacent crystallographic domains contributes to the hysteresis. Finally, the sample was rotated by 90° and the MR vs. field measurement was repeated in both orientations (J ∥ H and J ⊥ H). The MR behavior obtained was identical to that shown in Fig. 3(b), not differing by more than 0.5% at any field. This is due to the equivalence of the two orthogonal directions within the plane.

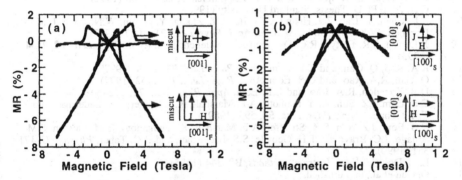

Fig. 3   Magnetoresistance vs. applied magnetic field for a SrRuO$_3$ thin film on (a) 2° miscut (001̄) SrTiO$_3$ substrate and (b) exact (001) LaAlO$_3$ substrate.

CONCLUSIONS

In summary, we have studied the domain structure dependent anisotropy in the magnetization and magnetotransport properties of epitaxial SrRuO$_3$ thin films. Single domain films on 2° miscut (001) SrTiO$_3$ substrates display a strong anisotropy in the magnetization and magnetotransport properties that reflect the inherent magnetocrystalline anisotropy of the material. In contrast, multi domain films on (001) LaAlO$_3$ substrate show identical properties in two orthogonal in-plane directions due to the presence of 90° domains within the plane. These films show large negative magnetoresistance effects (~10%) at low temperatures and high fields when the applied field and current are parallel.

ACKNOWLEDGMENTS

This work was supported by the ONR Grant No. N00014-95-1-0513, and the NSF Grant No. DMR 9421947, the NSF Young Investigator Award (CBE) and the David and Lucile Packard Fellowship (CBE).

REFERENCES

1. S. Jin, T.H. Tiefel, M. McCormack, R.A. Fastnacht, R. Ramesh, and L.H. Chen, *Science*, **246**, 413 (1994).
2. A. Kanbayashi, *J. Phys. Soc. Japan*, **44**, 108 (1978); G. Cao, S. McCall, M. Shepard, and J.E. Crow, *Phys. Rev. B*, **56**, 321 (1997).
3. A. Callaghan, C.W. Moeller and R. Ward, *Inorg. Chem.*, **5**, 1572 (1966); J.M. Longo, P.M. Raccah and J.B. Goodenough, J. Appl. Phys., 39, 1327 (1968).
4. L. Klein, J.S. Dodge, C.H. Ahn, G.J. Snyder, T.H. Geballe, M.R. Beasley, and A. Kapitulnik, *Phys. Rev. Lett.*, **77**, 2774 (1996).
5. A. Kanbayashi, *J. Phys. Soc. Japan*, **41**, 1879 (1976).
6. L. Klein, J.S. Dodge, C.H. Ahn, J.W. Reiner, L. Mieville, T.H. Geballe, M.R. Beasley, and A. Kapitulnik, *J. Phys: Condens. Matter*, **8**, 10111 (1996).
7. C.B. Eom, R.J. Cava, R.M. Fleming, J.M. Phillips, R.B. van Dover, J.H. Marshall, J.W.P. Hsu, J.J. Krajewski, and W.F. Peck, *Science*, **258**, 1766 (1992).
8. C.H. Ahn, Ph.D. Thesis, Stanford University (1996).
9. C.L. Chen, Y. Cao, Z.J. Huang, Q.D. Jiang, Z. Zhang, Y.Y. Sun, W.N. Kang, L.M. Dezaneti, W.K. Chu, and C.W. Chu, *Appl. Phys. Lett.*, **71**, 1047 (1997).
10. X.D. Wu, S.R. Foltyn, R.C. Due, and R.E. Muenchausen, *Appl. Phys. Lett.*, **62**, 2434 (1993).
11. R.A. Rao, Q. Gan, and C.B. Eom, *Appl. Phys. Lett.*, **71**, 1171 (1997).
12. Q. Gan, R.A. Rao, and C.B. Eom, *Appl. Phys. Lett.*, **70**, 1962 (1997).
13. D.B. Kacedon, R.A. Rao, and C.B. Eom, *Appl. Phys. Lett.*, **71**, 1724 (1997).
14. C.B. Eom, J.Z. Sun, K. Yamamoto, A.F. Marshall, K.E. Luther, S.S. Laderman, and T.H. Geballe, *Appl. Phys. Lett.* **55**, 595 (1989).
15. C.B. Eom, J.Z. Sun, S.K. Streiffer, A.F. Marshall, K. Yamamoto, B.M. Lairson, S.M. Anlage, J.C. Bravman, T.H. Geballe, S.S. Laderman, and R.C. Taber, *Physica C*, **171**, 351 (1990).
16. L. R. Dolittle, *Nucl. Instrum. Methods*, **B9**, 344 (1985).
17. Q. Gan et al., unpublished.
18. P.B. Allen, H. Berger, O. Chauvet, L. Forro, T. Jarlborg, A. Junod, B. Revaz, and G. Santi, *Phys. Rev. B* , **53**, 4393 (1996).
19. S.C. Gausepohl, M. Lee, K. Char, R.A. Rao, and C.B. Eom, *Phys Rev. B*, **52**, 3459 (1995).
20. L. Klein, A.F. Marshall, J.W. Rainer, C.H. Ahn, T.H. Geballe, M.R. Beasley, and A. Kapitulnik, preprint (1997).
21. A.F. Marshall, presented at the Materials Research Society - 1997 Spring Meeting : *Epitaxial Oxide Thin Film Symposium* (1997).

THE LOCAL ATOMIC STRUCTURE OF $La_{1-x}Sr_xCoO_3$: EFFECTS INDUCED BY THE SPIN-STATE AND NON-METAL TO METAL TRANSITIONS

DESPINA LOUCA, J. L. SARRAO, G. H. KWEI
Los Alamos National Laboratory, Condensed Matter and Thermal Physics Group, MST-10, MS K764, Los Alamos, NM 87545.

ABSTRACT

The pair density function (PDF) used in the analysis of pulsed neutron diffraction data of $La_{1-x}Sr_xCoO_3$ revealed new structural effects which are correlated to the susceptibility and transport transitions. The transition in the spin configuration of the Co ions from the low-spin (LS) to the high-spin (HS) state in the Co perovskite oxides can potentially induce structural distortions due to the coupling of the spin to the lattice and charge. The ground state of the pure compound, $LaCoO_3$, is in the LS state and is non-magnetic. A transition occurs to the HS state at ~ 50 K as indicated from the susceptibility measurements due to the thermal excitation of electrons to the e_g level. The $Co_{LS}O_6$ octahedra associated with the Co ions in the LS configuration are distinguished from the $Co_{HS}O_6$ octahedra with the Co in the HS configuration because the Co_{LS}-O bond length is shorter than the Co_{HS}-O distance due to the different size of the corresponding Co ions. Such bond lengths are clearly identified in the local structure between 15 - 300 K. This finding is in contrast to the average structure which shows only one type of bond length in this temperature range but two types of bond lengths are suggested at considerably higher temperatures. This suggests that whereas the LS and HS CoO_6-octahedra coexist, they are randomly distributed in the crystal lattice at lower temperatures and become ordered at higher temperatures. The introduction of charge carriers in the structure does not eliminate the coexistence of both the LS and HS states, indicating that with the transition to the ferromagnetic metallic state, the spin configuration is not entirely of the HS character and structural inhomogeneities are present.

INTRODUCTION

Crystals with the perovskite structure have been extensively studied in part due to their superconducting and magnetoresistive properties. The coupling of the charge to the spin and lattice degrees of freedom yields interesting phenomena such as the colossal magnetoresistance (CMR) [1] and high temperature superconductivity (HTSC) [2] where the underlying mechanisms are still under investigation. The Co perovskite system is yet another example that exhibits several magnetic and transport transitions. In addition, doped cobalt compounds of the $La_{1-x}A_xCoO_3$ composition where A is a divalent atom, are commonly used as cathode materials in solid oxide fuel cells (SOFC) and as a catalyst for controlling the emission of carbon monoxide in automotive combustion engines [3]. The transport is brought about by mixed oxygen and electron conduction.

The pure $LaCoO_3$ is a non-magnetic insulator with an average rhombohedral structure ($R\bar{3}C$). The low spin (LS) state (S = 0) is the ground state of this system with the Co ions in the nominal $3d^6$ configuration. Raising the temperature induces a spin transition to a high-spin (HS) configuration of the Co ions. The temperature at which this occurs is about 50 K as indicated by magnetic neutron scattering measurements [4] and susceptibility measurements.

The high temperature phase undergoes a structural as well as a transport transition. At about 400°C, a transition to an $R\bar{3}$ symmetry occurs [5, 6] and at 940°C, a semiconductor-metal transition occurs [6]. It was proposed early on by Raccah and Goodenough [5] that the octahedra corresponding to the LS and HS Co ions coexist and begin to order along the [111] plane above 375°C. This calls for a short range ordering below this temperature. In addition, an intermediate spin (IS) transition is suggested by photoemission [7-9]. The IS state could induce a structural distortion of the Jahn-Teller (JT) type because of the single occupancy of the e_g level. The temperature at which the IS state appears is controversial.

Hole doping in the LS ground state of the pure compound is believed to lead to the formation of localized magnetic polarons with unusually high spin numbers (S = 10-16) [10]. The doped holes induce a local transition from the LS to HS state and create a high spin polaron. The transition from an LS to HS state through doping has been confirmed by NMR studies of ^{59}Co and ^{139}La [11]. The FM state is presumably composed of mobile high-spin polarons. The formation of these magnetic polarons creates a precursor state to the doping-induced FM metallic state which occurs at about x = 0.18 in $La_{1-x}Sr_xCoO_3$. It is possible that ferromagnetism in the hole-doped Co oxides is mediated by the double exchange (DE) interaction similar to the CMR manganites via the coupling of the carriers to the local spins. However, the cobaltates can be more complicated because of the transition between the LS and HS states in addition to the stronger hybridization of the Co 3d orbital with the oxygen 2p orbital in comparison to the manganites. All these effects strongly suggest a strong lattice contribution which could change with doping and temperature.

The cobaltates are related to the manganites with regard to their properties but the Co system does not exhibit the CMR effect. In the case of the manganites, it has been shown that the lattice is actively involved in the mechanism that results in the magnetic and transport transitions [13-15] whereas the role of the lattice in cobaltates is not well defined. In the present work, by using pulsed neutron diffraction, we investigated how the structure responds to the change from a non-magnetic to a magnetic state, either paramagnetic (PM) or ferromagnetic (FM), and from an insulating to metallic conductivity. These preliminary results indicate a strong lattice response that is coincident with the change in the electrical and magnetic properties. The local atomic structure of the $La_{1-x}Sr_xCoO_3$ perovskite was studied as a function of temperature using the atomic pair density function (PDF) analysis.

EXPERIMENT

The $La_{1-x}Sr_xCoO_3$ powder samples were prepared by a standard ceramic method from La_2O_3, Co_3O_4 and $SrCO_3$, fired several times at 1100°C to achieve a single phase with intermediate grinding. Their final firing occurred at 1300°C in air for a day and were quenched from this temperature. The pure sample was additionally annealed in nitrogen at 900°C for a day. The samples were characterized by XRD and magnetic measurements. The neutron diffraction data were collected using the Glass Liquid and Amorphous Materials Diffractometer (GLAD) at the Intense Pulsed Neutron Source (IPNS) of the Argonne National Laboratory. Data were collected from room temperature (RT) down to 15 K. The structure function, S(Q), was determined up to 33 $Å^{-1}$ and was corrected for background, absorption, incoherent scattering, multiple scattering as well as for inelastic scattering (Placzek correction [16]). The multiple scattering correction procedure for the GLAD spectrometer was carefully calibrated so that it gives a correct PDF for crystalline Ni and $SrTiO_3$ powders [17]. The PDF, $\rho(r)$,

(Eqn. 1) is a real space representation of atomic density correlation and is obtained by Fourier transforming the structure function as follows:

$$\rho(r) = \rho_0 + \frac{1}{2\pi^2 r} \int_0^{\infty} Q[S(Q) - 1] \sin(Qr) dQ.$$ (1)

ρ_0 is the average number density of the material and Q is the momentum transfer. The PDF analysis provides direct information with regard to the local structure without a requirement of long range structural periodicity. Its application extends beyond the amorphous materials, to crystalline solids as well [18, 19].

RESULTS

In Fig. 1, the magnetic susceptibility for LaCoO$_3$ is plotted as a function of temperature and is measured at two fields where data were collected on warming after being cooled in almost zero field (1 Gauss). The top panel shows the high field measurement where a transition begins to occur at ~ 25 K and is complete by 80 K. This corresponds to the excitation from the LS to the HS configuration of the Co ions and is thermally activated in this system. The negative slope of the susceptibility at high temperature is an indication of the paramagnetic state of this system. Below 25 K, an upturn in the magnetization is observed which is present in the low field measurement as well (bottom panel). The origin of the low-temperature increase in the susceptibility is unclear at present. It is suggested that this could be due to frozen-in ferromagnetic domains which are trapped at lower temperatures. Another possibility would be due to the presence of impurities or due to a non-stoichiometry of the sample. However, annealing the sample in an inert atmosphere, N$_2$, did not change the results. Further investigation is under way to understand this phenomenon.

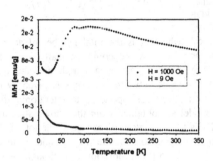

Figure 1: The susceptibility as a function of temperature at two fields, 1000 and 9 Oe. In the high field case, the magnetization increases between 25 and 80 K in going from the low-spin to the high-spin state. Above this temperature, the spin ordering becomes paramagnetic. Below 25 K, an upturn in the susceptibility is observed in both fields which could be due to frozen-in ferromagnetic domains.

The PDF analysis of the LaCoO$_3$ compound revealed interesting structural changes with temperature that have not been observed in previous crystallographic studies. The local atomic structure of LaCoO$_3$ changes as a function of temperature in the way shown in Fig. 2 for the data collected at 15 and 300 K. Note that the local structure determined at the two temperatures changes significantly in the short range region, particularly in the Co-O distance, the shortest distance in the crystal structure. The changes in the Co-O peak are a reflection of the changes of the octahedron. These are in turn associated with the transition from the LS to HS state. As seen from the PDF, two Co-O peaks are present. These correspond to two types of bond lengths, one associated with the LS Co ion (Co$_{LS}$) and the other with the HS configuration of the Co ion (Co_{HS}). The Co_{HS} ion is of a larger radius and the average Co_{HS}-O distance is 2.01 Å while the Co_{LS}-O distance is 1.95 Å [20]. However, the local structure shows that the actual longest bond length is at 2.15 Å and the shortest at 1.9 Å with possibly an even shorter distance at 1.75 Å. These results are in agreement with the results of Raccah and Goodenough [5] for the high temperature ordered state of the octahedra.

A model PDF calculated from the $R\bar{3}C$ structure of $LaCoO_3$ is also shown in Fig. 2 (solid line) and is compared to the PDF determined from the experiment. The model PDF is calculated in the following way:

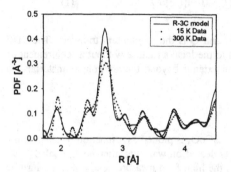

Figure 2: The PDF's of the pure $LaCoO_3$ determined at two temperatures is compared to the model PDF for the known crystallographic structure with R-3C symmetry. Note that although the intermediate local structure resembles that of the average structure, the short range structure differs significantly. At 15 K two types of Co-O distances are observed.

$$\rho(r) = \frac{1}{N} \sum_{i,j=1}^{N} \frac{b_i b_j}{\langle b \rangle^2} \delta\left(r - r_{i,j}\right), \quad (2)$$

where N is the total number of atoms, $<b>$ is the compositionally averaged neutron scattering length and the distance between the atoms is given by $r_{i,j}$ = $|r_i - r_j|$. The model PDF is constructed of δ-functions corresponding to interatomic distances in the structure. The parameters for the model are obtained from the known structure of the compound [5]. The sum of the partial PDF's for each atom gives rise to the total PDF of the crystal which is convoluted with a gaussian function to simulate thermal and quantum zero point vibrations. Although deviations from this model are observed at both temperatures, the PDF determined from the 300 K data is in closer agreement with the model PDF at the first nearest neighbor distance which is the Co-O distance. The Co-O peak in the model is very wide. The 15 K data, on the other hand, are actually quite different from the model, particularly in the first and second nearest neighbor distances. The experimental PDF clearly shows the presence of two peaks, whereas only one peak is evident in the model for the average structure. The two peaks establish the presence of both HS and LS Co ions. This is consistent with the magnetic measurements performed on this sam-

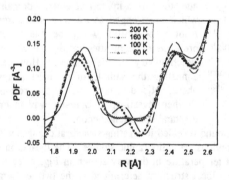

Figure 3: The temperature dependence of the Co-O PDF peaks in $LaCoO_3$. Note how both peaks change position with temperature. The shift in the peak position could be dynamic in origin.

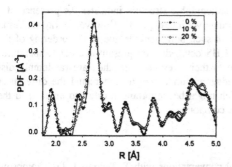

Figure 4: The PDF's of La$_{1-x}$Sr$_x$CoO$_3$ at x = 0, 0.1 and 0.2. The data were collected at 15 K. At this temperature, the peak at 2.15 Å is present at all compositions but is reduced at 20 % of doping.

ple which indicate that the spin state changes with temperature in this system.

The temperature dependence of the Co-O peaks of LaCoO$_3$ for intermediate temperatures is shown in Fig. 3. As the temperature is raised which corresponds to the transition from the LS to the HS state, the local structure shows a lot of changes. In particular, the peak at 2.15 Å is present at all temperatures. The nature of this peak could be dynamic due to the thermally excited spin fluctuations. An intermediate spin (IS) state has been proposed possibly at temperatures higher than room temperature [5, 7]. Such an IS state, which resembles a Jahn-Teller type of distortion, induces a bond length of longer than the average Co-O bond length of ~ 2.1 Å, longer than the average Co-O bond (1.95 Å). More experiments are under way to investigate the details of this state.

Doping with Sr changes the local structure as shown in Fig. 4. At 15 K and with 10% Sr, it has been proposed [21] that the system is in a spin-glass state, whereas with 20% of Sr, a ferromagnetic transition occurs. Although the average structure is rhombohedral for all compositions, the local structure changes with doping as indicated by the comparison of the 10 and 20% samples to the pure compound. At this temperature, it is expected that with 20% Sr, the Co ions would all be in the HS state but two Co-O peaks are observed, suggesting that the ferromagnetic metallic state is not made of a uniform HS state.

A similar temperature dependence of the PDF is observed in the doped samples as in the pure compound. The local structure of La$_{0.8}$Sr$_{0.2}$CoO$_3$ changes with temperature as shown in Fig. 5. The long Co-O peak is present in the local structure, similar to LaCoO$_3$, indicating that both LS and HS octahedra are present in the insulating and metallic states of this compound.

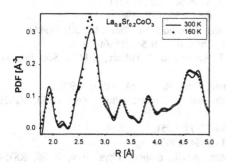

Figure 5: The PDF's of La$_{.8}$Sr$_{.2}$CoO$_3$ at two temperatures, 160 and 300 K. Note that even at this composition, two types of Co-O bonds are present as in the undoped compound.

DISCUSSION

The PDF analysis has shown that two types of Co-O bonds are present in the local structure despite the fact that the average crystallographic structure indicates the presence of a single type of Co-O bond in the temperature range we investigated. However, x-ray studies at higher temperature demonstrate the existence of such bond lengths due to the ordering of the octahedra [5]. This suggests that the LS and HS octahedra are present even at temperatures where average techniques become insensitive to their presence, and they are randomnly distributed in the crystal structure. Further studies are underway to understand the differences between the local and average structures which might be important in order to understand the mechanism of the spin-state transition in these compounds.

ACKNOWLEDGMENTS

The authors would like to acknowledge useful conversations with T. Egami, J. D. Thompson, R. H. Heffner and M. F. Hundley. They particularly acknowledge J. D. Thompson for the magnetic measurements. Work at the Los Alamos National Laboratory is performed under the auspices of the U.S. Department of Energy under contract W-7405-Eng-36.

REFERENCES

1. S. Jin, T. H. Tiefel, M. McCormack, R. A. Fastnacht, R. Ramesh and L. H. Chen, Science **264**, 413 (1994).
2. J. G. Bednorz and K. A. Müller, Z. Physik B **64**, 189 (1986).
3. S. R. Sehlin, H. U. Anderson and D. M. Sparlin, Phys. Rev. B **52**, 11681 (1995).
4. K. Asai, O. Yokokura, N. Nishimori, H. Chou, J. M. Tranquada, G. Shirane, S. Higuchi, Y. Okajima and K. Kohn, Phys. Rev. B **50**, 3025 (1994).
5. P. M. Raccah and J. B. Goodenough, Phys. Rev. **155**, 932 (1967).
6. G. Thornton, B. C. Tofield and A. W. Hewat, J. Solid State Chem. **61**, 301 (1986).
7. S. Yamaguchi, Y. Okimoto and Y. Tokura, Phys. Rev. B **55**, R8666 (1997).
8. S. Yamaguchi, H. Taniguchi, H. Takagi, T. Arima and Y. Tokura, J. Phys. Soc. Jap. **64**, 1885 (1995).
9. R. H. Potze, G. A. Sawatzky and M. Abbate, Phys. Rev. B **51**, 11501 (1995).
10. S. Yamaguchi, Y. Okimoto, H. Taniguchi and Y. Tokura, Phys. Rev. B **53**, R2926 (1996).
11. M. Itoh and I. Natori, J. Phys. Soc. Jap. **64**, 970 (1995).
12. D. Louca and T. Egami, J. Appl. Phys. **81**, 5484 (1997).
13. D. Louca, T. Egami, E. Bosha, H. Röder, A. R. Bishop, Phys. Rev. B **56**, R8754 (1997).
14. D. Louca and T. Egami, Physica B, in press (1997).
15. D. Louca and T. Egami, submitted Phys. Rev. B (1997).
16. G. Placzek, Phys. Rev. **86**, 377 (1952).
17. D. Louca, PhD thesis, University of Pennsylvania, 1997.
18. T. Egami and S. J. L. Billinge, Prog. Mater. Sci., **38**, 359 (1994).
19. B. H. Toby and T. Egami, Acta Cryst. A **48**, 336 (1992).
20. R. D. Shannon, Acta Cryst. A **32**, 7519 (1976).
21. S-W. Cheong, presented at the Spin-Charge-Lattice Dynamics Workshop at the Los Alamos National Laboratory, Los Alamos, NM, 1997.

ANALYSIS OF CATION VALENCES AND OXYGEN VACANCIES IN MAGNETORESISTIVE OXIDES BY ELECTRON ENERGY-LOSS SPECTROSCOPY

Z.L. WANG[a], J.S. YIN, Y. BERTA, J. ZHANG*

School of Material Science and Engineering, Georgia Institute of Technology, Atlanta, GA 30332-0245 USA. [a] e-mail: zhong.wang@mse.gatech.edu.
*Advanced Technology Materials, Inc., 7 Commerce Drive, Danbury, CT 06810; Currently at: Motorola, Inc., 3501 Ed Bluestein Boulevard, MD: K-10 Austin, TX 78721.

ABSTRACT

Magnetic oxides of $(La,A)MnO_3$ and $(La,A)CoO_3$ have two typical structural characteristics: cations with mixed valences and oxygen vacancies, which are required to balance the charge introduced by cation doping. The consequences introduced by each can be different, resulting in different properties. It is important to quantitatively determine the percentage of charges balanced by each, but this analysis is rather difficult particularly for thin films. This paper has demonstrated that electron energy-loss spectroscopy (EELS) can be an effective technique for analyzing Mn and Co magnetic oxides with the use of intensity ratio of white lines, leading to a new technique for quantifying oxygen vacancies in functional and smart materials.

INTRODUCTION

$(La,A)MnO_3$ and $(La,A)CoO_3$ (A = Ca, Sr, or Ba) are important magnetic oxides and their unique properties are determined by three structural characteristics [1]. First, the ferromagnetically ordered Mn-O (or Co-O) layers in the perovskite unit cell are isolated by non-magnetic La(A)-O layers, forming an intrinsic spin-coupling, which is the origin of the magnetic properties of the materials. Secondly, the electric conductivity of the materials is owing to the substituted divalent A^{2+}. The compounds with extreme values x = 0, 1 are neither ferromagnetic nor good electrical conductors. Only compounds with intermediate values of x are ferromagnetic. This is the most interesting characteristics of this type of materials. Finally, the co-existence of the cations with mixed valences and oxygen vacancies are the key factors for determining their unique properties.

The partial substitution of trivalent La^{3+} by divalent element A^{2+} is balanced by the conversion of Mn valence states from Mn^{3+} to Mn^{4+} (or Co^{3+} to Co^{4+} for Co) and the creation of oxygen vacancies. The ionic structure of $La_{1-x}A_xMnO_{3-y}$ is

$$La_{1-x}^{3+} A_x^{2+} Mn_{1-x+2y}^{3+} Mn_{x-2y}^{4+} O_{3-y}^{2-} V_y^O \qquad (1)$$

where V stands for the fraction of oxygen vacancies. Whenever a Mn^{3+} and a Mn^{4+} are on neighboring Mn sites, there exists the possibility of conductivity by electrons hopping from Mn^{3+} to Mn^{4+} with the assistance of an oxygen anion. That, this hopping current should be spin polarized was required for a process of two simultaneous electron hops (from Mn^{3+} onto O^{2-} and from O^{2-} onto Mn^{4+}, thus interchanging Mn^{4+} and Mn^{3+}), called *double exchange* [2]. Therefore, the residual charge introduced by cation substitution can be balanced by either valence conversion of the transition metal element or the creation of oxygen vacancies. If there is no valence conversion, the double exchange may not occur, leading to no or very minimal electrical conductivity. On the other hand, the increase of oxygen vacancies may increase the ionic conductivity.

From the analysis above, quantification of Mn valence and oxygen deficiency is the key for understanding the roles played by each in determination the properties and performance of the magnetic oxide. However, this analysis is a challenge to existing microscopy techniques particularly for thin films, because of the influence from the substrate surface, interface mismatch dislocations and defects in the film. A technique with high spatial resolution is required for this analysis. In this paper, electron energy-loss spectroscopy (EELS) in a transmission electron microscope (TEM) is introduced as a tool for quantifying the valence state of Mn (or Co) and the oxygen deficiency. The details of the technique are given, and its application in the two magnetic oxides containing Mn and Co, respectively, will be illustrated.

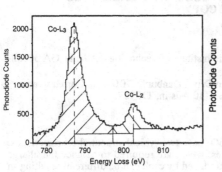

Figure 1. EELS spectrum acquired from $La_{0.67}Ca_{0.33}MnO_{3-y}$ showing the L_3 and L_2 white lines of the Mn ionization edges. The procedure of background subtraction is illustrated. This procedure must be followed consistently for all of the spectra.

Figure 2. (a) A comparison of single-scattering EELS spectra of Co-$L_{2,3}$ ionization edges acquired from $CoSi_2$, $CoCO_3$, $CoSO_4$, Co_3O_4, and $La_{0.5}Sr_{0.5}CoO_{3-y}$. The spectra are displayed by normalizing the heights of the L_3 white lines.

EXPERIMENTAL APPROACH

In TEM, the interaction of the incident electron with the crystal electrons can generate various inelastic excitation processes [3]. One of the processes is the ionization of atomic inner shell bounded electrons, resulting in the ionization edge observed experimentally. In EELS, the L ionization edges of transition-metal, rare-earth and actinide compounds usually display sharp peaks at the near edge region (Figure 1). These threshold peaks are known as *white lines*. For transition metals with unoccupied 3d and 4d states, the white lines are observed. The unoccupied 3d states form a narrow energy band, the transition of a 2p state electron to the 3d levels leading to the formation of white lines observed experimentally. Thus, the atomic state changes from $2p^63d^{(m)}$ to $2p^53d^{(m+1)}$ after the excitation of a 2p electron, where m stands for the number of unoccupied 3d states. More specifically, the L_3 and L_2 lines are the transition of $2p^{3/2} \rightarrow 3d^{3/2}3d^{5/2}$ and $2p^{1/2} \rightarrow 3d^{3/2}$, respectively. EELS experiments have shown that the change in valence states of cations introduces significant changes in the ratio of the white lines, leading to the possibility of identifying the occupation number of 3d or 4d electrons (or cation valence states) using the measured white line intensities in EELS [4-6].

To establish the relationship between white line intensity and the number of unoccupied d electrons states, the white lines must be isolated from the background intensity. The EELS data must be processed first to remove the gain variation introduced by the detector channels and the multiple scattering effect via deconvolution. A low-loss valence spectrum and the corresponding core-shell ionization edge EELS spectrum were acquired from the same specimen region. The energy-loss spectrum was used to remove the multiple-inelastic-scattering effect in the core-loss region using the Fourier ratio technique, thus, the presented data are for single scattering. The background intensity was modeled by step functions in the threshold regions [5]. A straight line over a range of approximately 50 eV was fit to the background intensity immediately following each white line. This line was then extrapolated into the threshold region and set to zero at energies below that of the white-line maximum. The L_2 white line was further isolated by smoothly extrapolating the L_3 background intensity under the L_2 edge. For a case in which two white lines are not widely separated (Figure 1), a straight line is fit to the background immediately following the L_2 white line over a region of approximately 50 eV and is then extrapolated into the threshold region. This line was then modified into a double step of the same slop with onsets occurring at the white-line maxima. The ratio of the step heights is chosen as 2:1 in accordance with the multiplicity of the initial states.

In the magnetic oxides with mixed valences and if the oxygen is ordered, the oxygen deficiency is directly correlated to the fraction of the mixed valences. From Eq. (1), the mean valence state of Mn is

$$\langle Mn \rangle_{vs} = 3 + x - 2y. \tag{2}$$

The $\langle Mn \rangle_{vs}$ can be determined using EELS as described above. Therefore, the content of oxygen vacancies y can be obtained. This is a new approach for studying oxide materials [7,8].

EXPERIMENTAL RESULTS

EELS analysis of valence state is usually carried out in reference to the spectra acquired from standard specimens with known cation valence states. If a series of EELS spectra are acquired from several standard specimens, an empirical plot of these data may serve as the reference for determining the valence state of the element present in a new compound. This is the basis of our analysis on the $La_{1-x}Sr_xCoO_{3-y}$ (LSCO) and $La_{1-x}Ca_xMnO_{3-y}$ (LCMO) magnetic oxides to be described below. The thin films were grown by metal-organic chemical vapor deposition (MOCVD) [9] and their crystal structures have been investigated previously [10].

$La_{0.5}Sr_{0.5}CoO_{3-y}$

$La_{1-x}Sr_xCoO_{3-y}$ has a perovskite-related structure, in which the mixed valence of the cations plays a vital role in determining the properties of the material. In the literature, Co has been believed to have valences 3+ and 4+ in this compound. The substitution of trivalent La^{3+} by divalent Sr^{2+} is balanced by creating oxygen vacancies as well as the conversion of Co^{3+} into Co^{4+} (or Co^{2+} into Co^{3+}). Thus, the ionic structure of $La_{1-x}Sr_xCoO_{3-y}$ can be either

$$La^{3+}_{1-x} Sr^{2+}_x Co^{3+}_{1-x+2y} Co^{4+}_{x-2y} O^{2-}_{3-y} V^O_y \quad \text{(for } y \leq x/2\text{)} \tag{3}$$

or

$$La^{3+}_{1-x} Sr^{2+}_x Co^{3+}_{1+x-2y} Co^{2+}_{2y-x} O^{2-}_{3-y} V^O_y \quad \text{(for } x \leq 2y\text{)} \tag{4}$$

depending on the concentration of the anion deficiency. In practice, the valence state of Co must be measured experimentally to determine the ionic configuration of the compound.

Figure 2 shows a comparison of the processed single-scattering EELS spectra of Co-$L_{2,3}$ ionization edges acquired from $CoSi_2$ (with Co^{4+}), Co_3O_4 (with $Co^{2.67+}$), $CoCO_3$ (with Co^{2+}) and $CoSO_4$ (with Co^{2+}) and $La_{0.5}Sr_{0.5}CoO_{3-y}$. The first four compounds are chosen as the standard specimens with known Co valences, and the last one is the specimen that we want to determine its Co valence state. It is apparent that the shape of $CoSi_2$ (with Co^{4+}) is dramatically different from the rest not only for its high L_2 edge but also for its broaden shape, simply because of its highest Co valence state. The other four specimens have the same line width and intensity except that Co_3O_4 (with $Co^{2.67+}$) has a higher L_2 edge, this is because of its larger Co valence. The two standard specimens showing the same shape of Co-$L_{2,3}$ edges have Co^{2+}. These spectra clearly establish the experimental basis of using the white line intensities for determination the Co valence in a new compound (Figure 3), which clearly shows that $I(L_3)/I(L_2)$ is very sensitive to the valence state of Co. From the empirical fitting curve, the valence of Co in $La_{0.5}Sr_{0.5}CoO_{3-y}$ can be determined from its $I(L_3)/I(L_2)$ value (= 5.05). The corresponding horizontal axis is approximately 1.93, which means the valence of Co in $La_{0.5}Sr_{0.5}CoO_{3-y}$ is 2+ with consideration of the experimental error. Therefore, the ionic structure of LSCO is described by Eq. (4) with y = $(x+1)/2 = 0.75$, which is

$$La^{3+}_{0.5} Sr^{2+}_{0.5} Co^{2+} O^{2-}_{2.25} V^O_{0.75}.$$

Chemical microanalysis using energy dispersive x-ray spectroscopy and EELS microanalysis and the in-situ EELS experiments [11] as well have confirmed this result. Based on this ionization formula, the crystal structure of a new rhombohedral phase of $La_{0.5}Sr_{0.5}CoO_{2.25}$ (or $La_8Sr_8Co_{16}O_{36}$) has been determined using the information provided by high-resolution transmission electron microscopy and electron diffraction [8]. The unit cell is made out of two-types of fundamental modules and it is composed of a total of 8 modules. Each module is a c-axis stacking of the anion deficient $SrCoO_{3-z}$ and $LaCoO_{3-\delta}$ basic perovskite cells. The unit cell preserves the characteristics of perovskite framework and it is a superstructure induced by oxygen vacancies.

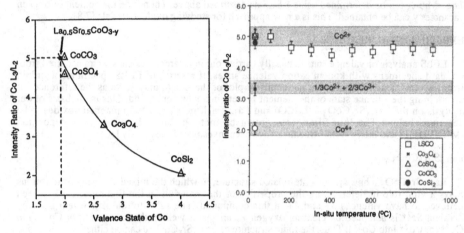

Figure 3. A plot of the intensity ratio of $I(L_3)/I(L_2)$ calculated from the spectra shown in Figure 2 for different compounds. The Co valence state in $La_{0.5}Sr_{0.5}CoO_{3-y}$ is obtained from the empirical fitting curve in reference to the known Co valences of the standard specimens.

Figure 4. The intensity ratio of Co $I(L_3)/I(L_2)$ measured from LSCO as a function of the specimen temperature in TEM, showing the stability of the white line intensity. The specimen decomposes at 900 °C.

To check the reliability of the white line intensity for valence state measurement, the LSCO specimen was annealed in-situ from room temperature to 900 °C, until the film is decomposed to lower oxides. Figure 4 is a plot of the intensity ratio of Co $I(L_3)/I(L_2)$ as a function of the specimen temperature. It is striking that the intensity ratio of $I(L_3)/I(L_2)$ (= 4.5-4.9) remains almost constant up to 900 °C, above which the specimen started to decompose. At 1000 °C, the decomposed film is polycrystalline, but the valence state of Co is still 2+, proving the stability of white line ratio.

$La_{0.67}Ca_{0.33}MnO_{3-y}$

LCMO is an important group of materials that have been found to exhibits the colossal magnetoresistive effect [12,13]. To determined the average valence of Mn, Figure 5 shows a plot of the experimentally measured intensity ratio of white lines $I(L_3)/I(L_2)$ for several standard specimens with known Mn valences. It is apparent that the intensity ratio of $I(L_3)/I(L_2)$ strongly depends on the valence state of Mn. This curve serve as the standard for determining the valence state of Mn although it is non-linear. The $I(L_3)/I(L_2)$ ratio for LCMO is 2.05 - 2.17, thus, the average valence state of Mn is 3.2 to 3.5. Substituting this value into Eq. (3), yields y ≤ 0.065, which is equivalent to less than 2.2 at.% of the oxygen content [7].

RELIABILITY AND ACCURACY OF THE MEASUREMENT

To determine the reliability and accuracy of the EELS measurement on valence state, a standard MnO_2 specimen was used in TEM to observe its in-situ reduction process as the specimen temperature is increased [16]. The chemical composition of the specimen is continuously determined using the integrated intensity of the ionization edges. A plot of composition, n_O/n_{Mn} and white line intensity, Mn $I(L_3)/I(L_2)$ is given in Figure 6, the shadowed bands indicate the white line ratios for Mn^{2+}, Mn^{3+} and Mn^{4+} as determined from the standard specimens of MnO, Mn_2O_3 and MnO_2, respectively. The reduction of MnO_2 occurs at 300 °C. As the specimen temperature increases, the O/Mn ratio drops and the $I(L_3)/I(L_2)$ ratio increases, which indicates the valence state conversion of Mn from 4+ to lower valence states. At T = 400 °C, the specimen contains the mixed valences of Mn^{4+}, Mn^{3+} and Mn^{2+}. As the temperature reaches 450 °C, the specimen is dominated by Mn^{2+} and the composition has O/Mn = 1.3 ± 0.5, slightly higher than that in MnO, which is consistent with the mixed valence of Mn cations and implies the uncompleted reduction of MnO_2.

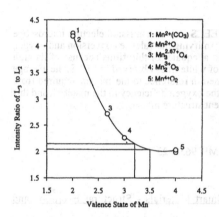

Figure 5. Plot of white line intensity ratio $I(L_3)/I(L_2)$ verses the valence state of Mn for several standard specimens of $MnFe_2O_4$, $MnCO_3$, MnO, Mn_2O_3 and MnO_2. A nominal fit of the experimental data is shown by a solid curve. The valence state of the LCMO film is obtained from the measured $I(L_3)/I(L_2)$ data.

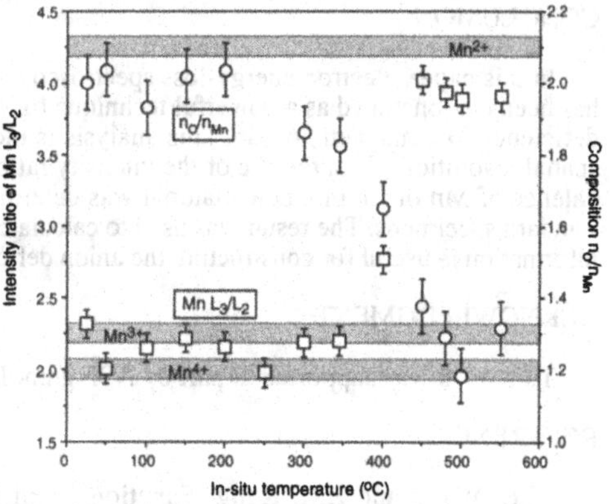

Figure 6. An overlapped plot of the white line intensity ratio of Mn $I(L_3)/I(L_2)$ and the corresponding chemical composition of n_O/n_{Mn} as a function of the in-situ temperature of the MnO_2 specimen based on EELS spectra, showing a continuously change in Mn valence state and oxygen composition.

DISCUSSIONS

It has been demonstrated by several authors [5,6] that the 3d and 4d electron occupations need to be determined with the use of the normalized white line intensity and the continuous spectrum at 50 to 100 eV above the edge threshold. This technique can give a linear fitting between the normalized white line intensity $[I(L_3) + I(L_2)]$ with the d state occupation, but it has two major shortcomings. One this linear curve is a good approximation for the entire occupation range from 0 to 10, but it is not a fair representation of the valence state if one is interested only in one element, such as Mn, whose 3d occupation can be 3, 2 or 1. The linear curve deviates largely from the experimental data if the 3d occupation is 1-3, leading to a significant inaccuracy in the measured result. With the use of the white line ratio $I(L_3)/I(L_2)$, the experimental data are steady and the difference between different valence state is significantly large, allowing more accurate determination of the valence state. The other one, in practical EELS, the intensity at a region 50-

100 eV above the edge threshold may be affected by the deconvolution and spectrum background subtraction procedures particularly when the noise level and gain variation are significant, resulting in a large error in the evaluation of the continuous part depending on specimen thickness. In contrast, the intensity ratio of $I(L_3)/I(L_2)$ has little dependence on the specimen thickness and its value is a steady number.

It has also been pointed out in the literature that the $I(L_3)/I(L_2)$ ratio is approximately the ratio of the electrons in the $j = 5/2$ and $j = 3/2$ states, thus, the white line intensity may be sensitive to the spin distribution [14,15]. From the EELS spectra of the four standard specimens, $CoCO_3$, $CoSO_4$, $CoSi_2$ and Co_3O_4 (with $Co^{2.67+}$) (see Figure 2), the former two with Co^{2+} show almost an identical $Co-L_{2,3}$ shape, while the last two with Co^{4+} and $Co^{2.67+}$ show a distinct difference in $Co-L_2$. A small difference in $I(L_3)/I(L_2)$ ratio between $CoCO_3$ and $CoSO_4$ in Figure 2 might be due to the spin effect, but this small fluctuation cannot significantly affect the measurement. This indicates that, at least in our case, the electron distribution in spin states, if any, plays a negligible role. The in-situ EELS analysis of valence conversion in transition metal oxides has shown the sensitivity and reliability of valence state measurement using white lines in Mn and Co oxides [16].

CONCLUSION

In this paper, electron energy-loss spectroscopy (EELS) in a transmission electron microscope has been demonstrated as a powerful technique for quantifying the valence conversion and oxygen deficiency in a magnetic oxide. This analysis is most adequate for thin films because of its high spatial resolution. With the use of the intensity ratio of white lines observed in EELS, the average valence of Mn or Co in a new material was determined in reference to the spectra acquired from standard specimens. The result was used to calculate the oxygen deficiency in the material, and this information is useful for constructing the anion deficient structure model.

ACKNOWLEDGMENT

This work was supported in part by NSF grant DMR-9632823.

REFERENCES

1. Z.L. Wang and Z.C. Kang, Functional and Smart Materials - Structural Evolution and Structure Analysis, Plenum Press, New York, 1997.
2. C. Zener, Phys. Rev. 82, 403 (1951).
3. R.F. Egerton, Electron Energy-Loss Spectroscopy in the Electron Microscope, 2nd ed., New York: Plenum Press, 1996.
4. J.H. Rask, B.A. Mine and P.R. Buseck, Ultramicroscopy 32, 319 (1987).
5. D.H. Pearson, C.C. Ahn and B. Fultz, Phys. Rev. B 47, 8471 (1993).
6. H. Kurata and C. Colliex, Phys. Rev. B 48, 2102 (1993).
7. Z.L.Wang, J.S. Yin, Y.D. Jiang and J. Zhang, Appl. Phys. Lett. 70, 3362 (1997).
8. Z.L. Wang and J.S. Yin, Phil. Mag. B, in press (1997).
9. J. Zhang, R.A. Gardiner, P.S. Kirlin, R.W. Boerstler and J. Steinbeck, Appl. Phys. Lett. 61, 2884 (1992).
10. Z.L. Wang and J. Zhang, Phys. Rev. B 54, 1153 (1996).
11. J.S. Yin and Z.L. Wang, Microscopy and Microanalysis 3, Suppl. 2, 599 (1997).
12. J. Jin, T.H. Tiefel, M. McCormack, R.A. Fastnacht, R. Ramech and L.H. Chen, Science, 264, 413 (1994).
13. R. Von Helmolt, J. Wecker, B. Holzapfel, L. Schultz and K. Samwer, Phys. Rev. Lett., 71, 2331 (1994).
14. S.J. Lloyd, G.A. Botton and W.M. Stobbs, J. Microsc. 180, 288 (1995).
15. J. Yuan, E. Gu, M. Gester, J.A.C. Bland and L.M. Brown, J. Appl. Phys. 75, 6501 (1994).
16. Z.L. Wang, J.S. Yin, W.D. Mo and J.Z. Zhang, J. Phys. Chem. B 101 (No. 35), 6793 (1997).

NEGATIVE MAGNETORESISTANCE IN $(Bi,Pb)_2Sr_3Co_2O_9$ LAYERED COBALT OXIDES

I. TSUKADA, T. YAMAMOTO, M. TAKAGI, T. TSUBONE, K. UCHINOKURA
Department of Applied Physics, The University of Tokyo, 7-3-1 Hongo, Bunkyo-ku, Tokyo
113, Japan, tsukada@ap.t.u-tokyo.ac.jp

ABSTRACT

Transport and magnetic properties of layered cobalt oxide $(BiPb)_2Sr_3Co_2O_9$ are investigated in detail under magnetic field up to 8 T. Parent compound, $Bi_2Sr_3Co_2O_9$, is a typical
band insulator with Co ions being in a low-spin 3+ state because of the well-separated
$d\epsilon$ and $d\gamma$ levels possibly due to a strong crystal field. We have tried to introduce holes
mainly by Pb substitution for Bi. The hole-doped sample shows metallic behavior in a
resistivity measurement between 300 and 30 K. Below 30 K, however, the resisitivity increases. Under the magnetic field the resistivity is strongly suppressed in this region. We
observed more than 30% resistivity drop at 2 K under $H = 8$ T, which is comparable to
insulating $(La,Sr)CoO_3$ system. We discuss the mechanism of hole doping and the origin of
negative magnetoresistance with tranport and magnetic properties, and point out that the
conventional double-exchange mechanism cannot be applied to this system. This means
that some new mechanism is necessary to explain this phenomenon.

INTRODUCTION

Since the discovery of high-temperature superconductors, various kinds of oxide materials have been discovered. Many of them were, however, synthesized only to search
for new superconductors, and their properties have not been studied sufficiently. Bi-Sr-
Co-O system is one of such materials [1, 2]. From a structural veiwpoint, it has a two-
dimensional structure with pseudosquare CoO_2 plane, and has a homologous series described by $Bi_2M_{n+1}Co_nO_y$ ($n = 1$, 2, and $M = $ Ca, Sr, Ba), as in BSCCO superconductors.
The $n=2$ phase (Co232) has been reported as an insulator for $M = $ Ca, Sr and as a metal
for $M = $ Ba [2]. These metallic Co232 compounds with highly two-dimensional transport
properties are good reference materials of high-temperature superconductors. The main
difference between Co221 and Co232 is the valence of Co ions; the former has basically
Co^{2+} state, while the latter has Co^{3+} state. This makes the physical properties of Co221
and Co232 completely different. We have tried to grow these compounds both as a bulk single crystal and a thin film. The latter was prepared by molecular beam epitaxy technique
(MBE) [3, 4], and has been tried as a barrier material to fabricate Josephson junctions
[5]. With single crystals, we have studied Pb substitution effect of $Bi_2Sr_3Co_2O_9$. In this
Proceedings, we report a large negative magnetoresistance in $(Bi,Pb)_2Sr_3Co_2O_9$.

EXPERIMENT

We show two samples without (Sample A) and with (Sample B) Pb in this paper. They
were grown by floating zone method with the growth condition similar to that for BSCCO

Table 1: Sample specifications. Chemical composition analyzed by ICP-AES is normalized as Co = 2.00.

| | chemical composition | | | | c-axis length [Å] |
	Bi	Pb	Sr	Co	
Sample A	2.02	0	2.64	2.00	29.861 ± 0.001
Sample B	1.73	0.59	2.61	2.00	30.030 ± 0.001

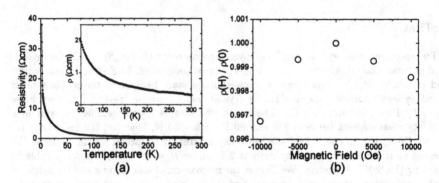

Figure 1: (a) Temperature dependence of the resistivity of sample A. (b) The normalized resistance of sample A at 5 K under magnetic field.

superconductors. Bi_2O_3, Pb_3O_4, $SrCO_3$, Co_3O_4 were mixed in nonstoichiometric ratio as Bi : Sr : Co = 2 : 2 : 2. If we start from the stoichiometric ratio (Bi : Sr : Co = 2 : 3 : 2), only Co221 phase was grown. The mixture of powders was presintered at 800°C in air for 24 h followed by regrinding and cold-press processes. The growth rate at floating zone process was kept around 0.2 mm/h, which is slightly faster than that of BSCCO.

Sample specifications are summarized in Table 1. Chemical compositions were analyzed by inductively coupled plasma-atomic emission spectroscopy (ICP-AES). It should be noted that Sr is naturally deficient in both samples. The Sr deficiency causes the doping of a certain amount of holes. As-grown samples were cut as a rectangular shape with typical dimension of $1 \times 4 \times 0.08$ mm³. The thinnest direction is parallel to the c axis. They were characterized by x-ray diffractometer to determine the c-axis length. Sample A has the same c-axis length as that previously reported [2]. Sample B shows the slightly longer c-axis length, but it is still far shorter than that of $Bi_2Sr_2CaCu_2O_8$. We did not measure the in-plane lattice constants. Resistivity and magnetization measurements were carried out with AC resistance bridge (PPMS 9T system: Quantum Design, Co. Ltd.) and SQUID magnetometer (chi-MAG 7T system: Conductus, Co. Ltd.).

RESULTS

The temperature dependence of the resistivity of sample A is shown in Fig. 1 (a). Sample A shows insulating behavior in the whole temperature range down to 2 K. We measured

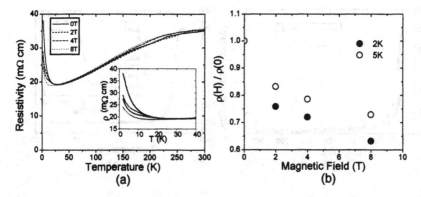

Figure 2: (a) Temperature dependence of the resistivity of sample B under various magnetic fields. The inset shows a low-temperature regions. (b) The normalized resistance of sample B at 2 K and 5 K under magnetic field.

magnetoresistance at 5 K with the magnetic field H parallel to the c axis as is shown in Fig. 1 (b). The observed magnetoresistance is quite small, which is only 0.15% under $H = 1$ T. This is a typical behavior of the samples without Pb.

On the other hand, the effect of magnetic field is remarkable in sample B as shown in Fig. 2. When $H = 0$ T, sample B took minimum resistivity around 30 K, and then showed steep increase down to 2 K. Application of a magnetic field along the c axis drastically changes the resistivity in this region. The resistivity at 2 K is approximately 38 mΩcm being comparable to the room temperature value. With the application of the magnetic field, it decreases and reaches 24 mΩcm under $H = 8$ T. The normalized magnetoresistance defined as $\Delta\rho(H) = (\rho(H) - \rho(0))/\rho(0)$ is estimated as -37%, which is larger than that of metallic $(La_{1-x}Sr_x)CoO_3$ $(x \geq 0.2)$ [6, 7]. Our results are rather comparable to insulating $(La_{0.93}Sr_{0.07})CoO_3$ [7], which shows -40% magnetoresistance at 50 K under 6 T.

Magnetization measurement revealed that the large negative magnetoresistance in sample B might be associated with a magnetic ordering. We show the temperature dependence of the susceptibility under 3 T in Fig. 3 (a). It is noticed that the susceptibility is far smaller than that of $LaCoO_3$ in the whole range below the room temperature [8, 9, 10]. The susceptibility was fitted by Curie-Weiss law, and we obtained quite small effective number of Bohr magneton (p) from the Curie constant; $p = 0.532$ and 0.643 for sample A and B, respectively. If we use a simple assumption that only Co ions contribute to the Curie term, we can estimate the averaged S as 0.0663 and 0.0943, respectively. We did not observe any sign of low spin-high spin [8, 9, 10] or low spin-intermediate spin [11, 12] crossover. Our data indicates that almost all Co^{3+} ions are in the low-spin state even up to room temperature, which means that the energy separation between $d\epsilon$ and $d\gamma$ levels is larger than that of $LaCoO_3$. Since we do not have data of high temperature susceptibility, we have no idea of the exact value of this gap, but it should be more than 300 K.

Despite small μ_{eff} sample B shows ferromagnetic transition below 4 K. Figure 3 (b) shows low temperature magnetization of sample B measured under $H = 100$ Oe. The saturation of magnetization appears below 4 K. The presence of ferromagnetic phase is also confirmed from M-H characteristic as shown in the inset of Fig. 3 (b). At 2 K,

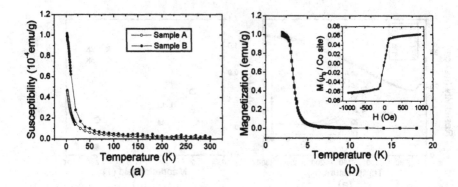

Figure 3: (a) Temperature dependence of the susceptibility of sample A and B under H = 3 T applyled along the a axis. (b) Temperature dependence of the magnetization of sample B under H = 100 Oe. The field direction is along the c axis. The inset shows M-H characteristics of sample B at 2 K.

magnetization rapidly increases up to H = 150 Oe, where the magnetization seems to be saturated. It continues, however, to increase and finally is saturated at H = 3 T, where the magnetization reaches approximately $0.17\mu_B$ / Co site. This small value is qualitatively consistent with the small μ_{eff} deduced from the Curie-Weiss fit in paramagnetic phase.

DISCUSSION

We believe that the negative magnetoresistance observed in sample B is strongly related with its ferromagnetic transition. It is well known that in a perovskite manganese oxides a double exchange interaction between the neighboring Mn ions stabilizes a ferromagnetic phase, and simultaneously, electrical conduction revives, which is the simplest explanation of the giant magnetoresistance in manganese oxides. The ferromagnetism in $(La,Sr)CoO_3$ is also explained in the same framework [6]. Roughly speaking, the double exchange scenario requires both $d\epsilon$ and $d\gamma$ electrons at a single ion site. The magnetization measurement of $(Bi,Pb)_2Sr_3Co_2O_9$ systems indicates that almost all Co^{3+} ions stay in a low-spin state and have no $d\gamma$ electrons. The estimated S for both samples are far smaller than those obtained if we assume that Co^{3+} ions take high-spin ($S = 2$) or intermediate-spin ($S = 1$) states. The stability of low-spin Co^{3+} ions in the parent compound was reported by Tarascon et al. [2].

The origin of such a stable low-spin state is due to the crystal field. Co232 structure can be regarded as a member of layered perovskite oxides as far as Co is concerned. The Co-O distance shows a significant difference between $LaCoO_3$ and Co232. The in-plane lattice constants of Co232 is quite small. For an M = Ca phase, a = 4.89 and b = 5.06 Å were reported [2], and we estimated $a \sim b \approx 5.0$ Å for thin film $Bi_2Sr_3Co_2O_9$ [3]. These values are the in-plane distance between the second-nearest-neighbor Co ions, and we can estimate the in-plane Co-O distance approximately 3.5 Å in M = Ca and Sr phases. On the other hand, Co-O distance in $LaCoO_3$ is 3.826 Å [13]. This difference suggests the presence of strong crystal field in Co232. The Co-O distance along the c axis is also suggested to be

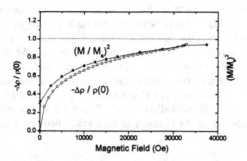

Figure 4: Field dependence of the square of normalized magnetization and the normalized magnetoresistance. The difference between them is remarkable, especially below 5000 Oe.

small. Tarascon *et al.* reported that the c-axis length of $Bi_2M_3Co_2O_9$ is 29.28 and 29.85 Å for M = Ca and Sr, respectively [2]. These values are smaller than that of $Bi_2Sr_2CaCu_2O_8$ superconductors, where the Cu-O distance is close to the Co-O distance in $LaCoO_3$. Thus we conclude that both the in-plane and out-of-plane Co-O lengths are smaller in Co232 than in $LaCoO_3$, which causes the large crystal-field splitting between $d\epsilon$ and $d\gamma$ levels in Co232. That is a main reason why Co^{3+} ions take low spin state even up to room temperature.

The magnetization of sample B indicates that hole is doped in $d\epsilon$ level. In $(La,Sr)CoO_3$, a rapid increase of magnetization in the temperature range typically below 100 K is attributed to the formation of a magnetic polaron with a high spin number around the doped hole [14]. In contrast to it, a doped hole in $(Bi,Pb)_2Sr_3Co_2O_9$ does not seem to be dressed with a magnetic moment of such a high spin number. The change of the average S from sample A to B is quite small. The saturated magnetization of sample B at 2 K indicates that only 10% of Co ions is in a low spin state Co^{4+} ion with $S = 1/2$. The average valence of Co ions estimated from the chemical composition in sample B is +3.2 corresponding to that 20% of Co^{3+} ions change to Co^{4+} states. The origin of this deviation is unclear because we do not know the exact oxygen content in sample B.

The breakdown of the double-exchange scenario is particularly remarkable in a low-field region. Figure 4 shows the temperature dependence of the square of normalized magnetization $((M/M_s)^2)$ and the normalized magnetoresistance $(-\Delta\rho/\rho(0))$ as functions of magnetic field. It is obvious that the magnetization and the magnetoresistance do not scale each other, especially in the low-field region. According to the theory based on the double-exchange interaction [15], $-\Delta\rho/\rho(0)$ is proportional to $(M/M_s)^2$. Our system does not follow this scenario. At the present stage, we have no definite answer to the mechanism of ferromagnetism and negative magnetoresistance in the $(Bi,Pb)_2Sr_3Co_2O_9$ system, which should be made clear in future.

SUMMARY

We have synthesized an oxide system of $(Bi,Pb)_2Sr_3Co_2O_9$. The parent compound is insulating and basically nonmagnetic. Most of the Co^{3+} ions are in a nonmagnetic low-spin state $(S = 0)$ because of the presence of a strong crystal field, which makes this system unique among the perovskite cobalt-based oxides. Hole doping by Pb substitution induces metallic conduction down to 30 K, but steep increase of the resistivity is observed in the lower temperature region. This resistivity behavior is related to the ferromagnetic transition with quite small saturation magnetization. Application of the magnetic field drastically reduces the resistivity in the lower temperature region. We observed typically over 30% negative magnetoresistance at 2 K under 8 T field, which is, to our knowledge, one of the largest negative magnetoresistances among cobalt-based metallic oxides.

We appreciate S. Konno's collaboration at the early stage of this study. We also thank I. Terasaki and Y. Tokura for fruitful discussion. This work is partially supported by Grant-in-Aid for COE Research of the Ministry of Education, Science, Sports and Culture.

REFERENCES

1. J. M. Tarascon, P. F. Miceli, P. Barboux, D. M. Hwang, G. W. Hull, M. Giroud, L. H. Greene, Y. LePage, W. R. McKinnon, E. Tselepis, G. Pleizer, E. Eibscutz, D. A. Neumann and J. J. Rhyne, Phys. Rev. B **39**, 11587 (1989).

2. J. M. Tarascon, R. Ramesh, P. Barboux, M. S. Hedge, G. W. Hull, L. H. Greene, M. Giroud, Y. LePage, W. R. McKinnon, J. V. Waszczak, and L. F. Schneemeyer, Solid State Commun. **71**, 663 (1989).

3. I. Tsukada, I. Terasaki, T. Hoshi, F. Yura and K. Uchinokura, J. Appl. Phys. **76**, 1317 (1994).

4. I. Tsukada, M. Nose and K. Uchinokura, J. Appl. Phys. **80**, 5691 (1996).

5. M. Nose, I. Tsukada and K. Uchinokura, Czech. J. Phys. **46**, 1323 (1996).

6. S. Yamaguchi, H. Taniguchi, H. Takagi, T. Arima and Y. Tokura, J. Phys. Soc. Jpn. **64**, 1885 (1995).

7. R. Mahendiran and A. K. Raychaudhuri, Phys. Rev. B **54**, 16044 (1996).

8. R. P. Heikes, R. C. Miller, and R. Mazelsky, Physica **30**, 1600 (1964).

9. P. M. Raccah and J. B. Goodenough, J. Appl. Phys. **39**, 1209 (1968).

10. M. Itoh, I. Natori, S. Kubota and K. Motoya, J. Phys. Soc. Jpn. **63**, 1486 (1994).

11. M. A. Korotin, S. Y. Ezhov, I. V. Solovyev, V. I. Anisimov, D. I. Khomskii and G. A. Sawatzky, Phys. Rev. B **54**, 5309 (1996).

12. S. Yamaguchi, Y. Okimoto and Y. Tokura, Phys. Rev. B **55**, R8666 (1997).

13. K. Asai, O. Yokokura, N. Nishimori, H. Chou, J. M. Tranquada, G. Shirane, S. Higuchi, Y. Okajima and K. Kohn, Phys. Rev. B **50**, 3025 (1994).

14. S. Yamaguchi, Y. Okimoto, H. Taniguchi and Y. Tokura, Phys. Rev. B **53**, R2926 (1996).

15. N. Furukawa, J. Phys. Soc. Jpn. **63**, 3214 (1994).

SURFACE MORPHOLOGY AND LATTICE MISFIT IN YIG AND LA:YIG FILMS GROWN BY LPE METHOD ON GGG SUBSTRATE

Duk-yong Choi, Su-jin Chung
Dept. of Inorg. Mat'ls Eng., Seoul National Univ., Seoul 152-742, Korea

ABSTRACT

$Y_3Fe_5O_{12}$(YIG) and La-doped YIG films were grown on the {111} GGG substrate using the $PbO-B_2O_3$ flux system. Pb, La incorporation and lattice misfit and annealing behaviors were studied. In the case of LPE growth of YIG film, lead ions from flux are substituted inevitably, and they play an important role in controlling film misfit. For a complete lattice matching, high supercooling is necessary in pure YIG growth, but this induces high defect concentration. In this experiment, La ions were added in the solution to sufficiently increase lattice parameter of the film grown under low supercooling. The concentration of substituted Pb and La were increased as the growth temperature was lowered and growth rate increased. The effective distribution coefficient of La was about 0.2 at a supercooling of 30℃. The optimum growth conditions which bring about very small misfit were determined by measuring the misfit by double crystal diffractometer. Strain distributions of pre-annealed and annealed samples were investigated by triple crystal diffractometer.

INTRODUCTION

The lattice misfit of YIG films grown on GGG substrates by LPE method can be controlled by various techniques. Using the PbO-based flux system, the dependency of Pb incorporation level on the growth temperature and growth rate, and the substitution of rare earth ions for Y are widely used for controlling the lattice misfit. For magneto-optic and magneto-static applications, YIG films should be grown with a thickness of more than a few tens microns. But in the growth of garnet films, misfit dislocations which relax the stress resulting from the lattice parameter differences are not generated[1,2]. Therefore, it is needed to reduce the misfit to avoid crack on the films. P.Görnert et al.[3] obtained experimentally the maximum lattice misfit needed for the growth of crack-free films. According to their results, misfit lower than 0.02% is necessary to grow film with 100μm thickness.

The misfit can be measured by double crystal diffraction(DCD)[2,3,4]. Especially, the lattice distortion form can be deduced by the diffraction of asymmetrical planes which are not parallel to the substrate. In the case of garnet films on (111) GGG substrates, it was shown that lattice type of film is changed from cubic to rhombohedral[5]because misfit component is absent parallel to the substrate and present perpendicular to that[4,5]. The misfit and FWHM can be measured precisely by virtue of recent development in x-ray equipment. Especially triple crystal diffraction (TCD) can measure the lattice distortion by mapping around the reciprocal lattice points[6].

It was known that garnet crystals exhibit growth-induced magnetic anisotropy and this anisotropy results from the site-preferences of cations. To stabilize the resonant frequency in the microwave devices, magnetic anisotropy should be annihilated. It was reported that this anisotropy can be annihilated by thermal annealing above 1100℃[7]. Plaskett et al.[8] reported that hillocks were formed on the film surface and the misfit was reduced by annealing the garnet film with high Pb incorporation. And these phenomena were supposed to occur by the reduction of Pb^{4+} ion to Pb^{2+} at high temperature[9].

In our work, the lattice distortion of films with positive, null and negative misfits were investigated by DCD and TCD mappings. The changes of lattice distortion and surface morphology of annealed samples were observed by TCD mappings , electron and optical microscope. From these studies, we can predict the optimum growth conditions and the incorporation mechanism of Pb ions.

Mat. Res. Soc. Symp. Proc. Vol. 494 © 1998 Materials Research Society

EXPERIMENT

YIG and La-doped YIG were grown by LPE method on (111) GGG substrates with 500 μm thickness. The composition of solution was as follows(R in molar ratios):

$$R1 = [Fe_2O_3]/[Y_2O_3] = 20.1 \tag{1}$$

$$R3 = [PbO]/[B_2O_3] = 6.5 \tag{2}$$

$$R4 = ([La_2O_3]+[Y_2O_3]+[Fe_2O_3])/([PbO]+[B_2O_3]+[La_2O_3]+[Y_2O_3]+[Fe_2O_3]) = 0.13 \tag{3}$$

$$[Y_2O_3]/[La_2O_3] = 14 \text{ or } \infty \tag{4}$$

By increasing the supercooling of solution, the incoporation level of Pb ions increased. As a result, the lattice parameters of grown films were enlarged. But in the case of pure YIG, it was difficult to increase the lattice parameter of film exceeding that of GGG by lowering the growth temperature. Thus we grew the films by substituting of La for Y ion to overcome this problem. The saturation temperature of solution was about 875℃. When the supercooling was 30℃, the misfit was lower than 0.01% and the effective distribution coefficient of La was around 0.2. Samples with various misfits were grown at the supercooling range of 5~67℃. The thickness of grown films were 3~5 μm and there was no surface-defect.

The extent and form of lattice distortions were observed by DCD and TCD mappings of symmetric (444) and asymmetric (642) planes. Fig. 1 shows the schematic drawing of x-ray diffractometer. CuKα radiation and the maximum output power of 2kW was used. The monochromator consists of two silicon channel-cut crystals(CCC) arranged in a (+,-,+,-) setting in such a way that x-ray undergoes a (022) reflection at each crystal faces. In TCD mode we placed another Si CCC before the detector. TCD mapping can be obtained by measuring the diffraction intensities around 444 and 642 reciprocal points in θ-2θ scan modes at various ω values. The number of data points were about 10,000 and the integration time was 2 seconds at each point.

Samples were annealed at 1150℃ in air. The lattice distortion and surface morphology of annealed samples were investigated by DCD, TCD mappings and optical microscope. To investigate the lattice distortion mechanism we observed the samples before and after annealing by transmission electron microscope.

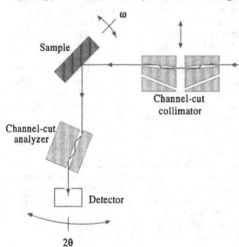

Fig. 1 Schematic drawing of x-ray diffractometer.

RESULTS

Lattice Distortion Before Annealing

The films were grown on both sides of substrates in order to suppress the bending of sampies. The radii of curvature of bending were measured by DCD and was found to be larger than 20m. On both sides of films there was no difference in the extent of misfit and its variation with position. This means that films were grown uniformly across the whole area of substrate. Pre-annealed two samples with positive and negative misfit were investigated by TCD mappings.

Fig. 2 shows the TCD mapping of two samples around (642) reciprocal

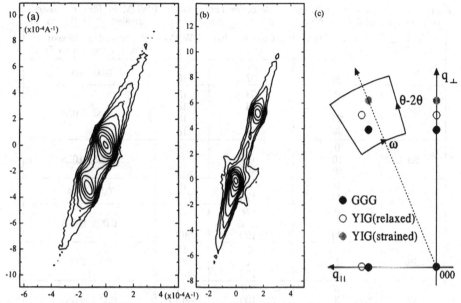

Fig. 2 TCD mappings of samples with positive(a) and negative(b) $\Delta\theta$ and the relative positions of two peaks with the extent of relaxation(c).

points. (642) diffracting plane is inclined with film surface by an angle 22.2°. The directions of abscissa and ordinate are in accordance with the reciprocal coordinate parallel and perpendicular to (642) plane. High intensity peaks are due to the diffraction by the film. The surface streaks are seen on an extension line along the two peaks. The components of separations between two peaks along $q_\perp$ and $q_\parallel$ indicate $\Delta\theta_B$ and $\Delta\varphi$. As shown in fig. 2-(c), the extent of relaxation can be deduced from the relative positions of two (642) peaks. The angle between the extension line along the two peaks and $q_\perp$ decreases as the extent of relaxation increases. The measured angles in fig. 2-(a) and(b) were about 22° and this indicates that films grew pseudomorphically regardless of the types(tensile or compressive) of stress.

Lattice Distortion After Annealing

Table I lists the changes of Bragg angle difference($\Delta\theta_{444}$) measured in (444) reflection, full-width at half maximum(FWHM) of film, estimated or measured Pb and La content(in formular unit) when samples were annealed at 1150℃ in air for different time. (T) designates the value was measured by TCD method. Pt incorporation from the crucible was lower than 0.01(in f.u.). In previous works, we had obtained the relationships between the lattice parameters of films and Pb, La content measured by electron probe micro-analysis(EPMA) and the supercooling of solution. When Pb and La were added, the increase of lattice parameter was around 0.1Å/f.u. So we could estimate Pb and La contents if supercooling and $\Delta\theta_{444}$ were known. And the estimated values agreed well with the measured ones.

When $\Delta\theta$ had positive values(S1,S2), $\Delta\theta$ did not change and FWHM increased silghtly by annealing. This result is different from the previous report[9] that the misfit was reduced by annealing due to the reduction of Pb^{4+} to Pb^{2+}. Due to the fact that cracks were not observed on the films we could know that these samples before

Table I The changes of Bragg angle difference($\Delta\theta$ 444), FWHM of film, estimated or measured Pb and La content(in f.u.) when samples were annealed at 1150℃ in air.

Sample #	annealing time (hrs)	$\Delta\theta_{444} = \theta_f - \theta_s$ (sec)	FWHM (sec)	Estimated or Measured Pb, La content(in f.u.)
S1	0	87	14	0.03, 0.00
	40	87	18	
S2	0	63	13	0.033, 0.013
	1	63	14	
	20	61	16	
	40	61	18	
S3	0	0	14	0.053, 0.035
	10	0	42	
	60	0	35(T)	
S4	0	-59	16	0.09, 0.04
	40	-120	30(T)	
S5	0	-69	16	0.095, 0.045
	20	-107	-	
	70	-109	28(T)	
S6	0	-97	14	0.11, 0.05
	1	-111	28(T)	
	40	-135	28(T)	
S7	0	-107	14	0.115, 0.055
	3	-105	22(T)	
	53	-122	23(T)	
S8	0	-115	15	0.04, 0.13
	1	-120	16	
	20	-100	30	

(T) designates FWHM of the film measured by TCD method.

annealing exerted small tensile stress. Also, the surface morphology was not changed by annealing. S1 sample which didn't contain La ion showed a similar annealing behavior with S2, though it had the largest $\Delta\theta$. S3 had no misfit because it was grown at suitable supercooling and La content. By annealing $\Delta\theta$ of S3 did not change similar to S1 and S2. But FWHM increased the more with the annealing time. These results indicated that the structure of film was changed by annealing so that the distribution of interplanar distances of (444) became broader. Samples of S4~S7 had negative $\Delta\theta$ because they were grown at high supercooling so that Pb and La contents were high. Facetting induced by the compressive stress was not observed on the film surface. FWHM of these samples incresed by annealing, and the increments were independent on $\Delta\theta$ of pre-annealed samples. As the annealing time increased, $\Delta\theta$ reduced so that the misfit was enlarged.

Fig. 3-(a),(b) and (c) show the TCD mappings of (444) reflection when S6 was annealed for 0, 1 and 40 hours, respectively. These mappings were performed at constant x-ray power. In the mapping of (642) reflection, we could not make a distinction between the film and the substrate. In fig. 3 the intensity difference of each contour line is 2.3 times. As explained in the previous section, the surface streaks are seen on an extension line along the two peaks. The other streak(in fig. 3-(a)) tilted by 30° with $q_\perp$ is due to the analyser crystal. As the annealing time increased, the maximum intensity of film was reduced and the detectable area of x-ray was broader. As a result, FWHM of film increased. This means that the structure of (444) planes was changed a bit. Especially FWHM changed more along the $q_\parallel$-direction. Due to he fact that the separation of two peaks became farther by annealing, it can be concluded that the interplanar distance of (444) increases.

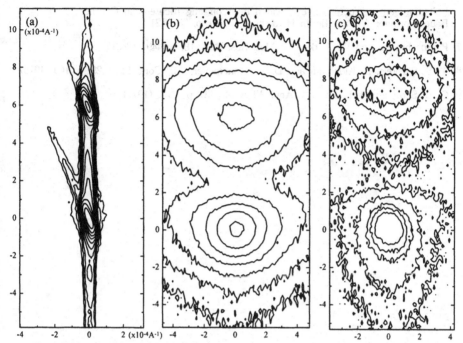

Fig. 3 TCD mappings of (444) reflections when S6 was annealed for 0(a), 1(b), 40hours(c)
(The intensity difference of each contour line is 2.3times).

CONCLUSIONS

The lattice distortions of films were investigated by TCD mapping. YIG ar d La:YIG films grew pseudomorphically on (111) GGG substrates regardless of the sign of misfits. When the films were annealed at high temperature, $\Delta\theta$ and FWHM changed differently according to the signs of $\Delta\theta$. In the case of positive $\Delta\theta$, FWHM changed slightly but $\Delta\theta$ did not. FWHM increased with annealing time in the film with nearly zero misfit. When the film had negative $\Delta\theta$, FWHM increased and misfit was enlarged.

REFERENCES

1. Howard L. Glass, Proc. of the IEEE **76** (2), p. 151 (1988).

2. R. Hergt, H. Pfeiffer, P. Görnert, M. Wendt, B. Keszei and J. Vandlik, Phys. Stat. Sol. (a) **104**, p. 769 (1987).

3. P. Görnert, R. Hergt, E. Sinn, M. Wendt, B. Keszei and J. Vandlik, J. Crystal Growth **87**, p. 331 (1988).

4. H. L. Glass and M. T. Elliott, J. Crystal Growth **27**, p. 253 (1974).

5. S. Isomae, S. Kishino and M. Takahashi, J. Crystal Growth **23**, p. 253 (1974).

6. Alois Krost, Günther Bauer and Joachim Woitok, in Optical Characterization of

Epitaxial Semiconductor Layers, edited by Wolfgang Richter and Günther Bauer(Springer-Verlag, Berlin Heidelberg, 1996), p. 287-421.

7. P. Röschmann and W. Torksdorf, Mat. Res. Bull. **18**, p. 449 (1983).

8. T. S. Plaskett, E. Klokholm and D. C. Cronemeyer, AIP Conf. Proc. **29**, p. 109 (1975).

9. G. B. Scott and J. L. Page, J. Appl. Phys. **48** (3), p. 1342 (1977).

MAGNETOTRANSPORT IN THIN FILMS OF $La_{n-nx}Ca_{1+nx}Mn_nO_{3n+1}$ (n=2,3, and ∞)

H. Asano, J. Hayakawa, and M. Matsui
Dept. of Crystalline Materials Science, Nagoya University, Furo-cho, Chikusa-ku, Nagoya, 464-01, Japan, asano@numse.nagoya-u.ac.jp

ABSTRACT

With a use of the epitaxial a-axis thin films of perovskite series $La_{n-nx}Ca_{1+nx}Mn_nO_{3n+1}$ (n=2,3, and ∞) with fixed carrier concentration (x=0.3), the transport properties of the series compounds have been examined to be associated with the difference in the number of the MnO_2 layers. Results have indicated that a reduction in the number of layers results in systematic changes in the various features. These include an increase in resistivity, a decrease in resistivity peak temperature T_c^{ρ} corresponding to the metal-insulator transition, an enhancement of the maximum MR near T_c^{ρ}, and an increase in low temperature intrinsic MR. In order to explain the variation in these features with the number of MnO_2 layers, it is necessary to take both anisotropic c-axis transfer interaction and two-dimensional spin fluctuation into account.

INTRODUCTION

The observation of the colossal magnetoresistance (CMR) effect[1-6] in the mixed valence manganese perovskites has revived interest in the physical properties of this class of compounds. Several studies on the pseudocubic perovskites $La_{1-x}M_xMnO_3$ (M=Ca, Sr) have shown that materials with an optimized carrier concentration $x\sim0.3$, which undergo a transition from the paramagnetic insulator state to the ferromagnetic metal state upon cooling, exhibit CMR in a narrow temperature range around the Curie temperature. There is evidence that distortion of the Mn-O-Mn bond, which modifies the one-electron bandwidth, is a crucial parameter governing spin-charge coupling, hence the magnitude of the CMR.

Recent attention has focused on layered perovskite series of Ruddlesden-Popper compounds $(La-M)_{n+1}Mn_nO_{3n+1}$, since CMR was observed in the n=2 compounds (M=Sr, Ca).[7,8] In contrast to pseudocubic perovskites (n= ∞), the layered versions of $(La-M)_{n+1}Mn_nO_{3n+1}$, consisting of perovskite blocks, n MnO_6 octahedra thick, offset along the c-axis and with an intervening layer of (La-M)O ions, possess a two-dimensional and anisotropic character. The layered manganites with n=2,3 and $x\sim0.3$ are found to be ferromagnets exhibiting CMR effects. Studies[7-11] of the n=2 compound have shown that they exhibit remarkable features including MR ratio enhancement, anisotropy in charge transport, magnetization, and magnetostriction, existence of two-dimensional ferromagnetic ordering in a certain temperature range, the characteristic low-temperature intrinsic MR effect. However, the microscopic origin of several features remains unclear, and a unified interpretation explaining the variation in the features with the number of MnO_2 layers in the perovskite family is not yet available.

In this paper, we present results on the transport properties of $La_{n-nx}Ca_{1+nx}Mn_nO_{3n+1}$ thin films for n=2,3, and ∞ together with a discussion of the variation in various features with different numbers of MnO_2 layers. Based on a comparison of the magnitude of these features among the compounds we argue that the reduced c-axis transfer interaction as well as the two dimensional spin fluctuation, which result from the anisotropic (double) exchange interaction, play an important role in enhanced CMR and other related properties.

EXPERIMENT

A magnetron sputtering technique was used for preparing the thin film samples of $La_{n-nx}Ca_{1+nx}Mn_nO_{3n+1}$ (n=2,3, and ∞). The preparation methods and conditions have been reported elsewhere.[8,12] The film thickness was typically 150 nm, and the substrates were MgO (001). Energy dispersive X-ray microanalysis (EDX) indicated that the compositions of the films were nearly identical to the nominal one. X-ray diffraction analysis showed that all the films were single phase and had an a-axis (100) normal orientation to the substrate surface. The ϕ-scan (in-plane rotation) X-ray analysis on the n=2,3 thin films indicated that the a-axis films were ordered in the film plane and consisted of two domains, rotated at $90°$ to each other, in the plane (the so-called mosaic structure). The diffraction data including the ϕ-scan mode gave us lattice parameters of a_0=0.3864 nm for the n=∞ films, a_0=0.3864 nm, c_0 =1.920 nm, and c_0/a_0=4.970 for the n=2 films, and a_0=0.3867 nm, c_0 =2.680 nm, and c_0/a_0=6.930 for the n=3 films.

RESULTS

The top panel in Fig. 1 shows the temperature dependence of the resistivity ρ in a zero field for thin film samples of $La_{n-nx}Ca_{1+nx}Mn_nO_{3n+1}$ with n=2,3, and ∞. All three samples display a maximum in the ρ-T curve at T_c^ρ with a metallic behavior below and a semiconducting behavior above the temperature. When comparing these thin film data (n=2, ∞) with those of polycrystalline bulk samples with identical n and carrier concentration values (no available data for n=3 bulk), the present thin films exhibit T_c^ρ values, which are in good agreement with the bulk values (140 K for n=2, and 240 K for n=∞).

The slight reduction ($\sim$10K) in the T_c^ρ values in the thin films might be due to strain induced by the substrate or slight deviation of the carrier concentrations. The thin films exhibited reasonably low resistivity values at 4.2 K (about two orders of magnitude lower than those of the

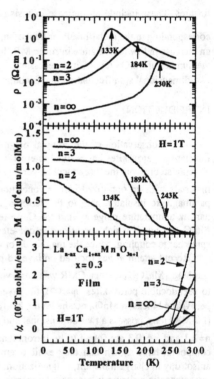

Fig. 1. Temperature dependence of resistivity ρ, magnetization M and inverse susceptibility $1/\chi$ for thin films with n=2,3, and ∞.

bulk samples), indicating a film epitaxial quality free from granularity.

The middle panel in Fig. 1 shows the temperature dependence of the magnetization M in a magnetic field of 1 T for thin film samples with n=2,3, and ∞. The ferromagnetic Curie temperatures T_c are determined by the conventional M^2-T method. All the thin film samples possessed the T_c values which are correlated well with the T_c^{ρ} values. However, the three samples exhibit different behavior in the M-T curves, especially at temperatures higher than the T_c. The inverse susceptibility $1/\chi$ -T curves at the high temperatures for the three samples are shown in the bottom panel in Fig. 1. A linear temperature dependence of $1/\chi$ is observed above $\sim$ 260 K, and the temperatures Θ_p, extrapolated from the Curie-Weiss plot, are about 250 K for all the samples. Although the Θ_p value is close to the T_c value for the n= ∞, the Θ_p value is much higher than the T_c values for the layered compounds with n=2 and 3. The large deviation between the T_c and the Θ_p values has been also reported for single crystals of the layered (n=2) compounds[10]. This could be interpreted as being the result of the anisotropic electron transfer and exchange interaction, which is due to the different Mn-O bond configurations in the a-axis and c-axis directions. Therefore, the upper transition temperature, Θ_p, can be ascribed to the stronger a-axis interaction, and the lower one, T_c, to the weaker c-axis interaction.

The temperature dependence of the negative MR ratio $-\Delta\rho/\rho(0)$ (H=1 T) for the three samples is shown in Fig. 2. The MR ratio is defined as $\Delta\rho/\rho(0)=(\rho(H)-\rho(0))/\rho(0)$, where $\rho(H)$ and $\rho(0)$ are the resistivity in an applied magnetic field and the zero field resistivity, respectively. It is clear that all the samples exhibited maximum magnetoresistance near T_c^{ρ}. For the n=∞ compound, the relatively large MR associated with T_c^{ρ} is restricted to a narrow temperature range, and MR ratio decreases with decreasing temperature and becomes negligible in the low temperature range. By contrast, for the layered compound with n=2, 3, we can see a broad maximum in the MR-T curve and observe an appreciable MR effect over a wide temperature range from low temperature to around T_c^{ρ}. It is shown that a reduction in n resulted in remarkable enhancement of both the maximum and low temperature MR.

We now turn to the origin of the relationship between number of MnO_2 layers, n and the electrical transport in $La_{n-nx}Ca_{1+nx}Mn_nO_{3n+1}$. As a consequence of the structural change from perovskite (n=∞) down to n=2, which is brought about by insertion of the insulating $(La,Ca)_2O_2$ layer into the perovskite layers, a two-dimensional character is introduced. This is expected to produce an anisotropic reduction of the one-electron (e_g) bandwidth. The bandwidth is the critical parameter governing the CMR and related properties in perovskite manganites. Within the framework of a simple double exchange theory, the exchange interaction is proportionally related

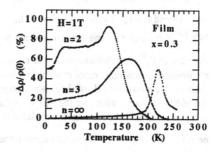

Fig. 2.   Temperature dependence of MR ratio ($-\Delta\rho/\rho(0)$) for thin films with n=2,3, and ∞.

to the transfer interaction t , and the ferromagnetic transition is associated with the M-I transition. Therefore, the M-I and ferromagnetic transition temperature is expected to be proportional to the effective bandwidth (W). The reduction in W accompanied by a reduction in n was also demonstrated in the magnitudes of the electrical resistivities. The increase in electrical resistivity ρ with decreasing n is attributed to a narrowing of the one-electron bandwidth. This is expected from the basic concept of the double exchange model: both T_c and ρ reflect real charge motion determined by the transfer interaction t, via the one-electron bandwidth.

When discussing the n dependence of the MR effect, we first consider the effect of the bandwidth W of the e_g electron. The role of the bandwidth on the magnitude of MR have been suggested from the previous results on three-dimensional compounds with the different average ionic radius of the La-site. It is shown that increasing the bond distortion enhances the MR response and reduces the ferromagnetic transition temperature T_c. Several reports on the three dimensional compounds have shown a general trend that the MR response is nearly inversely

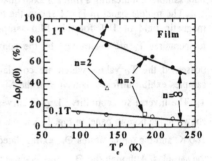

Fig. 3. MR ratio (-Δρ/ρ(0)) as a function of the T_c^{ρ} values for thin films in the present study (n=2,3, and ∞) and previous studies (Ref. 2-4) (n=∞). The closed symbols denote data in H=1 T, and the open symbols denote data in H=0.1 T. The (open and closed) circles (indicated by arrows) denote present data for n=∞, and the other circles denote reported data (Ref.1-3) for n=∞. The squares denote data for n=3, and the triangles denote data for n=2.

proportional to the ferromagnetic transition temperature T_c. Within the framework of the recently developed Kondo lattice model [13], the narrowing of the bandwidth results in both the lowering of the T_c and enhanced coupling (J_H/W) between itinerant carriers and localized spins. To investigate the role of W in the relationship between the MR response and T_c and related issues in the present compounds, we plot in Fig. 3 the relationship between magnitude of the maximum MR (at 0.1T and 1T) and the T_c^{ρ} values obtained in the present study (n=2,3 ∞) and the previous studies (n= ∞).[2-4] It is clear that the MR magnitude in the n=2 compounds increases more rapidly with a decrease in T_c^{ρ}, and no (inverse) linear relationship can be seen between them. The deviation from the linear relationship is more evident in a low magnetic field (0.1T). This indicates that we cannot attribute the enhanced MR of these two-dimensional compounds solely to the narrower bandwidth. An additional factor affecting the MR response is the spin correlation inherent to two-dimensional compounds. This is because spin-correlated fluctuation scattering has a dominant effect on the MR near T_c^{ρ}. To obtain an insight into these issues, we examined the field dependence of normalized resistivity ρ(H)/ρ(0) for the three thin film samples at T_c^{ρ} and the results are shown in Fig. 4. In the ρ(H)/ρ(0)-H curves, a steep decrease in ρ(H)/ρ(0) is observed in the low magnetic field range below 0.5 T, particularly for the n=2 compound. The observed field dependence of the MR response shows that in the two dimensional materials the spin-correlated fluctuation can be easily suppressed by a low external field. Recent studies suggest a two-dimensional and/or short-range spin correlation above T_c^{ρ} for the n=2 compound.[9,14]

Below T_c^{ρ}, the spin correlation becomes long range. The two-dimensional spin ordering in the layered compound possibly originates from the anisotropic exchange interactions along the a-b axis (in-plane) and the c-axis (out-of plane) directions. Larger anisotropy expected for a smaller n compound might lead to more enhanced two-dimensional fluctuation. The broad MR peak with an asymmetric shape seen in the MR-T curve only for the n=2,3 compounds (Fig. 2) may suggest that two-dimensional spin correlated fluctuations survive down to lower temperatures. The existence of two-dimensional spin correlated fluctuations even in the ferromagnetic state may be associated with the observation of enhanced Jahn-Teller distortion upon charge delocalization for the double layer manganites.[15]

The most distinctive difference between the features of the three- and two- (layered) dimensional compounds is in their low temperature MR behavior. The field dependence of normalized resistivity $\rho(H)/\rho(0)$ and magnetization M for the three thin film samples at low temperatures are shown in Fig. 5, and indicative of apparent hysteresis behavior for the n=2,3 compounds. We found that the magnetic fields leading to the resistivity peak in the $\rho(H)/\rho(0)$-H curve correspond to the fields at which zero magnetization is obtained in the M-H curve. The observed MR with hysteresis behavior for the n=2,3 compounds can be understood by the intragrain (intrinsic) spin-polarized tunneling through the insulating $(La,Ca)_2O_2$ layer. Moreover, it is found that the magnitude of the tunneling MR depends critically on the number (2 or 3) of MnO_2 layers. This means that the magnitude of the tunneling MR is not solely determined by the spin polarization, since nearly complete spin polarization is also expected for the perovskite manganites at low temperatures.

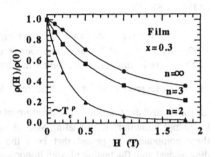

FIG. 4. Field dependence of normalized resistivity $\rho(H)/\rho(0)$ at near T_c^{ρ} for the thin films of with n=2,3, and ∞. The solid lines are drawn only as a guide to the eyes.

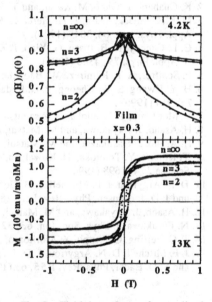

Fig. 5. Field dependence of normalized resistivity $\rho(H)/\rho(0)$, and magnetization M at low temperatures for the thin films with n=2,3, and ∞. These data were obtained after the initial application of a magnetic field of +1.8T.

CONCLUSIONS

We have reported results on the magnetotransport properties of $La_{n-nx}Ca_{1+nx}Mn_nO_{3n+1}$ (n=2,3, and ∞, x=0.3) thin films. Our results demonstrate that a reduction in the number of MnO_2 layers in the unit cell resulted in systematic changes in various features. These changes include an increase in resistivity, a decrease in resistivity peak temperature T_c^ρ corresponding to the metal-insulator transition, an enhancement in the maximum MR near T_c^ρ, and an increase in the low temperature intrinsic tunneling MR. A comparison of the magnitude of these features among these compounds suggested that both the reduced c-axis transfer interaction and the two-dimensional spin fluctuation play an important role in determining the CMR and related properties in layered manganites.

REFERENCES

1. R. von. Helmolt, J. Wecker, B. Holzapfel, L Schultz, and K. Samwer, Phys. Rev. Lett. 71, 2331 (1993).
2. K. Chahara, T. Ohno, M. Kasai, and Y. Kozono, Appl. Phys. Lett. 63, 1990 (1993).
3. M. McCormack, S. Jin, T. H. Tiefel, R. M. Fleming, J. M.Phillips, and R. Ramesh, Appl. Phys. Lett. 64, 3045 (1994).
4. C. L. Canedy, K. B. Ibsen, G. Xiao, J. Z. Sun, A. Gupta, and W. J. Gallagher, J. Appl. Phys. 79, 4546 (1996).
5. P. Schiffer, A. P. Ramirez, W. Bao, and S. W. Cheong, Phys. Rev. Lett. 75, 3336 (1995).
6. H. Y. Hwang, S. W. Cheong, P. G. Radaelli, M. Marezio, and B. Batlogg, Phys. Rev. Lett. 75, 914 (1996).
7. Y. Moritomo, Y. Tomioka, A. Asamitsu, Y. Tokura, and Y. Matsui, Nature 380, 141(1996).
8. H. Asano, J. Hayakawa, and M. Matsui, Appl. Phys. Lett. 68, 3638 (1996).
9. H. Asano, J. Hayakawa, and M. Matsui, Phys. Rev. B56, 5395 (1997).
10. T. Kimura, Y. Tomioka, H. Kuwahara, A. Asamitsu, M. Tamura, and Y. Tokura, Science 274, 1698 (1996).
11. D. N. Argyriou, J. F. Mitchell, J. B. Goodenough, O. Chamaissem, S. Short, and J. D. Joegensen, Phys. Rev. Lett. 78, 1568 (1997).
12. H. Asano, J. Hayakawa, and M. Matsui, Appl. Phys. Lett. 71, 844 (1997).
13. N. Furukawa, J. Phys. Soc. Jpn. 63, 3214 (1994).
14. T. G. Perring, G. Aeppli, Y. Moritomo, and Y. Tokura, Phys. Rev. Lett. 78, 3197 (1977).
15. J. F. Mitchell, D. N. Argyriou, J. D. Joegensen, D. G. Hinks, C. D. Potter, and S. D. Bader, Phys. Rev. B55, 63 (1997).

MICROMORPHOLOGY, MICROSTRUCTURE AND MAGNETIC PROPERTIES OF SPUTTERED GARNET MULTILAYERS

R. MARCELLI°, G. PADELETTI, N. GAMBACORTI, M.G. SIMEONE, D. FIORANI
°Istituto di Elettronica dello Stato Solido del CNR, via Cineto Romano 42, 00156 Roma, Italy
Istituto di Chimica dei Materiali del CNR, via Salaria km. 29.5, 00016 Monterotondo Stazione (RM), Italy

ABSTRACT

The growth technique, the micromorphological and microstructural characterization by means of atomic force microscopy (AFM) and secondary ions mass spectrometry (SIMS) as well as the magnetic properties of a novel class of magnetic multilayers, based on radio frequency (RF) sputtered thin amorphous garnet films, are presented. One, three and five thin film multilayers composed by amorphous pure yttrium iron garnet (a:YIG) and amorphous gadolinium gallium garnet (a:GGG) have been grown on GGG single crystal substrates. The multilayer interfaces have been found to be comparable in both, the three and five-layers structure. Low field susceptibility measurements, showed a paramagnetic behaviour for the single layer YIG film. For the three and five layers samples, irreversibility effects were observed, giving evidence of magnetic clusters at the interface YIG/GGG.

INTRODUCTION

Magnetic multilayers are of growing interest in magnetism for the unusual basic properties of the layered structures and for applications in magnetic devices. The multilayers present some typical novel effects, with respect to single-layer films, such as the giant magnetoresistence (GMR) and the oscillatory exchange interaction between the magnetic layers composing the structure. Moreover, the higher magnetic anisotropy increase make them appealing for applications in magnetic field sensors as well as in magnetooptical recording media [1].

Magnetic garnets are well known ferrimagnetic materials utilized in microwave and magnetooptical applications, whose response to magnetic resonance and optical signal propagation is generally well understood [2]. From a compositional point of view, with exception for the keystone composition yttrium iron garnet ($YIG = Y_3Fe_5O_{12}$), all the other substituted garnets, obtained from exotic combinations of the ions substituting Y and Fe, are currently studied looking for optimized microwave filtering response [3] and/or low optical absorption and high Faraday rotation characteristics [4]. From a structural point of view, a wide literature is currently available on the growth and characterization of crystalline garnets obtained by bulk or epitaxial growth techniques [5-7]. On the other hand, less informations have been obtained on the properties and possible applications of non-crystalline garnets. Amorphous (or vitreous) garnet films have been prepared in different ways, by directly depositing the film in non-crystalline state or destroying the crystalline order by means of a laser annealing or fast neutrons irradiation [8].

The sputtering technique is important when integrated structures (thin films and multilayers) must be deposited onto heterogeneous substrates not having the same composition nor lattice parameter of the film which has to be grown, and without care for the single crystal phase. This solution is convenient when the polycrystalline material has a physical response comparable to that of a single crystal one, as, for instance, in magnetooptical recording applications at well defined light wavelengths [9].

The amorphous state obtained during the sputtering deposition of garnet films has been mostly considered in the past few years just as an intermediate or undesidered result, but not extensively analyzed. In this paper, the magnetic properties of amorphous garnet films and the change introduced in the magnetic properties by the multilayer structure will be stressed. The aim of this work is to describe the growth technique, the interface behaviour, the surface quality, and the magnetic properties of thin amorphous garnet films and multilayers, realized by directly depositing amorphous layers and not by amorphizing a crystalline structure as described in previous papers [9].

EXPERIMENT

One (1L), three (3L) and five-layers (5L) structures of amorphous pure yttrium iron garnet and gadolinium gallium garnet (GGG) films have been grown alternatively by RF sputtering on <111> oriented, single crystal GGG wafers, heating the substrate up to 300 °C. Ceramic, polycrystalline, stoichiometric targets of pure YIG and pure GGG have been prepared by sintering of oxide powders. The targets were 2 inch in diameter and 4 mm thick. By simply rotating the sputtering matching head, the two different targets were placed over the GGG substrate. The partial heating helps the crystallization process if a thermal post-annealing is used, but it does not allow for an *in situ* crystallization of the films. By imposing 200 watt of RF power and 5 min of deposition time, each layer grown was 50 nm thick. After each deposition, the structures were annealed at 850 °C before the successive growth. Low level inclusions of Ar have been also recorded.

A reactive high energy electron diffraction (RHEED) analysis was used to test the crystalline status of the samples, and the stoichiometric compositions were measured by means of electron probe microanalysis (EPMA).

The morphological structure was characterized by AFM technique. The measurements were carried out in air, using a Nanoscope III - Digital Instruments microscope, equipped with an optical deflection system in combination with Si_3N_4 tips and cantilevers. Images were acquired using the instrument in tapping mode (Digital patent). With this technique, the cantilever is oscillated near its resonant frequency as it is scanned over the sample surface. By this method it is possible to achieve high resolution without inducing destructive frictional forces. The frequency range is from 50 to 500 kHz. Topographic images have been recorded over scanned area ranging from 500 x 500 nm^2 up to 5 x 5 μm^2, each with a resolution of 256 x 256 data points. The surfaces were characterized by means of the maximum excursion pick-valley registered in the scanned area (Zr) and by means of the mean roughness (Ra) of the surface relative to a reference center plane.

The SIMS analysis was carried out with the high resolution magnetic sector spectrometer Fisons Instruments (VG Isotech) - MicroSIMS. To minimize the microtopography effect and to reduce the matrix effects [133] Cs^+ primary ions were used and $[CsM]^+$ were detected. The [133] Cs^+ primary ion was fixed at an energy value of 10 keV and the $[CsM]^+$ secondary ions were extracted at an energy of 6 keV. The primary ions current was fixed at 10 nA and the beam was scanned across an area of 100 x 100 μm^2. The secondary ions were collected from a circular area (diameter=10 μm) in the center of the rastered area. To avoid electrical charge build up the samples were covered with Au and electron-gun neutralization was used; the base pressure in the analysis chamber during the depth profile was lower than 1.1 x 10^{-9} Torr.

Low field susceptibility measurements have been carried out by a commercial SQUID magnetometer in the temperature range 4.2 - 250 K, applying a field of 5 Oe perpendicular to the film plane.

RESULTS AND DISCUSSION

By using the RF sputtering technique without a proper heating of the substrate it is not possible to get the crystalline order even if a crystalline substrate is used. Only after a thermal or laser post-annealing the films are crystallized (single phase, polycrystals) and the calibration of the annealing process depends on the garnet composition [10]. The stoichiometry of the films can be mainly controlled through the RF power and the target composition. The RHEED analysis (not shown) reveals that the as grown films were almost completely amorphous. The annealed films and multilayers exhibit partially a crystalline phase, and they are almost entirely amorphous close to the surface. Moreover, very thin amorphous films and multilayers, in the order of tens of nm, do not crystalize easily due to the lattice mismatch between YIG and GGG. In fact, we have verified that, at the same growth and post annealing conditions, films and multilayers having a thickness close to 1 μm become polycrystalline. This is in agreement with the improving of crystal quality generally achieved in mismatched systems by means of increased layer thickness or by the use of a "superlattice-like" structure. Almost stoichiometric compositions, exception done for a small Y excess, have been obtained, as measured by EPMA. It can be deduced that the results obtained by sputtering and annealing are comparable with that ones obtained on single-crystal garnets amorphized by using fast neutrons; In fact, in both cases, only the crystalline quality is destroyed, without alteration of the chemical composition [11-13].

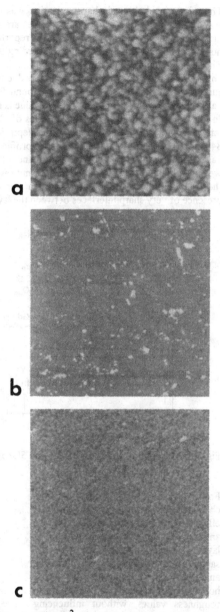

Fig. 1: 2x2 μm² AFM images obtained on 1L (a), 3L (b) and 5L (c) samples respectively. Maximum brightness corresponds to highest heigth. (For Z-scale refers to Tab. 1)

139

The micromorphological characterization was carried out by means of AFM technique: the textures observed for the different samples are shown in Fig. 1 (a,b,c). The Ra and Zr values measured for the three samples are reported in Tab.1. The sample 1L presents an homogeneous texture, but granular, according to the effects generally produced by sputtering. The sample 3L, that undergone annealing cycles, is granulous, homogeneous, with finer grains with respect to the previous one, probably due to the annealing effect. The most homogeneous surface of the samples series has been found for the 5L sample, as confirmed by the lowest value recorded for Zr parameter; its Ra value is comparable with that one in 3L sample.

In Fig. 2a and 2b the SIMS depth profiles of 3L and 5L samples are shown respectively. The interface width has been defined as the depth interval over which the intensity changes from 84% to 16 %. Concerning the Gd depth-profile signal of the upper GGG layer, the 3L sample has a leading edge (LE) width of 11 nm and a trailing edge (TE) of 14 nm. For the 5L sample we have obtained LE=11 nm and TE=19 nm respectively. The inferred values are very close to the best depth resolution obtained with our experimental conditions, thus confirming the presence of very sharp interfaces between the layers, in both samples.

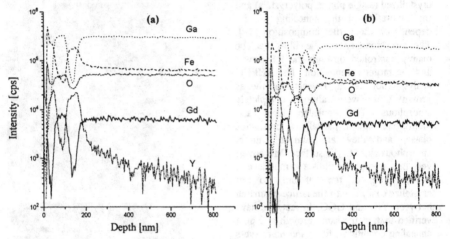

Fig.2: SIMS depth-profiles for 3L (a) and 5L (b) samples respectively.

From the AFM and SIMS characterizations it turns out that the effect of the thermal annealing is more and more important when the number of layers increases, going from the 1L sample to the 5L one. In particular, the annealing is responsible for more homogeneous surfaces and lower mean roughness values, without influencing the interface widths with interdiffusion phenomena.

Sample #	R_a (nm)	Z_r (nm)
1L	9.1	71
3L	1.1	28
5L	1.5	20

Tab.1: Mean roughness (Ra) and pick-valley excursion (Zr) values obtained on 2x2 μm^2 AFM images

The results of magnetization measurements performed after cooling the samples in zero field (ZFC) as well as after cooling them in a magnetic field (FC, 5 Oe), are shown in Fig.3, where the inverse of the susceptibility is shown for the one, three and five layers configurations. No irreversibility (i.e. difference between the ZFC and FC measurements) and an almost linear behaviour are observed for the 1L configuration, indicating a paramagnetic behaviour. On the other hand, irreversibility is observed for both the 3L and 5L configurations above 50 K and 80 K respectively. This give evidence of cluster formation at the YIG/GGG interface. Presently it is under investigation how the number of interfaces and the annealing cycles play a decisive role in the difference measured between the FC and ZFC magnetization curves, which is enhanced when the multilayers are considered.

In our measurements, the magnetic response is not affected by rotations in the film plane, thus confirming the complete lack of crystalline order.

CONCLUSIONS

In this paper the micromorphological and microstructural properties as well as the magnetization behaviour as a function of the temperature for single and multiple layers of amorphous garnets grown by RF sputtering have been investigated. The flattest surface was observed for the sample that undergone a larger number of annealing cycles. In both the three and five multilayer configurations the interface widths are comparable between them, thus suggesting that interdiffusion phenomena are not present. The presence of magnetic clusters at the interface YIG/GGG has been evidenced by magnetization measurements.

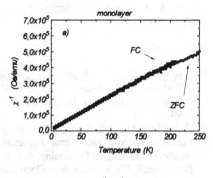

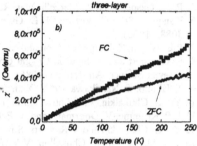

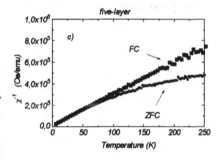

Fig. 3: Inverse of the susceptibility as a function of temperature for the one (a), three (b), and five (c) layers configurations.

ACKNOWLEDGEMENTS

We would like to thank Mr. G. Petrocco and Mr. L. Scopa of CNR-IESS for their assistance in the sputtering growth and EPMA characterization and Prof. Vitali and Dr. M. Rossi of

Università di Roma "La Sapienza" - Dipartimento di Energetica, for the RHEED measurements.

REFERENCES

1. Magnetic Thin Films Multilayers and Superlattices, A. Fert, G.Guntherodt, B.Heinrich, E.E.Marinero, M.Maurer Editors, E-Mat. Res. Soc. Proc. **16**, Amsterdam, 1991. - Magnetic Multilayers, L.H. Bennett and R.E. Watson Editors, World Scientific, London, 1994.
2. G. Winkler, Magnetic Garnets, F. Vieweg & Sohn, Braunschweig/Wiesbaden, 1981,
3. Circuits, Systems and Signal Processing, special Issue on Magnetostatic Waves and Applications to Signal Processing, **4**, (1985).
4. D.D. Stancil, IEEE J. Quantum Electronics, **27**, 61, (1991).
5. P. De Gasperis and R. Marcelli, Mat. Res. Bull. **22**, 235, (1987).
6. Kunquam Sun, C. Vittoria, H.L. Glass, P. De Gasperis, R. Marcelli, J. Appl. Phys. **67**, 3088, (1990).
7. P.E. Wigen, Z. Zhang, L. Zhou and M. Ye, J. Appl. Phys. **73**(10), 6338, (1993).
8. P. Goernert and E. Sinn, in Crystal Growth of Electronics Materials, edited by E. Kaldis, Elsevier Science, Amsterdam (1985), p. 81-101.
9. T. Hirano, T. Namikawa and Y. Yamazaki, IEEE Trans. on Magnetics **28**, 3237, (1992).
10.Yu. G. Chukalkin, V.R. Shtirz and B.N. Goshchitskii, Phys. Stat. Sol. A **112**, 161, (1989).
11.J.-P. Krumme, V. Doormann and R. Eckart, IEEE Trans. on Magnetics **20**, 983, (1984).
12.Yu. G. Chukalkin and V.R. shirts, Sov. Phys. Sol. State **31**(7), 1215, (1989).
13.V.R. Shirts, Yu. G. Chukalkin, V.V. Petrov and B.N. Goshitskii, Sov. Phys. Solid State **29**(3), 509, (1987).

IMPROVEMENT OF THERMAL STABILITY OF METAL/OXIDE INTERFACE FOR ELECTRONIC DEVICES

Yo Ichikawa, Masayoshi Hiramoto, Nozomu Matsukawa, Kenji Iijima and Masatoshi Kitagawa
Central Research Laboratories, Matsushita Electric Industrial Co. Ltd., Seikacho, Kyoto 619-02, Japan

ABSTRACT

The nano-meter controlled iron/iron-oxide multilayer materials have been successfully obtained by the pulse reactive sputtering method with high deposition rate. These multilayer demonstrated a good thermal stability of its structure and magnetic properties up to 500℃ when a small amount of Si was doped in the structure, whereas the non-doped multilayer degraded at above 300℃. The difference of the oxidation energy between Fe and Si increases the thermal stability of the interface between Fe and Fe-O layer.

INTRODUCTION

A precise control of the interface between a metal and a ceramic is important to develop a new material for advanced electronics devices. Especially preparation of a magnetic material having a nano-meter scale interlayer of ceramics having good thermal stability is crucial to fabricate micro-electronics devices, such as a magnetic head for a new type of recording device and a thin film device using tunneling effect of electron. Recent progress of the data processing equipment or the mobile tool is pushing the research in this field.

There are two points of view for the research; one is material research and the other one is processing research. There have been many reports for about Fe/SiO_2, Fe/SiC, Fe/FeN and Fe/FeC multilayers[1-5]. For the Fe/SiO_2 or Fe/SiC systems, the thermal stability of the interface structure is investigated from the view point of surface and interface energy and chemical potential of the elements. However, a material which has a good magnetic and thermal stability has not been obtained. There are few report about the iron and nitride or carbide compound system because of its complexity. On the other hand, the iron oxide has many phases such as hematite, magnetite, wustite, etc., and those have many interesting magnetic and electrical properties. Therefore, multilayers made of iron and iron oxide are expected to provide a noble material which will open a path in the field of electronics component.

Material processing study is another view point of the research for multilayer. A pulse reactive sputtering (PRS) method is one of the selection for processing technology. In the PRS method, mulatilayers are prepared using a single target and an intermittent incorporation of pure oxygen gas to grow thin oxide layers between metal layers. The PRS realizes very high deposition rate with nano-meter scale thickness control and it is about 100 times faster than that of MBE or multi-target sputtering method. Such high deposition rate has a merit to avoid the contamination of the surface of a thin film and a target during the growth, in addition to the industrial advantages.

In this study, we investigated the preparation of nano-meter controlled Fe/Fe-O multilayers to obtain a new material which has a good thermal stability of the structure and magnetic properties.

EXPERIMENT

Samples were prepared by conventional rf-magnetron sputtering apparatus in which the electric micro-valve was installed to incorporate the oxygen gas intermittently into the sputtering chamber. The interval of oxygen incorporation was controlled precisely by the micro-valve and a 32-bit personal computer. The sputtering targets were iron disc of 3 inches in diameter with the purity of 99.9% and iron-silicon (11.5 wt% Si) alloy disc with the same purity. The sputtering gas was Ar with the pressure of 4Pa and the fluctuation of gas pressure due to the oxygen incorporation is controlled less than 5% by using additional evacuation system. The deposition rate was typically 20nm/min.. The substrates were optical polished TiMgNiO ceramic wafers with the thickness of 0.6mm. Thermal treatment of the samples were carried out in a evacuated furnace for 30 minutes at various temperatures.

The samples were characterized by X-ray diffraction and TEM observation. Coercive force, Hc was measured by a B-H loop tracer at 60Hz. Permeability, μ, was obtained from the impedance change of a ferrite yoke with and without sample.

RESULT AND DISCUSSION

1.Characterization of the films

Figure 1 shows the X-ray diffraction pattern of the multilayer of [Fe(4.6nm)/Fe-O(0.4nm)]$\times$ 800layer. The total thickness of this sample was about 4 μm and it was obtained only 3 hours deposition which is unattainable by conventional MBE or multi-target sputtering. In a lower angle region, the diffraction peaks from the superstructure designed for the period of 5nm (=4.6nmFe+0.4nmFeO) are observed. We can clearly observe up to the 10th diffraction peak of superstructure and the artificial modulation length calculated from two theta-values of these diffraction peaks was 5.3nm which is in a close agreement with the settled value in the deposition. In a higher angle region, only strong and sharp *200* diffraction peak of α-Fe is observed. It indicates that Fe layer in the multilayer is composed by [100] oriented α-Fe with high crystallinity.

In Fig. 3(a), typical cross sectional TEM photograph of Fe/Fe-O multilayer is shown. In this photograph, we can observe that thin oxide layers separating the thick metal layer introduced periodically and a fine multilayer is successfully obtained by this method. These results guaranteed that the PRS can produce a nano-meter thickness controlled multilayer material with high deposition rate.

2.Thermal stability of the multilayer material

Figure 2 shows the low angle X-ray diffraction pattern for Fe/Fe-O and silicon doped Fe:Si/Fe:Si-O multilayers before and after the thermal treatment at various temperature up to 600℃. For the Fe/Fe-O, the change of fourth and fifth diffraction peaks is shown in the figure. In the case of non-doped Fe/Fe-O multilayer, the intensity of the peaks decreased remarkably after the 400℃ annealing indicating the artificially introduced layer structure destroyed at this annealing temperature. The higher angle X-ray diffraction indicated that the full width at half maximum of *002* reflection of α-Fe did not changed up to 400℃ and it decreased after the 500℃ annealing. It indicates that the degradation or break down of the interface of iron and oxide layer is preceded before the grain growth of α-Fe in metal layer. On the other hand, Si doped Fe:Si/Fe:Si-O mutilayer showed an excellent resistance for thermal treatment up to 500℃. In Fig. 2(b), superstructure

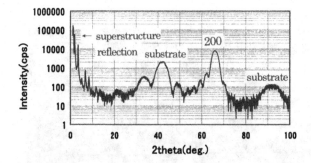

Fig. 1 X-ray diffraction pattern of Fe/Fe-O multi-layer of [Fe(4.6nm)/Fe-O(0.4nm)]×800.

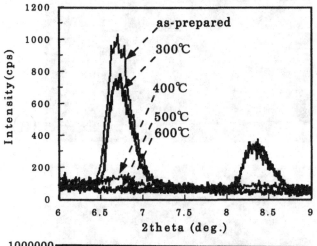

Fig. 2 Temperature dependence of low angle superstructure reflection for the multilayer of Fe/Fe-O (a) and Si doped Fe:Si/Fe:Si-O (b).

(a)

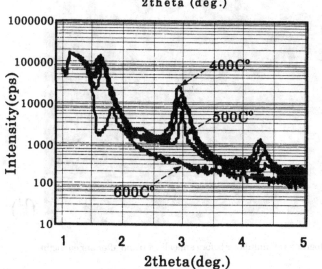

(b)

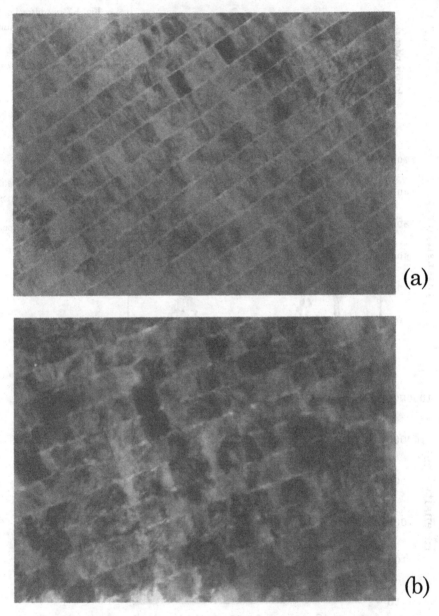

(a)

(b)

Fig.3 TEM photographs of multilayer before annealing(a) and after annealing(b).

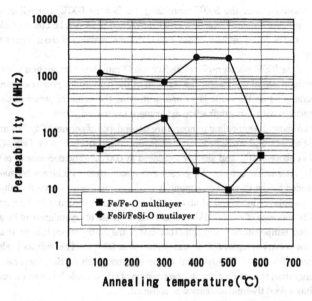

Fig. 4 Annealing temperature vs. permeability of the multilayer.

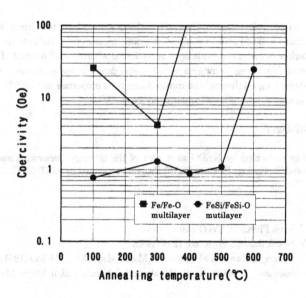

Fig. 5 Annealing temperature vs. coercivity, Hc of the multilayer.

reflections can be observed after the 500℃ annealing. After the 600℃ annealing, although the second and third peaks are smeared, the first order diffraction peak can be observed clearly. This result indicates that the Si doped multilayer has a good thermal stability of the interface between iron and iron oxide layer.

Figure 3 shows the TEM photographs of Fe:Si/Fe:Si-O multilayer before and after the thermal treatment. After the thermal treatment, the contrast of the thin oxide layer is smeared and some grains of Fe is overlapping on the oxide layer and destroying the artificial period of the multilayer. These result is consistent with X-ray diffraction experiment mentioned above.

Figures 4 and 5 show the annealing temperature dependence of permeability, μ, and coercivity, Hc, for both the Fe/Fe-O and Fe:Si/Fe:Si-O multilayers, respectively. μ of Fe/Fe-O multilayer decreased sharply at above 300℃ and slightly increased at 600℃. The decrease of μ is due to the change of magnetic interaction between Fe grains and also between Fe layers because multilayer structure was degraded by the thermal treatment at 400℃, as mentioned above. Simultaneously, the Hc of the Fe/Fe-O multilayer increased at above 300℃ remarkably because of the same reason for the change of μ. The increase of μ at 600℃ is due to the increase of crystallinity of Fe grains by the annealing at elevated temperature. On the other hand, the Si doped multilayer is keeping high permeability and low coercivity up to 500℃ and degraded at 600℃. This results is also consistent with those of X-ray diffraction. The addition of the element, such as Si, which has lower oxide formation free energy than that of Fe promote the creation of stable oxide layer and consequently the multilayer which has a good thermal stability was obtained.

CONCLUSION

The nano-meter controlled multilayers in the Fe/Fe-O system have been successfully obtained by PRS. X-ray diffraction, TEM observation and magnetic measurement indicate that the thermal stability of the interface of metal and oxide layer is reinforced when a small amount of Si, which has lower oxide formation free energy compared with Fe, was doped in the structure. Consequently, those multilayer showed a good thermal endurance of magnetic properties up to 500℃ whereas non-doped material degraded at lower annealing temperature of 300℃.

ACKNOWLEDGEMENT

This work was supported by NEDO as a part of the Synergy Ceramics Project under the Industrial Science and Technology Frontier (ISTF) Program promoted by AIST, MITI, Japan.

REFERENCES

1. T. Kobayashi , J. Appl. Phys., 73, 858 (1993)
2. M. Senda and Y. Nagai, J. Appl. Phys., 65, 1238 (1989)
3. H. Sakakima, K. Osano, K. Ihra and M. Satomi, J. Magn. Magn. Mat., 93, 349 (1991)
4.. H. Fujimori, N. Hasegawa, N. Kawaoka, S. Nagata and S. Yamaguchi, J. Magn. Magn. Mat., 121, 42 (1993)
5. R. R. Ruf and R. J. Gambino, J. Appl.. Phys., 55, 2628 (1984)

ROOM TEMPERATURE MAGNETORESISTIVE RESPONSE IN CMR PEROVSKITE MANGANITE THIN FILMS

MICHAEL A. TODD, CHARLES SEEGEL AND THOMAS H. BAUM
Advanced Technology Materials, Inc., Advanced Delivery and Chemical Systems Division
7 Commerce Dr., Danbury, CT 06810

ABSTRACT

Perovskite-structured $La_xSr_yMnO_3$ thin-films have been deposited onto $LaAlO_3$ substrates via liquid delivery chemical vapor deposition (LD-CVD) using metal(ß-diketonato) precursors, $M(thd)_x$ [where M= Ca, Sr, La and Mn, thd = 2,2,6,6-tetramethyl-3,5-heptanedionato and x = 2-3]. Thin films were deposited at temperatures between 500 and 700 °C and subsequently annealed at 1000 °C under O_2. These films possess stoichiometries that are: i) vastly different from the $La_{0.67}Sr_{0.33}MnO_3$ compositions commonly reported in the literature and ii) display high temperature, low field responses that may be technologically important. Resistance versus temperature measurements revealed a metal to semiconductor transition at room temperature and above. Hall measurements on a film of $La_{0.35}Sr_{0.24}MnO_3$ displayed a magnetoresistive response (MR) of –10% at 57 °C in a fixed magnetic field of 780 Oe. Based upon our research, the observed film properties are directly related to the deposited film stoichiometry and the best results were observed at Sr / La ratios between 0.30 and 1.0 for A-site deficient $La_xSr_yMnO_3$ thin-films after thermal annealing.

INTRODUCTION

Magnetoresistance (MR) is defined as the change in the resistance of a material under an applied magnetic field (R_H), relative to the original resistance of the material in the absence of a magnetic field (R_0), as given in Equation 1:

$$MR = \Delta R / R_0 = (R_H - R_0) / R_0 \qquad (1)$$

where R is the resistance and H is the applied magnetic field strength

MR in permalloy ($Ni_{0.80}Fe_{0.20}$) thin-films is extremely well documented; the 2% resistance change in permalloy has enabled commercial products to be developed with widespread applicability. More recently, giant magnetoresistance (GMR) changes (7-8%) were demonstrated in thin-film, metallic multilayers [1]; improved signal-to-noise ratios and significant gains in magnetic recording densities can be realized with GMR thin-film designs [2]. This same phenomenon is being explored for high-density, magnetic random access memories (MRAMs) in conventional integrated circuits [3]. Although GMR appears to be commercially viable, the deposition technique requires ultra-high vacuum (UHV) hardware and precise control (1Å) of the individual metal layers, which are from 2 to 15 nm thick. Several advantages of the GMR metallic multilayers include room temperature operation, large magnetoresistance response in small magnetic fields and the deposition of well characterized metallic films.

Perovskite-structured $LaMnO_3$ oxides are increasingly important for technological applications based upon their ability to be both ferromagnetic conductors and anti-ferromagnetic insulators [4]. Of particular interest, lanthanum manganites doped with Group II elements, $La_xM_yMnO_3$

149

(where M is a Ca, Sr or Ba dopant), are being scrutinized for their unique CMR responses. The ferromagnetic conductor (metal-like) to anti-ferromagnetic insulator (semiconductor-like) transition is strongly influenced by the sample temperature and applied magnetic field. Historically, CMR responses are only observed in large magnetic fields (1 – 12 Tesla) and at temperatures well below room temperature (i.e., 200 K) [5]. Both of these undesirable physical characteristics have resulted in CMR materials being a research curiosity with questionable commercial potential.

Doped manganite materials have been synthesized in bulk crystalline form by classical solid-state ceramic techniques [6]. Thin-films have been deposited by pulsed laser deposition (PLD) [7], molecular beam epitaxy (MBE) [8], wet, chemical techniques, such as sol-gel [9] and by chemical vapor deposition (CVD) [10]. Although these materials present a significant opportunity for both scientific understanding and commercial applications, they currently suffer from low temperature transitions and require large magnetic field strengths to induce the CMR response. Recently, however, magnetoresistive responses were observed in A-deficient lanthanum manganite materials [4] which provide a potential pathway to higher density recording media under ambient conditions.

In this research, we have utilized a liquid delivery chemical vapor deposition (LD-CVD) process to develop CMR materials that exhibit room temperature (300 K) magnetoresistance responses in relatively small applied fields (i.e., 780 Oe). The control of film stoichiometry is critical to the observed thin-film properties, including the transition temperature and magnetic field response. The focus of our research was the growth of $La_xSr_yMnO_3$ thin films with stoichiometries that are unique from those typically reported in the literature. These films exhibit room temperature MR responses, a large thermal dependence and low applied magnetic field responses. Perovskite CMR materials, therefore, exhibit commercial potential for use in magnetic sensors, thermal switches, solid oxide fuel cell electrodes, thin-film recording heads and magnetic random access memory (MRAM) devices.

EXPERIMENT

The thin films described in this work were deposited using a LD-CVD process in an inverted low-pressure reactor, shown schematically in Figure 1. Films were typically deposited by delivering an organic solution containing the precursor reactants, metal(thd)$_x$ (where metal = La, Sr, Ca and Mn; thd = 2,2-6,6-tetramethyl-3,5-heptanedionato ligand), to a heated vaporizer (T_{vap} = 245 °C). The reactants were "flash" vaporized and decomposed on pre-cleaned $LaAlO_3$ and MgO substrates held at a pedestal temperature between 500 and 700 °C. The precursors were introduced into the reactor using a nitrogen carrier gas and were oxidized using reactive gases such as oxygen, nitrous oxide and ozone, alone or in combination. The gases were introduced to the reactor system via mass flow controllers with flow rates varying from 250 sccm to 1.0 slm. The films were deposited, cooled under flowing oxygen, analyzed, and subsequently annealed at 1000 °C under flowing oxygen for four hours.

The films were characterized prior to and after annealing using XRD, EDS and 4-point probe resistance versus temperature measurements. Selected samples were also measured using cross-sectional SEM to determine film thickness, used to calculate average film growth rates. Hall measurements were performed using a heated sample stage at fixed magnet currents (eg. at fixed fields and constant magnet temperature) on both as-deposited and annealed samples. Magnetic measurements were also made in collaboration with Princeton University and Honeywell in order to corroborate our experimental findings.

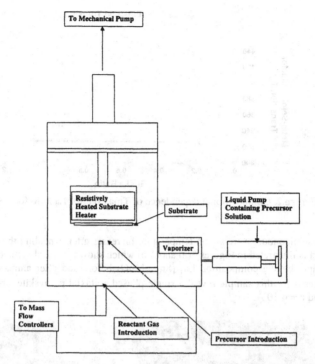

Figure 1. Liquid delivery CVD reactor system used to deposit $La_xSr_yMnO_3$ thin films.

RESULTS

Films having composition in the range $La_xSr_yMnO_3$ (where $x = 0.21$ to 0.50 and $y = 0.08$ to 0.50) were deposited at temperatures ranging from 500 to 700 °C with film thickness ranging from 1,000 Å to 3,500 Å. Under typical deposition conditions, average growth rates were 20 Å/min to 60 Å/min. These films exhibited metal to semiconductor transitions at temperatures ranging from 270 to 350 K as-deposited and at temperatures ranging from 310 to 460 K after annealing in oxygen. The thin-films also exhibited a compositional dependence towards T_{max}, the temperature (T) where the measured film resistance is maximized. In general, the greater T_{max}, the greater the metal-semiconductor transition temperature. The compositional dependence of T_{max} as a function of Sr / La ratio within the thin-films measured is shown in Figure 2.

Clearly, the highest values of T_{max}, and thus, the highest metal-semiconductor transition temperatures, are occurring for Sr / La ratios between 0.30 and 1.0. The maximum temperature appears to be realized near 0.48. These observations are important for the following reasons; (1) the measured MR response is observed over a temperature range corresponding to the metal-semiconductor transition region (in annealed samples) and indicates that the two phenomena are physically inter-related and (2) the Sr / La ratios are nearly equal to, or increased relative to the Sr /La ratio in $La_{0.67}Sr_{0.33}MnO_3$, a composition that is widely reported in the literature.

The metal to semiconductor transition temperature also exhibits a dependence on the crystallinity and / or degree of oxidation of the films, as can be deduced from the observation that

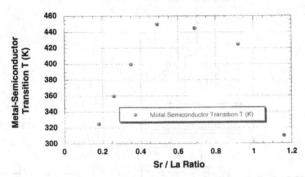

Figure 2. Compositional dependence of Tmax on Sr / La ratio for LSMO films.

the metal-semiconductor transition temperature increases after annealing the samples in oxygen. This effect is illustrated in Figures 3(a) and 3(b) which show the metal-semiconductor transition for a sample having composition $La_{1.1}Sr_{0.37}MnO_3$ before and after annealing in oxygen. The XRD patterns for this sample reveal a single-phased epitaxial perovskite structure with a = b = 3.86 Å and c = 4.10 Å.

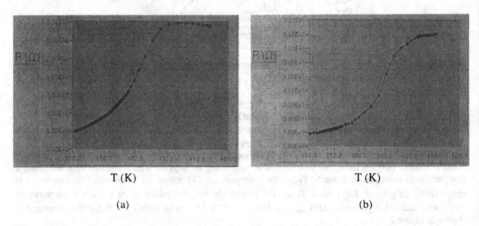

(a)   (b)

Figure 3. R vs. T plots for a $La_{1.1}Sr_{0.37}MnO_3$ thin film (a) the metal-semiconductor transition is observed at ~ 320 K prior to annealing and (b) the transition temperature is shifted upward to ~ 350 K after annealing in O_2 at 1000 °C.

Resistivity measurements on a sample having a composition of $La_{0.35}Sr_{0.24}MnO_3$ are shown in Figure 4. The resistance of the sample in no applied field and as a function of three fixed magnetic fields (780, 4100 and 6200 Oe) applied to the sample are displayed as a function of the sample temperature. This film exhibits a negative MR of 10% at 57 °C in an applied field of 780 Oe. To our knowledge, this represents the highest MR observed for a LSMO sample at elevated temperatures. The MR response is also observed over the entire temperature regime of the metal

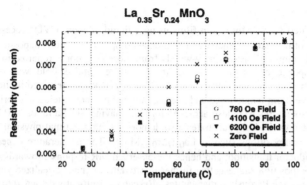

Figure 4. Resistivity measurements for a $La_{0.35}Sr_{0.24}MnO_3$ sample at zero applied field, 780, 4100 and 6200 Oe. The maximum MR response observed at 780 Oe is –10 % and occurs at 57 °C. The MR respsonse is nearly saturated at 780 Oe, indicating that the film may exhibit useful MR responses at lower applied magnetic fields.

MR (%) As a Function of Temperature

No Applied Field	Resistivity (Ω cm)	780 Oe Applied Field	Resistivity (Ω cm)	MR (%)
37 °C	0.004012	37 °C	0.003837	-4
47	0.004763	47	0.004429	-7
57	0.006008	57	0.005378	-10
67	0.007050	67	0.006497	-8
77	0.007567	77	0.007301	-4
87	0.007923	87	0.007750	-2

Table 1. MR data obtained for a $La_{0.35}Sr_{0.24}MnO_3$ thin-film deposited by liquid delivery CVD. The resistance was measured by 4 point probe as a function of temperature in a) no applied field and b) a fixed 780 Oe applied field. Also shown is the calculated MR % for each temperature. A useful MR response is observed over a wide temperature range in a relatively small magnetic field.

to semiconductor transition region, with MR's ranging from –4% at 37 °C to –2% at 87 °C. The tabulated MR data, as a function of temperature in a fixed field, is presented in Table I. It should also be noted that the peak MR response for this sample occurs at the temperature corresponding to the "center" of the metal to semiconductor transition, an observation consistently noted in this study. This observation implies that the MR response is directly related to the physical phenomenon responsible for the abrupt conductivity change (metal to semiconductor transition) and is a function of both the sample temperature and thin-film compositional stoichiometry.

The temperature dependence of the metal to semiconductor transition with film stoichiometric composition strongly suggests that the transition is a function of the film microstructure and/or electronic band structure. Both crystalline microstructure and band structure can be related to the elemental and stoichiometric composition of the deposited thin-films. Experiments are currently underway to determine the effects of bandgap and strain towards the MR properties in these Perovskite materials. Our desire is to optimize the elemental composition of the films, determine the optimum film stoichiometries and to elucidate the influence of substrate lattice matching on the properties of the manganite films.

CONCLUSION

We have deposited $La_xSr_yMnO_3$ thin-films using a liquid delivery CVD process that enables the direct control of film stoichiometry and elemental composition. Films deposited in this study exhibit elemental compositions that vary widely from those commonly reported in the literature. A peak MR response of -10% at 57 °C in a 780 Oe applied field was observed in a $La_{0.35}Sr_{0.24}MnO_3$ film. These results represent the best MR values reported in the literature for temperatures above room temperature and in small applied magnetic fields. Superior MR results and transition temperatures were also achieved for films having Sr / La ratios of 0.30 to 1.0. The thin-films deposited on $LaAlO_3$ substrates also demonstrate that optimal physical properties are not observed for compositions such as $La_{0.67}Sr_{0.33}MnO_3$. The temperature dependence of the metal to semiconductor transition, and hence the observed MR response, towards elemental and stoichiometric composition suggests that further improvements may be realized via bandgap and strain engineering of the Perovskite manganites. Research specifically designed to optimize film stoichiometries, crystalline orientation and microstructure are required to better understand the fundamental limits of doped lanthanum manganites for use in magnetic sensor applications.

ACKNOWLEDGMENTS

The authors wish to thank the Ballistic Missile Defense Organization for financial support under contract # NAS3-27809, Mr. Eric Clark of NASA Lewis Research Center for helpful discussions, Drs. P Ong and P. Matle of Princeton University for high field MR measurements, Dr. L. Ranganatham of Honeywell for low field MR measurements and Dr. Z. Wang of Georgia Tech. Institute for analytical results and transmission electron microscopy results. A special thanks is offered to Jiming Zhang, Dan Studebaker and Anna Raskova for their early film growth studies on LCMO films.

REFERENCES

1. K. Derbyshire and E. Korczynski, "Giant Magnetoresistance for Tomorrow's Hard Drives", *Solid State Technology*, **September**, 57 (1995).
2. P. Singer, "Read/Write Heads: The MR Revolution", *Semiconductor International*, **February**, 71 (1997).
3. S. S. P. Parkin, *Mater. Res. Soc. Symp. Proc.*, , (1997).
4. A. Gupta, T. R. McGuire, P. R. Duncombe, M. Rupp, J. Z. Sun, W. J. Gallagher and G. Wang, *Appl. Phys. Lett.*, **67** (23), 3494 (1995).
5. R. von Helmont, J. Wecker, B. Holzapfel, L. Schulz and K.Samwer, *Phys. Rev Lett.*, **71**, 2331 (1993).
6. G.H. Rao, J.R. Sun, J.K. Liang, W.Y. Zhou and X.R. Cheng. *Appl. Phys. Lett.*, **69**(3), 424 (1996).
7. Y. Tomioka, H. Kuwahara, A. Asamitsu, M. Kasai and Y. Tokura. *Appl. Phys. Lett.*, **70** (26), p. 3609 (1997). N. Sengoku and K. Ogawa, *Jpn. J. Appl. Phys.*, **35**(pt.1, 10), p. 5432 (1996). J.Y. Gu, K.H. Kim, T.W. Noh and K.S. Suh, *J. Appl. Phys.*, **78**(10), p. 6151 (1995).
8. I. Bozovic and J.N. Eckstein, *Appl. Surf. Sci.*, 4018 , p.1 (1997).
9. S.Y. Bae and S.X. Wang. Appl. Phys. Lett. **69**(1), p. 121 (1996).
10. J. Zhang, S. Pombrik, *J. Mater. Res.*, (1994). K. H. Dahmen and M.W. Carris, *Chem. Vap. Deposition,* **3**(1), p. 27 (1997).
11. G.J. Snyder, R.Hiskes, S. DiCarolis, M.R. Beasley and T.H. Geballe, *Phys. Rev. B*, 53(21), p. 14, 434 (1996).

Part III

Metallic Magnetic Oxide
Theory and Devices

THE MAGENTIC SUSCEPTIBILITY IN ULTRATHIN FILMS OF MAGNETIC MATERIALS

KAMAKHYA P. GHATAK,[*] P.K. BOSE[**] AND GAUTAM MAJUMDER[**]
Department of Electronic Science, University of Calcutta,
University College of Science and Technology,
92, Acharya Prafulla Chandra Road, Calcutta-700 009,INDIA.
Department of Mechanical Engineering,
Faculty of Engineering and Technology, Jadavpur University,
Calcutta-700 032, INDIA.

ABSTRACT

In this paper we have studied the dia and paramagentic susceptibilities of the holes in ultrathin films of magnetic materials in the presence of a parallel magentic field on the basis of a newly derived dispersion law for such systems. The numerical computations are performed taking $Hg_{1-x}Mn_xTe$ and $Cd_{1-x}Mn_xSe$ as examples. Both the susceptibilities increses with decreasing doping and film thickness respectively. It is important to note that not only the paramagnetic-to-diamagnetic susceptibility ratio for the present case deviates from (1/3) in conventional semiconductors, but also that is a critical region, where quenching of the diamagnetic occurs. The theoretical analysis is in agreement with the experimental datas as given elsewhere.

Introduction

The interest in epitaxial films of $Cd_{1-x}Mn_xTe$ arises from the novel magentic properties due to the Mn atoms and the close lattice match and chemical compatibility with $Hg_{1-x}Cd_xTe$, an ideal material for long wave length infrared detectors. It is worth remarking that the close lattice match and chemical compatibility of CdMnTe with HgCdTe are important because it is possible to monolithically integrate devices such as CdTe transistors (Hg,Cd,Mn) [1,2] the heterojunction lasers [3,5] and other optical and electronic devices with HgCdTe infrared focal plane arrays. Recent advances in molecular beam epitaxy techniques have made possible the controlled substitutional doping of CdTe [6], CdMnTe [7] and CdMnTe-CdTe superlattices [8]. The combined compatibility of controlled substitutional doping and heterostructure of dilute magnetic materials make possible magnetically tuned lasers, magnetic field sensor and magnetic field sensitive transistors [9]. Though considerable work has already been done nevertheless it appears from the literature that the dia and paramegnetic susceptibilities of the holes in ultrathin films of dilute magnetic materials have yet to be investigated. In this paper we shall study both the susceptibilities for the more interesting case which occurs from the presence of a parallel magnetic field, taking $Hg_{1-x}Mn_xSe$ and $Cd_{1-x}Mn_xSe$ as examples.

Theoretical Background

The hole energy spectra in bulk specimens of DMS are given by [10]

$$E_{\pm3/2}(k) = \hbar^2 a_1(k_x^2 + k_y^2) + \hbar^2 a_2 k_z^2 \pm 3A_o \qquad (1)$$

$$E_{\pm1/2}(k) = \hbar^2 b_1(k_x^2 + k_y^2) + \hbar^2 b_2 k_z^2 \pm A_o \qquad (2)$$

where $a_1 = \frac{1}{2}[(3/4m_{1h}) + (1/4m_{hh})]$, $a_2 = 1/2m_{hh}$

$b_1 = \frac{1}{2}[(1/4m_{1h}) + (3/4m_{hh})]$, $b_2 = 1/2m/h$ and the other notations are defined in [10]. In the presence of a parallel magnetic filed B along y-direction, the modified hole energy spectra can be written as

$$E_{\pm3/2}(k) = a_1\hbar^2 k_s^2 + a_2(\hbar n_1\pi/a)^2 + a_2 e^2 B^2 <x^2> \pm 3A_o \qquad (3)$$

and

$$E_{\pm1/2}(k) = b_1\hbar^2 k_s^2 + b_2(\hbar n_2\pi/a)^2 + b_2 e^2 B^2 <x^2> \pm A_o \qquad (4)$$

The hole concentration and free energy can, respectively, be expressed as

$$p_o = (\pi a_1\hbar^2)^{-1} \sum_{n_1=0}^{n_{1max}} F_o(\eta_1) + (\pi b_1\hbar^2)^{-1} \sum_{n_2=0}^{n_{imax}} F_o(\eta_2) \qquad (5)$$

and

$$F(B) = p_o E_F - (k_B T/\pi a_1\hbar^2) \sum_{n_1=0}^{n_{1max}} F_1(\eta_1) - (k_B T/\pi b_1\hbar^2) \sum_{n_2=0}^{n_{2max}} F_1(\eta_2) \qquad (6)$$

where $\eta_1 = (k_B T)^{-1}[E_F - E_1]$, $E_1 = a_2(\hbar n_1\pi/a)^2$

$+ a_2 e^2 B^2 <x^2> \pm 3A_o$; $\eta_2 = (K_B T)^{-1}[E_F - E_2]$,

$E_2 = b_2(\hbar n_2\pi/a)^2 + b_2 e^2 B^2 <x^2> \pm A_o$ and $F_j(\eta)$ is the one parameter Fermi-dirac integral.
The equation for diamagnetic susceptibility is

$$\chi_d = -\mu_o \partial^2 F(B)/\partial B^2 \qquad (7)$$

By combining (5) to (7) we can find χ_d. When spin splitting is considered, the free energy can be writhen as

$$f(B) = p_oE_f - (k_BT / 2\pi a_1\hbar^2) \sum_{n_1=0}^{n_{1max}} \left[F_1(\eta_{1+}) + F_1(\eta_{1,-}) \right]$$

$$-(k_BT / 2\pi b_1\hbar^2) \sum_{n_2=0}^{n_{2max}} \left[F_1(\eta_{2,+}) + F_1(\eta_{2,-}) \right] \tag{8}$$

where $p_o = (2\pi a_1\hbar^2)^{-1} \sum_{n_1=0}^{n_{1max}} \left[F_o(\eta_{1,+}) + F_o(\eta_{1,-}) \right]$

$$+(2\pi b_1\hbar^2)^{-1} \sum_{n_2=0}^{n_{2max}} \left[F_o(\eta_{2,+}) + F_o(\eta_{2,-}) \right]$$

$$\eta_{1,\pm} = (k_BT)^{-1}\left[E_f - E_1 \pm \frac{1}{2}g_1\mu B \right],$$

$\eta_{2,\pm} = (k_BT)^{-1}[E_f - E_2 \pm \frac{1}{2}g_2\mu B,]$, g_1 and g_2 are the magnitudes of the effective g

factors at the valence bands.

Thus by using $\chi_p = -\mu_o(\partial^2 f(B) / \partial B^2)$ \hfill (9)

and combining (8) and (9) we can determine the paramagnetic susceptibility.

Results and Discussion

Using the appropriate equations and taking the parameters as given in [10] we have
plotted χ_d and χ_p versus p_o and a in the electric quantum limit in ultrathin films of
$Hg_{1-x}Mn_xTe$ and $Cd_{1-x}Mn_xSe$ as shown in Fig. 1 and 2 where the circular plots in Fig.1
exhibit the experimental datas [11]. It appears from Fig.1 that in general, χ_d and χ_p
both decrease with increasing concentrations.The χ_d, which normally takes only negative
values, becomes positive below a certain critical value of p_o, i.e. the diamagnetism
disappears in this range. When an external magnetic field is applied to a DMS, the
holes (characterized by their effective masses) can move only in the quantum mechanical
orbits.These paths are in a plane perpendicular to the direction of the magnetic field,
the sense of rotation governed by the Lenz's law. This rotational motion gives rise

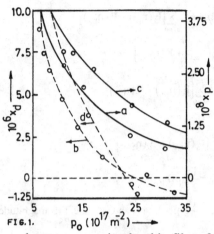

FIG.1.

Fig.1 : Variation of χ_p and χ_d versus p_o in ultrathin film of DMS in the presence of a parallel magnetic field a(χ_p) and b(χ_d) for $Hg_{1-x}Mn_xTe$; c(χ_p) and d(χ_d) for $Cd_{1-x}Mn_xSe$;

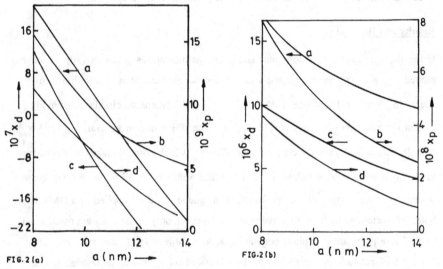

FIG.2 (a)   a (nm) ⟶   FIG.2 (b)   a (nm) ⟶

Fig.2 : Variation of χ_p and χ_d versus a for all the cases of fig.1 (a) $p_o = 2.5 \times 10^{17} m^{-2}$, while χ_d becomes negative only above a particular a (nearly equal to 10 nm.) (b) $p_o = 5 \times 10^{17} m^{-2}$ when there is a complete quenching of diamagnetism.

to a magnetic field which is opposite to the direction of this field. Thus the net magnetic field inside the system is less than that of applied externally. This is the well-known phenomena of diamagentism. In ultra thin films of DMS, the values of the magnetic field and the dimensions of the system are such that the holes can not move on complete orbits. They hit the boundary walls somewhere and get reflected to move in a curved path. The hole trying to move along a circular arc (clockwise according to Lenz's Law) hits the system somewhere at the boundary, gets reflected and repeats its course, crossing the initial position. This is equivalent to an anticlockwise rotation of the holes in the presence of magnetic field. This sense of rotation is opposite to that which gives rise to diamagnetism.

We wish to state that the individual arc contributes to diamagnetism to some extent. When the contribution to magentic moment, arising from the equivalent anticlokcwise rotation, exceeds the total contribution to the individual arcs. diamagnetism can not occur. The predominance of one contribution over the other is controlled by the individual arc lengths and the hole statistics of the allowed levels. For this reason the crawling motion of the holes in the less populated higher energy states, around the boundary of a 3D system, can not influence the diamagnetism which is due to the cyclic motion of the holes in the densely populated lower energy states. Thus the diamagnetism is not quenched for the 3D system.

It appears from the figures that the susceptibility ratio deviates from 1/3 unlike the case with parabolic semiconductors. In Fig. 2. the variation of the susceptibilities with the diemnsion (a) is shown. With the increase of a, the susceptibilities decrease. In Fig. 2., it may be seen that there is a critical value of a below which the diamagnetism disappears for the whole range of thickness as considered here. Thus there is a critical relation between the hole concentration and the dimensions of the ultrathin film to determine whether the diamagnetic susceptibility will appear or not.

Finally, it may be noted that though the mixing of holes and many body effects should be taken into account, this simple analysis exhibit the basic qualitative features of the magnetic susceptibilities in ultrathin films of dilute magnetic materials and theoretical results are in agreement with the experimental data as given elsewhere[11].

[1]   D.L. Dreifus, R.M. Koblous, K.A. Harris, R.N. Bicknell, N. C. Giles and J.F. Schetzina, Appl. Phys. Letts. 51, 931 (1987).
[2]   D.L. Dreifus, R.M. Kolbus, J.R. Tassitino, R.L. Harper, R.N. Bickness and J.F. Schetzina, J. Vac. Sci. Tech. 6A, 2722 (1988).
[3]   R.N. Bicknell, N.G. Giles-Taylor, J.F. Schetzina, N.G.

Anderson and W.D. Laidig, J. Vac. Sci. Tech. 4, 2126 (1986).

[4]   R.N. Bicknell, N.G. Giles, J.F. Schetzina, N.G. Anderson and W.D. Laidig, Appl. Phys. Letts. 46, 238 (1985).

[5]   P. Becla, J. Vac. Sci. Tech. 6A, 2725 (1988).

[6]   R.N.Bicknell,N.G.Gailes and J.F.Schetzina, Appl. Phys. Letts. 49,1095 (1986).

[7]   R.N. Bicknell, N.C. Giles, J.F. Schetzina and C. Hitzman, J. Vac. Sci. Tech. 5A, 3059 (1987).

[8]   R.L. Harpar, S. Hwang, N.C. Giles, R.N. Bickness, J.F. Schetzina, Y.R. Lee and A.K. Ramdas, J. Vac. Sci. Tech. 6A, 2627 (1988).

[9]   D.L. Dreifas and R.M. Kolbous, Appl. Phys. Letts. 53, 1279 (1988).

[10]   J.K. Furdyna, J. Appl. Phys. 64, 29 (1988).

[11]   A.L. Elfros and B. Kakanov, J. Exp. Theo. Phys. 96, 218 (1994).

POLARON FORMATION AND MOTION IN MAGNETIC SOLIDS

DAVID EMIN
Department of Physics and Astronomy, University of New Mexico, Albuquerque, NM 87131-1156

ABSTRACT

This paper addresses aspects of the theory of the formation and motion of polarons that appear relevant to understanding some metal-to-semiconductor transitions in oxides. First, the physical bases of both the long- and short-range electron-lattice interactions usually considered in polaron theory are described and contrasted with one another. Then the notion of self-trapping and the formal theory of polaron formation are presented. Using a scaling analysis of the nonlinear wave equation that lies at the heart of polaron formation, essential features of polaron formation are readily obtained for both types of electron-lattice interaction operating individually and in tandem. The theory is extended to apply to a carrier bound within a Coulomb potential.

Two distinct types of bound polaron state can exist. A "small" polaron's electronic carrier is confined to a single site. Alternatively, a "large" polaron's electronic carrier is distributed over multiple sites. When separated by an energy barrier, these distinct states can coexist. A "collapse" occurs when a continuous change of physical parameters produces an abrupt change of the groundstate from being large-polaronic to being small-polaronic.

To introduce magnetic effects, the scaling analysis is first applied to the formation of a large magnetic polaron, a charge carrier that moves freely within a large ferromagnetic cluster embedded within an antiferromagnet. The polaron is large enough that the predominant interactions are the exchange interactions of local magnetic moments among themselves and with the charge carrier.

The scaling analysis is then extended to describe the donor-state collapse that is thought to drive the metal-to-insulator transition that occurs in n-type EuO as this ferromagnet is heated toward its paramagnetic state. In this case, the metallic impurity conduction that dominates transport at low-temperatures is suppressed when the ferromagnet's large-radius donor states collapse to small-polaronic states upon approaching the paramagnetic regime. At appropriate doping levels, this transition is associated with a huge negative magneto-resistance.

This paper finally addresses small-polaronic hopping transport in p-type $LaMnO_3$. Attention is focused on the effects of compensating holes with electrons generated by oxygen vacancies. The Curie temperature is reported to be insensitive to this compensation. The low-temperature ferromagnetism is even unaffected when the hole density is reduced enough to eliminate metallic conductivity. These results imply that the ferromagnetism is not carrier-induced. Furthermore, the strong sensitivity of the high-temperature Seebeck coefficient to compensation suggests that the carriers hop amongst only a small subset of Mn sites. These cation sites may be associated with the divalent cation dopants. The observation of an n-type Hall effect is consistent with the notion that the hopping is a type of impurity conduction. Indeed, Hall effect sign anomalies are predicted and observed for the hopping of holes in disordered solids. In this view the transition from a ferromagnetic-metal to a paramagnetic-semiconductor in doped $LaMnO_3$ is similar to that of EuO, in that both transitions are associated with the collapse of carriers from extended states into small-polaronic impurity states as the temperature approaches the Curie temperature.

Mat. Res. Soc. Symp. Proc. Vol. 494 © 1998 Materials Research Society

INTRODUCTION

Charge carriers in common covalent semiconducting crystals are often viewed as being nearly free electronic quasiparticles whose motions are only occasionally impeded by scattering events. By contrast, charge carriers in oxides often interact so strongly with their surroundings as to alter the nature of the carrier itself. This type of situation is the concern of polaron theory.

Carriers become polarons when the states of these electrons are strongly affected by displacements of the ions that surround them. While generally less strong, the dependence of a carrier's energy on the alignments of nearby local magnetic moments can sometimes lead to a carrier inducing the alignment of these spins, thereby producing a magnetic polaron. These polaronic interactions tend to confine a carrier within regions about which ions are displaced and local magnetic moments are realigned. Such localizing interactions are generally opposed by the tendency of carriers to spread through a solid, governed by the electronic bandwidth.

Polaron effects not only induce a carrier's confinement but are enhanced when a carrier is confined. As a result, polaron formation is a feedback phenomenon. This feedback produces a synergy between different interactions that facilitate a carrier's localization. Thus, the localizing effect of disorder acts in tandem with polaronic interactions to foster carrier confinement. For this reason, traps tend to be seats at which polarons form.

Polarons are found in oxides because the interactions of the carriers with the surroundings are stronger, the electronic bands are often narrower, and the defect levels are much higher than those of conventional semiconductors. One type of polaron, a "small" polaron, has its electronic carrier confined to a single site. Small polarons generally move by phonon-assisted hopping. Small-polaron hopping has been established in MnO [1] and UO_2 [2]. Large polarons by contrast have electronic carriers that extend over multiple atomic sites. Large polarons generally move itinerantly with very large effective masses. Electrons in both forms of TiO_2 are believed to be large polarons [3,4]. In some solids small polarons only form within the confining potential of a Coulombic center. For example, the holes produced in NiO upon replacing some Ni cations by Li cations are believed to only form small polarons while adjacent to the Li sites [5]. In addition, V_2O_5, Ti_4O_7 are two oxides in which like-signed carriers are thought to pair together about a common site to form bipolarons [6,7].

In this paper, the fundamentals of atomic-displacement-polaron and magnetic-polaron formation will be discussed. Feedback phenomena via which carriers "collapse" into a severely localized small-polaronic state are emphasized. It is shown that the increasing fluctuations of the alignments of magnetic moments which occur as the temperature of a ferromagnetic semiconductor is raised can induce the collapse of a donor-state electron into a small-polaronic state. This phenomenon is suggested to be at the heart of the metal-to-insulator transition in doped EuO. Experiments implying similarities between EuO and doped $LaMnO_3$ are discussed.

ELECTRON-LATTICE INTERACTIONS

Alterations of the positions of a solid's atoms will change the potential experienced by an electronic carrier. In particular, when atoms nominally located at positions designated by u are shifted by $\Delta(u)$, the potential energy of a carrier at position r changes by:

$$V(r) = \int du \, Z(r - u) \, \Delta(u), \qquad (1)$$

where the dependence of Z(r-u) on r-u denotes the range of the interactions.

In ionic solids the strongest interactions are often Coulomb forces between a charge carrier and the ions. The net strength of these interactions is measured by $U/(1/\varepsilon_\infty - 1/\varepsilon_0)$, where U measures the Coulomb interaction energy and ε_0 and ε_∞ are the static and optical dielectric constants, respectively. The difference between the low- and high-frequency dielectric constants indicates the displacability of a solid's ions. The ions of many oxides are especially displaceable: $\varepsilon_0 \gg \varepsilon_\infty$.

Short-range electron-lattice interactions result because the energies of electrons in bonding, antibonding or non-bonding orbitals generally depend on the separations between the relevant atoms. In covalent solids the electron-lattice forces are typically about 2-3 eV/Å. Since the long-range electron-lattice interactions of ionic (polar) solids are usually much stronger than short-range electron-lattice forces, the quasiparticles produced by these forces were termed "polarons."

Especially strong short-range electron-lattice interactions occur if a carrier's presence has a qualitative effect on bonding. For example, the addition of a hole in KCl leads to the formation of a bond between two adjacent halogen ions. In addition, the electrons assigned to oxygen dianions have especially strong short-range electron-lattice interactions. This effect arises because an isolated oxygen atom cannot bind two additional electrons. It is only the positive charge of cations that surround an oxygen anion in a solid that bind the second electron in the vicinity of an oxygen anion, enabling it to be regarded as a dianion. Hence, an oxygen dianion's outermost electron is freed as the surrounding cations move away. The exceptionally strong short-range electron-lattice interaction of oxygen dianions indicates the sensitivity of their outer electron to the positions of the surrounding cations.

POLARON FORMATION: A NONLINEAR PROBLEM

A charge carrier can be bound within a potential well that is established by the displacements of atoms from their carrier-free equilibrium positions. The lowering of the energy of the carrier associated with its being bound within such a potential can exceed the strain energy required to produce the atomic displacements. Then it is energetically favorable for atoms to be displaced about a charge carrier so as to produce a bound state for it. Such a carrier is said to be "self-trapped" since the atomic displacements that create the potential well that traps the carrier are only stabilized when the carrier is bound within the potential well. The term "polaron" refers to the unit comprising the self-trapped electronic carrier and the atomic-displacement pattern that binds the carrier. The term "polaron" indicates the expectation of self-trapping in polar solids.

Self-trapping is a feedback phenomenon: the pattern of atomic displacements that produces the potential well that binds the electron itself depends upon the electronic state. This feature reflects itself in the nonlinearity of the wave equation for the self-trapped electron's groundstate:

$$\left[-\frac{\hbar^2}{2m}\nabla^2 - \frac{1}{S}\int dr' \, |\psi(r')|^2 \int du \, Z(r'-u) \, Z(r-u) \right]\psi = E\psi, \tag{2}$$

where S measures the stiffness of the deformable medium with respect to atomic displacements. It is evident from Eq. (2) that the nature of a self-trapped state generally depends on the range of the electron-lattice interaction, expressed by the function Z(r-u).

SCALING THEORY OF POLARON FORMATION

A scaling theory has been developed which permits extraction of essential physical information about self-trapped states without having to explicitly solve Eq. (2) for different models of the electron-lattice interaction [8]. This formulation of the theory of polaron formation considers E(R), the total energy of the groundstate (electronic energy plus strain energy) as a function of R, the radius of the self-trapped state in units of the radius of a carrier confined to just a single site, defined as $R \equiv 1$. The actual energy and radius of the groundstate are those which minimize E(R).

When the electron-lattice interaction is purely of long range, E(R) is:

$$E(R) = \frac{W}{2R^2} - \frac{U}{2R}\left(\frac{1}{\varepsilon_\infty} - \frac{1}{\varepsilon_0}\right),$$ (3)

where W is the electronic bandwidth and U is the Coulomb repulsion between elementary charges separated by the interatomic separation. The plot of this function in Fig. 1 indicates that this function always has a solitary minimum. The characteristic spatial extent of this state may be written as $a_B/(1/\varepsilon_\infty - 1/\varepsilon_0)$, where a_B is the Bohr radius when the electron mass is replaced by the effective mass appropriate to the electronic band of width W. Large polarons in ionic solids typically have radii of about 5 Å.

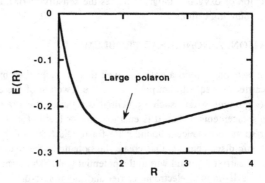

Fig. 1. E(R) from Eq. (3) is plotted versus R.

When the electron-lattice interaction is purely of short-range, E(R) is

$$E(R) = \frac{W}{2R^2} - \frac{E_b}{R^d},$$ (4)

where E_b is termed the small-polaron binding energy and d is the dimensionality of carrier motion. Figure 2 illustrates that E(R) of Eq. (4) has no minima between $R = 1$ and $R = \infty$ for a carrier that is able to move in all three dimensions, $d = 3$. Rather, the solitary maximum of Fig. 2 separates two distinct minima. The minimum at $R = 1$ corresponds to a self-trapped carrier that has "collapsed" to the smallest allowed radius. This self-trapped electronic state is that of a small polaron. The minimum at $R = \infty$ refers to a free carrier that extends over all available space.

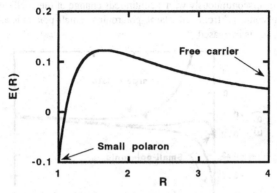

Fig. 2. E(R) from Eq. (4) is plotted versus R.

The small polaronic state is only the groundstate if $E_b > W/2$. While E_b is measured to be as large as several eV (for holes in alkali halides), E_b is usually estimated from experiment to be less than an electron volt. These results imply that small polarons would only form when the electronic bandwidth is relatively narrow (W < 1 eV). However, small polarons are frequently found when electronic bandwidths are estimated to very much larger, several eV. Thus interactions other than the short-range electron-lattice interaction, considered in seminal studies of small-polaron formation [9], are often crucial for small-polaron formation.

SMALL-POLARON COLLAPSE

Forces that drive carriers toward localization often trigger their collapse into small polarons. For example, small polarons' distinctive transport is frequently observed in covalent glasses [10], even though the glasses have wide electronic bands and far too mild electronic disorder to produce these severely localized states solely through Anderson localization [11]. In these instances modest disorder seems to trigger the collapse of carriers into small polarons [12].

Two characteristic properties of semiconducting oxide crystals promote the confining of electronic charge carriers. First, the long-range electron-lattice interaction is particularly effective since oxide's ions are often especially displaceable, $\varepsilon_0 >> \varepsilon_\infty$. Second, semiconducting oxides often contain significant densities of charged dopants and defects (e.g., vacancies).

The synergistic localizing effects of 1) the short-range electron-lattice interaction, 2) the long-range electron-lattice interaction and 3) an attractive Coulombic center are readily studied within the scaling approach. In particular, with these interactions the scaling energy has to form:

$$E(R) = \frac{W'}{2R^2} - \frac{E_b}{R^d} - \frac{V_c'}{R} , \qquad (5)$$

where the final term contains the carrier's attraction to a Coulombic center and the primes indicate that W is decreased to W' and V_c is increased to V_c' by the long-range component of the electron-lattice interaction. Figure 3 illustrates that the location of the absolute minimum of E(R) of Eq. (5) can change discontinuously with a continuous change of the coefficients. For example, the groundstate can "collapse" from being large-polaronic to small-polaronic as the strength of an attractive potential, V_c' is increased.

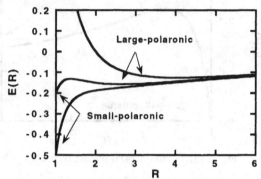

Fig. 3. As the strength of V_c' of Eq. (5) is increased, E(R) progressively changes from 1) having a single minimum at R > 1 to 2) having minima both at R > 1 and at R = 1 to 3) having a solitary minimum at R = 1. Minima at R > 1 corresponds to large-polaronic states while minima at R = 1 indicate small-polaronic states. The abrupt change of the absolute minimum from being at R > 1 to being at R = 1 is termed a small-polaron collapse.

MAGNETIC POLARONS

So far this discussion has only considered polaron formation produced by electron-lattice interactions. However, polaron effects also arise from exchange interactions between a carrier and the local moments of a magnetic solid. These magnetic interactions tend to dominate electron-lattice interactions if a polaron is very large. Then a polaron of purely magnetic origin can form

A magnetic polaron, depicted in Fig. 4, denotes a carrier confined to a ferromagnetically aligned region of local moments that is embedded within an antiferromagnet. The carrier's intra-site exchange interactions with the aligned local moments produces the well that binds the carrier.

Fig. 4. Schematic depiction of a magnetic polaron.

The intrasite exchange interaction, I, between a local moment (of net spin S) and a carrier (of spin 1/2) fosters a parallel (or antiparallel) alignment of the carrier's spin with that of the local moments the carrier contacts. In addition, the carrier has the desire (measured by its electronic bandwidth, W) to retain its spin alignment while spreading to geometrically equivalent sites of a crystal. It is energetically favorable for a carrier to maintain a universal spin alignment rather than to align with each local moment when W > IS [13].

The energy of the carrier's interaction with local moments and the carrier's transfer energy can both be minimized when the local moments are aligned with one another, as depicted in Fig. 4. In an antiferromagnet this parallel alignment of local moments comes at the cost of increasing the (intersite) exchange energy between local moments. In a pure magnetic polaron, the competition between a carrier's desire to spread out amongst aligned local moments and the resistance of an antiferromagnet's local moments to be aligned with one another defines the size of the ferromagnetically aligned region about a carrier in an antiferromagnet.

The formation of such a magnetic polaron has been addressed with a scaling formalism akin to that used in addressing atomic-displacement polarons [14]. The energy functional for a magnetic polaron formed by a universally aligned carrier in a three-dimensional antiferromagnet is:

$$E(R) = \frac{W}{2R^2} - IS/2 + c_3 JS^2 R^3, \tag{6}$$

where J is the intersite exchange energy and c_3 is the numerical constant appropriate for a three-dimensional antiferromagnet. Since the carrier extends over N ($\propto R^3$) local moments, its interaction with each of these moments is -IS/2N. The net intrasite exchange energy is thus -IS/2.

As illustrated in Fig. 5, E(R) has a minimum at $R_{min} = (W/3c_3JS^2)^{1/5}$. This energy minimum only corresponds to a stable magnetic polaron, E(R) < 0, if IS > $(5c_3JS^2/2)(R_{min})^3$. Thus, large-radius magnetic polarons form in only very weak antiferromagnets.

Large magnetic polarons occur in EuTe, where they are identified by their very large magnetic susceptibilities [15]. The magnetic susceptibility has a Curie-type contribution from magnetic-field induced alignment of the polaron and a contribution from magnetic-field-induced alteration of the polaron's size [14]. Increasing the density of magnetic polarons ultimately drives the host antiferromagnet to be a ferromagnetic metal with its carriers extending over the entire crystal [14].

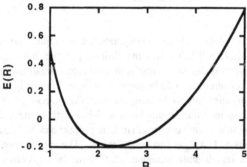

Fig. 5. E(R) for a three-dimensional magnetic polaron, Eq. (6), possesses a solitary minimum.

METAL-TO-SEMICONDUCTOR TRANSITION IN EuO

EuO is one of the few ferromagnetic insulators (T_{Curie} = 70 K). Direct ferromagnetic exchange between the S = 7/2 Eu^{2+} cations gives rise to its ferromagnetism [15]. The smaller direct Eu-Eu ferromagnetic exchange that results when the oxygen anions are replaced with much larger chalcogen anions accounts for EuS, EuSe and EuTe being antiferromagnets.

Oxygen vacancies or trivalent rare-earth cations produce shallow donors with 6s electrons. When the concentration of these large-radius donors is high enough, the donors overlap sufficiently to yield metallic impurity conduction [16-18]. However, the metallic conductivity falls rapidly as the temperature is raised above 50 K. Semiconducting behavior is observed above the Curie temperature. With optimum doping (about 1 %) the change in conductivity reaches 15 orders of magnitude at this metal-to-insulator-transition [16-18]. The transition shifts to higher temperature in an applied magnetic field (2 K/T) yielding a huge negative magneto resistance [19].

This transition has been attributed to the thermally induced abrupt shrinking of the large-radius donor states [20-21]. In particular, the collapsing of the donor states reduces the overlap between donors, thereby suppressing the metallic impurity conduction. The large negative magnetoresistance arises because the temperature of the thermally induced collapse, and hence the metal-to-insulator transition, increases with applied magnetic field [22].

Thus, the metal-to-insulator transition is driven by the thermally induced collapse of the donors. The temperature-dependent donor collapse, in turn, has its origin in the interaction between the donor electron and the localized spins. Namely, the presence of the donor electron stiffens the local moments with respect to deviations from their optimally aligned groundstate. However, this stiffening depends on the relationship between the coherence lengths of the spin deviations and the diameters of the donor state. In particular, deviations of the local moments' alignment (magnons in the spin-wave approximation) that have coherence lengths that exceed the diameter of the donor state experience less stiffening than the misalignments with shorter coherence lengths. As a result, the R-dependent portion of the free energy of the donor and the magnons is:

$$F(R,T) = \frac{W'}{2R^2} - \frac{E_b}{R^d} - \frac{V_c'}{R} - C\left(\frac{I}{k_B T_{Curie}}\right)^2 \left(\frac{k_B T}{R^2}\right), \qquad (7)$$

where C is a numerical constant.

The final term of Eq. (7) depends both on temperature and on the donor-state radius. Through this term's proportionality to R^{-2}, it drives the shrinking of the donor state. Although donor electrons generally provide stiffening against spin deviations, small-radius donors provide less stiffening against spin deviations than do large-radius donors. Thermally induced spin deviations produce this term's proportionality to the temperature. The inverse dependence of this term on the square of the Curie temperature results because the density of thermal spin deviations falls with increasingly strong interatomic exchange (in turn, proportional to T_{Curie}).

Figure 6 presents plots of F(R,T) at two temperatures. The upper curve, corresponding to the lower temperature, has its absolute minimum at a finite radius, corresponding to the large-polaronic donor state being stable. By contrast, the high-temperature curve has its absolute minimum at R = 1 indicating the stability of the small-polaronic donor state. These curves illustrate the thermally induced collapse of a donor in a ferromagnet.

The thermally induced donor collapse illustrated in Fig. 6 relies on the presence of two distinct effects. First, the short-range component of the electron-lattice interaction produces the dichotomy between large-radius and small-radius states. In particular, the energy barrier separating these states requires the presence of the short-range electron-lattice interaction. Second, the interaction of the donor electron with the ferromagnet's spin deviations provides a contribution to the free energy that increasingly favors small-radius states as the temperature is

170

raised. By itself, the electron-magnon interaction would produce a continuous change of the donor radius with temperature.

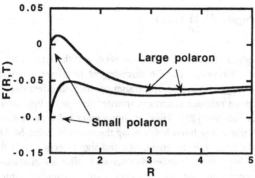

Fig. 6. The R-dependent portion of the free energy of a donor that interacts with the local moments of a ferromagnetic donor, from Eq. (7), is plotted at two temperatures (after Ref. [21]). At low temperatures the large-polaronic minimum is stable. However, at higher temperatures the small-polaronic minimum at R = 1 is stable. These curves illustrate the thermally induced collapse of a donor state in a ferromagnet.

SMALL-POLARON HOPPING

A self-trapped carrier can only move when the atoms that participate in the self-trapping alter their positions. Since atoms generally move slowly, polaronic motion is always very slow. The slowness of the carrier motion distinquishes polaronic transport from common transport.

In particular, a large-polaron moves slowly with an effective mass that is much larger than that of a nonpolaronic carrier, $m_{LP} \gg m_e$. However, since these massive quasiparticles are not easily scattered, $\tau_{LP} \gg \tau_e$, free large polarons have mobilities that are comparable to those of conventional nonpolaronic carriers, $e\tau_{LP}/m_{LP} \approx e\tau_e/m_e$ [23].

By contrast, small-polaron motion is not readily described in terms of a freely moving carrier that is occasionally interrupted by scattering events. Rather small polarons move by a succession of phonon-assisted hops. Small-polaron motion is characterized by low mobilities, < 1 cm²/V-sec, that rise with increasing temperature [24]. The mobility rises in a nonArrhenius manner at temperatures well below the characteristic phonon temperature and achieves an Arrhenius temperature dependence at higher temperatures [25,26].

The Hall effect of small-polaronic hopping has three distinguishing features. First, the Hall mobility due to small-polaronic hopping is low, < 1 cm²/V-sec. Thus, its magnitude is lower than those of conventional semiconducting carriers. Second, unlike the carriers in conventional transport, the temperature dependence of the Hall mobility generally differs from the mobility that enters into the dc conductivity, the "drift mobility." The Hall mobility rises more gently with rising temperature than does the drift mobility [27,28]. Third, the sign of the Hall effect is often anomalous. In particular, the Hall effect arises from the interference between different hopping paths that transport a carrier between a pair of sites. The Hall-effect sign depends on the sign of the carrier, q, the net number of steps in both hopping paths, n, and the product of transfer energies, $t_{j,j+1}$, linking adjacent sites of the closed loop defined by the interference [29]:

$$\text{Sign}(R_{Hall}) = \text{Sign}(q^{n+1} \prod_{j=1}^{n} t_{j,j+1}). \qquad (8)$$

For hopping along the edges of a cubic crystal, n is even, and the sign of the Hall effect indicates the sign of the carrier. However, for the three-legged interference processes that dominate hopping in irregularly arranged sites, $n = 3$, the sign of the Hall effect depends only on the sign of the product of the three relevant electronic transfer energies. For example, this product is negative for impurity conduction [29]. Thus, the hopping of holes gives an n-type Hall effect.

Indeed, the hopping of small polaron holes along the edges of cubic MnO gives a Hall effect that agrees in magnitude, temperature dependence and sign (p-type) with these predictions [2]. However, small-polaron hopping in disordered materials is often anomalously signed [10,11].

A small polaron's interaction with local moments of a magnetic semiconductor affects its mobility and Seebeck coefficients. While space is not available to discuss these phenomena, references to some papers on this subject are provided [30-32].

COMPARING THE METAL-TO-INSULATOR TRANSITIONS OF LaMnO₃ AND EuO

The origin of the ferromagnetism in doped LaMnO₃ has been debated since its discovery almost half a century ago [33]. By comparing the effect of different divalent dopants on the material's magnetic properties, Jonker concluded that doped LaMnO₃ is ferromagnetic because dopants thwart a low-temperature distortion that would make insulating LaMnO₃ antiferromagnetic [34]. Subsequently Tanaka found that removing carriers by compensating dopants with oxygen vacancies has little effect on the materials ferromagnetism [35]. These results suggest that the ferromagnetism of doped LaMnO₃ is produced by direct exchange between magnetic cations.

Volger reported that p-type doped LaMnO₃ underwent a metal-to-semiconductor transition near the Curie temperature [36]. He found that application of a magnetic field produced a large negative magnetoresistance peak in the vicinity of the transition. From attempts to measure the Hall effect, he concluded that the carriers had very low mobilities. Subsequent transport measurements identified the carriers as small polarons [35,37]. However, the Seebeck coefficient is much smaller than expected for the density of holes (Mn^{4+}) provided by doping if the holes are distributed among all Mn sites [38,39]. It was suggested that the density of holes is increased to about 50% that of the Mn sites by Mn^{3+} disproportionation: $2Mn^{3+} \rightarrow Mn^{2+} + Mn^{4+}$. This conjecture is at odds with the finding that the Seebeck coefficient rises markedly as the hole density is decreased by compensating holes with electrons provided by oxygen vacancies [35]. An alternative explanation of the Seebeck anomalies is that the small-polaronic holes only move between a subset of the Mn sites. This type of impurity conduction is consistent with the measurement of a n-type Hall effect in this p-type material [36,40].

This analysis of the literature on p-type doped LaMnO₃ suggests that 1) the low-temperature ferromagnetism is not induced by the carriers but results from direct Mn-Mn exchange and 2) the carriers in the semiconducting state are located at dopant-related sites. If these conclusions are correct, the metal-to-semiconductor transition in this material is analogous to that in doped EuO. Then the mechanism of the transitions may well be similar to one another.

SUMMARY

There are two distinct components of electron-lattice interaction in oxide semiconductors. The long-range electron-lattice interaction, produced by Coulomb interaction between a carrier and the solid's ions, is generally very strong. The short-range electron-lattice interaction, produced by alterations of bond lengths, is typically much weaker than the long-range component. The tendency for localization that comes from the long-range component of the electron-lattice interaction competes with carriers' tendency to spread out, measured by the electronic bandwidth, to produce a finite-sized "large" polaron. By contrast, the short-range electron-lattice interaction will by itself only permit a severely localized bound carrier, a small-polaron. However, forming a small-polaron in this way requires unusually narrow electronic bands.

Nonetheless, small polarons can reasonably be expected to form when the short-range electron-lattice interactions act in tandem with other interactions that foster localization. Two omnipresent localizing effects in oxide semiconductors are the long-range electron-lattice interaction and a carrier's attraction to available charge centers, e.g., dopants. Acting together these localizing effects permit both large- and small-polaronic states. These two types of polaron can even coexist. With a continuous change of parameters the groundstate can abruptly "collapse" from a large-polaronic state to a small-polaronic state. A small polaron formed by attraction toward a charge center is confined in its vicinity.

The intra-site exchange between local moments and a polaron's electronic carrier tends to ferromagnetically align the local moments within the polaron. When a host antiferromagnet is sufficiently weak, a carrier will form a local ferromagnetic region within the antiferromagnet, a magnetic polaron. In sufficient density, magnetic polarons induce global ferromagnetism.

As rising temperature increases the prevalence of deviations of local moments from ferromagnetic alignment, polaronic states can be induced to collapse. The collapsing of donor states in the ferromagnetic insulator EuO induce a metal-to-insulator transition by suppressing the metallic impurity conduction that dominates the low-temperature transport. A large negative magnetoresistance results from the shifting of the transition to higher temperatures in an applied magnetic field. Consideration of measurements on polycrystalline doped $LaMnO_3$ suggests that its metal-to-semiconductor transition is similar in many respects to that of EuO.

ACKNOWLEDGMENTS

This work supported by Sandia National Laboratories through BES/DOE Contract DE-AC04-94AL85000.

REFERENCES

1. J. Devreese, R. DeCominck and H. Pollak, Phys. Stat. Sol. 17, 825 (1966).
2. C. Crevecoeur and H. J. de Wit, J. Phys. Chem Solids 31, 783 (1970).
3. H. P. R. Frederikse, J. Appl. Phys. 32, 2211 (1961).
4. L. Forro, O. Chauvet, D. Emin, L. Zuppiroli, H. Berger and F. Levy, J. Appl. Phys. 75, 633 (1994).
5. A. J. Bosman and H. J. van Daal, Adv. Phys. 19, 1 (1970).
6. C. Schlenker, S. Ahmed, R. Buder and M. Gourmala, J. Phys. C. 12, 2503 (1979).

7. B. K. Chakraverty, M. J. Sienko and J. Bennerot, Phys. Rev. B **17**, 3503 (1979).
8. D. Emin and T. Holstein, Phys. Rev. Lett. **36**, 323 (1976).
9. T. Holstein, Ann. Phys. (N.Y.) **8**, 325 (1959).
10. D. Emin, C. H. Seager and R. K. Quinn, Phys. Rev. Lett. **28**, 813 (1972).
11. D. Emin in <u>Amorphous Thin Films and Devices</u> L. L. Kazmerski, editor (Academic Press, New York, 1980), pp. 17-57.
12. D. Emin and M.-N. Bussac, Phys. Rev. B **49**, 14 290 (1994).
13. P. W. Anderson and H. Hasegawa, Phys. Rev. **100**, 675 (1955).
14. D. Emin and M. S. Hillery, Phys. Rev. B **36**, 7353 (1987).
15. S. Methfessel and D. C. Mattis, <u>Magnetic Semiconductors</u>, Zeitschrift fur Physik **18a**, Springer-Verlag, Heidelberg 1968 pp. 389-562.
16. J. B. Torrence, M. W. Shafer and T. R. McGuire, Phys. Rev. Lett. **29**, 1168 (1972).
17. T. Penny, M. W. Shafer, aned J. B. Torrence, Phys. Rev. B **5**, 3669 (1972).
18. C. Godart, A. Mauger, J. P. Desfours and J. C. Achard, J. de Physique **41**, C5-205 (1980).
19. M. R. Oliver, J. O. Dimmock, A. L. McWhorter and T. B. Reed, Phys. Rev. B **5**, 1078 (1972)
20. D. Emin, M. Hillery and N. L. H. Liu, Phys. Rev. B, **33**, 2933 (1986).
21. D. Emin, M. Hillery and N. L. H. Liu, Phys. Rev. B **35**, 641 (1987).
22. M. S. Hillery, D. Emin and N. H. Liu, Phys. Rev. B **38**, 9771 (1988).
23. H.-B. Schüttler and T. Holstein, Ann. Phys. (N.Y.), **166**, 93 (1986).
24. T. Holstein, Ann. Phys. (N.Y.) **8**, 343 (1959).
25. D. Emin, Phys. Rev. Lett. **32**, 303 (1974).
26. D. Emin, Adv. Phys. **24**, 305 (1975).
27. L. Friedman and T. Holstein, Ann. Phys. (N.Y.) **21**, 494 (1963).
28. D. Emin and T. Holstein, Ann. Phys. (N.Y.) **53**, 439 (1969).
29. D. Emin, Phil. Mag. **35**, 1189 (1977).
30. N.-L H. Liu and D. Emin, Phys. Rev. Lett. **42**, 71 (1979).
31. D. Emin and N.-L. H. Liu, Phys. Rev. B **22**, 4788 (1983).
32. N.-L. H. Liu and D. Emin, Phys. Rev. B. **30**, 3250 (1984).
33. G. H. Jonker and J. H. Van Santen, Physica **XVI**, 337 (1950).
34. G. H. Jonker, Physica **XXII**, 707 (1956).
35. J. Tanaka, M. Umehara, S. Tamura, M. Tsukioka and S. Ehara , J. Phys. Soc. Japan, **51**, 1236 (1982)
36. J. Volger, Physica **XX** 49 (1954).
37. M. Kertesz, I. Riess, D. S. Tannhauser, R. Langpape and F. J. Rohr, J. of Solid State Chem. **42**, 125, (1982).
38. R. Raffaelle, H. U. Anderson, D. M. Sparlin and O. E. Parris, Phys. Rev. B **43**, 7991 (1991).
39. M. F. Hundley and J. J. Neumeier, Phys. Rev. B **55**, 11511 (1997).
40. M. Jaime, H. T. Hardner, M. B. Salamon, M. Rubenstein, P. Doresy and D. Emin, Phys. Rev. Lett. **78**, 951 (1997).

CALCULATED TRANSPORT AND MAGNETIC PROPERTIES OF SOME PEROVSKITE METALLIC OXIDES AMO$_3$

G. SANTI, T. JARLBORG
Département de Physique de la Matière Condensée, Université de Genève, 24, Quai Ernest-Ansermet, CH-1211 Genève 4, Switzerland

ABSTRACT

We study some compounds of the perovskite (or pseudo-cubic perovskite) series AMO$_3$, where M is a transition metal and A is Ca, Sr, or Nd, by LSDA self-consistent electronic structure calculations with the LMTO method. Transport and magnetic properties, as well as Fermi surfaces are calculated. These materials exhibit sharp density of states features in the vicinity of the Fermi level that strongly affect their transport and magnetic properties and make them very sensitive to structural deformation and stoichiometry. Calculated total energies are very close for anti-ferromagnetic and ferromagnetic solutions. This explains qualitatively the magnetoresistive anomalies shown by this family of compounds.

INTRODUCTION

The pseudo-cubic perovkite series of oxides of type AMO$_3$, where A is a divalent metal (or a rare earth) and M a transition metal, exhibit quite a wide range of magnetic and transport properties (depending on the elements A and M, as well as the doping). This, and their structural resemblance with the cuprates, has caused a revival of experimental and theoretical interest. Furthermore, many technical applications are appealing since thin films of these compounds can be grown relatively easily. Almost all these compounds have a magnetically ordered ground state. However, the transition temperatures and the different types of order can vary considerably from one system to another. The peculiar transport properties, such as the colossal magnetoresistance (CMR) for the Mn-based compounds are associated with a transition from a metallic ferromagnetic (FM) to an insulating anti-ferromagnetic (AF) configuration. Sometimes this transition is accompanied by a structural distortion (as for Nd$_{1-x}$Sr$_x$MnO$_3$ [1, 2]).

Here we will focus on the ternary compounds SrRuO$_3$ (SRO) and CaRuO$_3$ (CRO) for the ruthenates and CaMnO$_3$ (CMO) and Nd$_{1-x}$Sr$_x$MnO$_3$ (NSMO) for the manganites. SRO and CRO have been studied in [3, 4, 5, 6], while independent investigations were made on (La, Ca)MnO$_3$ (LCMO) [7, 8]. In this work we study mainly the Nd$_{1/2}$Sr$_{1/2}$MnO$_3$ system (NSMO) in order to compare with recent, detailed experimental information for this system [9]. Except for the Nd-f band, the main band features of all these compounds are very similar.

METHOD

We use the self-consistent, spin-polarized LMTO band method within the LDA for the electronic structure calculations. Details for the application to SRO and CRO systems can be found in ref. [4], where the calculations were done both with the idealised cubic structure and the real distorted orthorhombic one with 20 atoms per cell. For CMO and NSMO, we have used a double cubic cell with 10 atoms per cell in order to describe the

175

AF configuration as well as $Nd_{1/2}Sr_{1/2}MnO_3$. The calculations were done in the simple cubic structure for both CMO and NSMO, even if NSMO, unlike CMO, shows a slight orthorhombic distortion [10]. We took the lattice constants to be 3.73 Å for CMO [7] and 3.87 Å for NSMO, the latter corresponding to the value observed in thin films [11]. Indeed, NSMO has the particularity to adapt its cell dimensions (at constant volume) to the substrate up to a large thickness (about 1000 Å) without dislocations (the a parameter can vary from 3.75 to 3.9 Å) [11]. Self-consistent results are determined using 89 k-points in the irreducible Brillouin Zone. For CMO, we have considered both A and G-type AF configurations, and for half-doped NSMO, we have simulated the doping by using a rock-salt structure that corresponds to G-type AF configuration. Two sets of calculations were done for NSMO, where the Nd f-electrons were treated as itinerant valence or as core electrons. In the first case, we obtain an f-band occupancy of about 3.6, and a spin of about 3.2 electrons per Nd atom (for both FM and AF spin-polarized calculations). This is not too far from what is expected from a Hund's rule configuration of a Nd atom, although the spin is reduced a bit due to hybridization. The density-of-states (DOS) at ϵ_F is very high since ϵ_F is within the f-bands, and the DOS structures on Mn are affected by the hybridization. However, the f-bands are very flat in the f-bands and their conductivity is therefore limited. Guided by these results of the charge and spin configuration of the itinerant Nd f-electrons, we next proceeded to calculations where the f-electrons were treated as a core state containing 4 electrons. The goal is to see whether the AF and FM configurations are different in NSMO as is the case in CMO. In these cases the f-electrons cannot contribute to the DOS although their charge (and spin) can relax spacially during self-consistency. The f-charge is sufficiently localized so that imposing no or complete polarization of the 4 f-electrons, does not change the electronic structure of the rest of the system very much. The DOS is essentially identical, while the total energy is clearly lowest for the spin polarized case ($\approx$ 200 mRy/Mn). Furthermore, the configuration with the f-electrons in the core is easier to converge since the huge peak in the DOS at ϵ_F due to f-bands affects the numerical stability.

RESULTS AND DISCUSSION

Beside the Nd f-bands part, the electronic structures of these ternary oxides all show the same main features as can be seen from figs. 1 and 2 and refs. [4, 7, 6]: the DOS at ϵ_F shows a relatively large peak dominated by the (Ru,Mn)-d, and below an extended part dominated by the O-p. The lower part of the d-bands is formed from the t_{2g} sub-shells (the lowest degenerate bands of the d part at Γ on fig. 1), and the higher part by the e_g ones (almost at ϵ_F). However, it must be noted that the orthorhombic distorion is likely to wash out the splitting of the d-bands in t_{2g} and e_g as it does for SRO and CRO [4].

The calculations for CMO were done for non-magnetic (NM), FM, A-type AF and G-type AF configurations. The bands for the NM case are shown on fig. 1b. This system is known for having FM to AF transitions without structural changes, and this fact makes our total energy results more meaningful. The most stable solution is clearly spin-polarized and the Mn was carrying a magnetic moment of 2.70 μ_B for FM and 2.64 μ_B for AF in good agreement with Pickett and Singh [7]. But, from the total energies, we do not find the same order between the different configurations: the FM state is the lowest, but is almost indistinguishable ($<$ 1 mRy) from the G-type AF state, while the A-type AF state is about 14 mRy higher. However, some sensitivity of the total energies on the choice

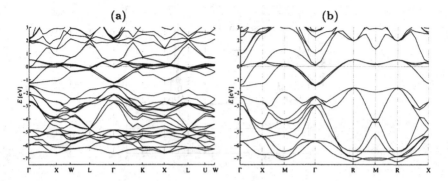

Figure 1: **(a)** Non-polarized bands for $Nd_{1/2}Sr_{1/2}MnO_3$ (Nd-f in the core, 10 atoms per cell). **(b)** Same for $CaMnO_3$ (5 atoms per cell).

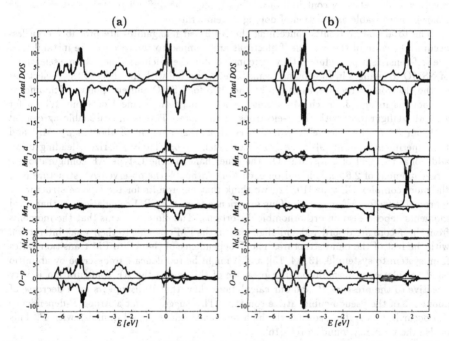

Figure 2: Total and partial DOS for **(a)** ferromagnetic and **(b)** anti-ferromagnetic $Nd_{1/2}Sr_{1/2}MnO_3$ (Nd-f is in the core).

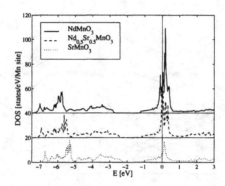

Figure 3: Comparison between the total NM DOS of $Nd_{1/2}Sr_{1/2}MnO_3$ and end-point stoichiometric compounds $NdMnO_3$ and $SrMnO_3$ (for a lattice constant of 3.85 Å and the Nd f-bands in the valence, causing the huge peak at ϵ_F).

of the linearisation energies is observed. In addition the T-dependence of the free energy involves several entropy contributions. Thus it might be difficult to state that one or the other is more stable function of doping or temperature.

The total energy results concerning the Nd-based manganites are probably even less accurate because of the way the f-electrons are confined to the core (or the instability of the self-consistent procedure if they were included in the valence). The extreme sensitivity of NSMO to changes in the doping seems to be connected with the effect of the f-electrons on the electronic structure, as shown by the DOS for NSMO and end-point stoichiometric compounds in fig. 3. Each Mn atoms carries a magnetic moment of about 1.3 μ_B for an FM configuration with the f-electrons in the core. This is in reasonable agreement with our "f in the valence" calculation (with a total moment of about 4 μ_B/Mn) and the experimental result [9] if the Nd f core shell is completely polarized, leading to an additional moment of 2 μ_B/Mn. On the other hand, in the G-type AF configuration, a larger moment of 2.6 μ_B is found on each Mn. Although the observed AF structure is of the more complex CE-type [10, 12], we think that our results for the G-type structure is representative for AF moments. Our experience from the SRO calculations is that local moments depend also on orthorhombic distortions. Therefore, it seems that the moments from the f-core are disordered and align only in the FM case due to their weak interaction with the rest of the band structure. The T-dependence of the AF to FM transition varies from system to system [9, 13, 14, 15], and it might be too delicate to describe by ab-initio band calculations. In addition, the presence of the gap in the AF configuration is very sensitive to the pressure P, and our calculations show that it disappears for a decrease of about 1% of the pseudo-cubic lattice constant. This suggests that a strong P-dependence can occur in this system, so that the high resistance values should diminish with P. This is also the tendency found in ref. [16].

The calculated properties of the different perovskites are summarised in table I. The conductivity has been estimated from the band structure through the Boltzmann approximation. The average conductivity is given by [4] $\langle \sigma \rangle = \frac{1}{3} \sum_\alpha \sigma_{\alpha\alpha}$, with $\sigma_{\alpha\beta} = e^2 \tau (n/m^*)_{\alpha\beta}$, and,

Compound	Magnetic configuration	DOS at ϵ_F [1/Ry/M-atom] ↑	↓	Total	μ [μ_B/M-atom]	$\langle\sigma\rangle$ [$\mu\Omega^{-1}m^{-1}$] ↑	↓	Total
SrRuO$_3$	NM	-	-	119	-	-	-	5.0
	FM	15	19	34	1.7	1.4	3.6	5.0
CaRuO$_3$	NM	-	-	88	-	-	-	2.0
CaMnO$_3$	NM	-	-	82	-	-	-	5.5
	FM	23	0	23	3.0	1.1	0.0	1.1
	AF (A)	13	13	26	0.0	0.4	0.4	0.8
	AF (G)	0	0	0	0.0	0.0	0.0	0.0
Nd$_{1/2}$Sr$_{1/2}$MnO$_3$	NM	-	-	93	-	-	-	2.6
	FM	95	37	132	1.3	1.3	1.6	2.9
	AF (G)	0	0	0	0.0	0.0	0.0	0.0
NdMnO$_3$	NM	-	-	232	-	-	-	0.6
SrMnO$_3$	NM	-	-	97	-	-	-	2.1

Table I: Calculated electronic structure, magnetic and transport properties for the different manganites and ruthenates in the cubic structure. μ is the cell moment per M atom. The relaxation times τ are 1.5×10^{-14} and 0.4×10^{-14} s for SRO and CRO respectively from a fit with experimental data [4] (thus, the total $\langle\sigma\rangle$ is the experimental value). For the other compounds, the same value as for SRO has been taken. Note that the DOS for NdMnO$_3$ includes f-electrons.

$$\left(\frac{n}{m^*}\right)_{\alpha\beta} = \frac{1}{\Omega_{cell}} \sum_{n,\mathbf{k}\in BZ} v_{n\mathbf{k}\alpha} v_{n\mathbf{k}\beta} \delta(\epsilon_F - \epsilon_{n\mathbf{k}}) \tag{1}$$

The τ for SRO and CRO was chosen to fit the experimental data [3, 4] at low T. For those compounds, that are not metallic at low T, we set $\tau = 1.5 \times 10^{-14}$ s (same as for cubic SRO) in order to give a rough idea of the comparative properties of the different compunds. It can be seen that the effect of the f-bands is not very important on the transport properties due to their flatness.

Thus, the band results confirm that a FM to AF transition is behind the CMR mechanism. Moreover, the small difference in total energy between FM and AF configuration explains qualitatively why, if the AF state (which is insulating or semi-metallic) is most stable for some doping and T, it is not very difficult to reach the FM state (which is a good metal in all cases) by application of a magnetic field. This has also been proposed in several studies (see for instance [1]). However, our results find no support for the assumption that Mn changes valence ("charge ordering") in this transition [9, 16, and references therein]. The local charges on Mn are very similar in our AF, FM and NM solutions, the charge within the Mn Wigner-Seitz sphere remaining the same (≈ 6.3 el./Mn).

CONCLUSIONS

The calculated band structures for the different perovskite compounds considered here show quite similar features. From our calculations on SRO and CRO in the orthorhombic structure, we find that the t_{2g} and e_g sub-bands hybridise thereby preventing any simple interpretation in terms of occupation of the d-subshells. The calculations for the different magnetic configurations in CMO shows that the insulating G-type AF and the metallic FM configuration are extremely close in energy. This explains qualitatively the field-induced transition observed experimentally. The role of the f-band is quite important in NSMO for determining the magnetic properties. We find that the f-shell has to be magnetically polarized (consistently with Hund's rules) to explain the magnetic moment observed in the FM phase. The polarization of the f-core does not seem to affect the band structure in the AF configuration. Therefore, we think that the f-moments are disordered except in the FM state where they are aligned through their weak interaction with the band structure. In these compounds, we have found no indication of any kind of "charge-ordering" in the Mn accompanying the AF transition. Finally, we find here a clear indication that the origin of the CMR is in the change of the electronic structure by the magnetic transition going from an FM metallic behaviour to a insulating AF behaviour.

ACKNOWLEDGEMENTS

We would like to thank here S. Reymond and E. Koller for stimulating discussions and for communicating their unpublished results prior to publication.

REFERENCES

1. H. Kuwahara et al., Science **270**, 961 (1995).

2. H. Kuwahara et al., Science **272**, 80 (1996).

3. P. B. Allen et al., Phys. Rev. B **53**, 4393 (1996).

4. G. Santi and T. Jarlborg, J. Phys.:Condens. Matter **9**, 9563 (1997).

5. D. J. Singh, Appl. Phys. Lett. **79**, 4818 (1996).

6. I. I. Mazin and D. J. Singh, Phys. Rev. B **56**, (1997).

7. W. E. Pickett and D. J. Singh, Phys. Rev. B **55**, 1146 (1996).

8. S. Salpathy, Z. S. Popović, and F. R. Vukajlović, J. Appl. Phys. **79**, 4555 (1996).

9. H. Kuwahara et al., Phys. Rev. B **56**, 9386 (1997).

10. H. Kawano et al., Phys. Rev. Lett. **78**, 4253 (1997).

11. S. Reymond, 1997, private communication.

12. E. O. Wollan and W. C. Koehler, Phys. Rev. **100**, 545 (1955).

13. J. Z. Liu et al., Appl. Phys. Lett. **66**, 3218 (1995).

14. P. Schiffer, A. P. Ramirez, W. Bao, and S.-W. Cheong, Phys. Rev. Lett. **75**, (1995).

15. P. G. Radaelli et al., Phys. Rev. Lett. **75**, 4488 (1995).

16. Y. Moritomo, H. Kuwahara, Y. Tomioka, and Y. Tokura, Phys. Rev. B **55**, 7549 (1997).

EXPERIMENTAL DETERMINATION OF THE KEY ENERGY SCALES IN THE COLOSSAL MAGNETORESISTIVE MANGANITES

D. S. DESSAU[1], T. SAITOH[1], C.-H. PARK[2], Z.-X. SHEN[2], Y. MORITOMO[3,*], Y. TOKURA[3]

[1]Department of Physics, University of Colorado, CO 80309

[2]Department of Applied Physics, Stanford University, Stanford, CA 94305

[3]Department of Applied Physics, University of Tokyo, Tokyo 113, Japan and Joint Research Center for Atom Technology (JRCAT), Tsukuba, Ibaraki 305, Japan

ABSTRACT

We have performed X-ray absorption (XAS) and angle-resolved photoemission (ARPES) on single crystals of both the layered and cubic colossal magnetoresistive manganites to determine the electronic structure and the relevant energy scales in the problem: the intra-atomic exchange energy J (~ 2.7 eV), the Jahn-Teller energy gain E_{J-T} ($<.25$ eV), the one-electron bandwidth W (>3 eV for layered compounds) and the lattice relaxation or polaron binding energy E_B (.65 eV ferromagnetic phase and .8 eV paramagnetic phase). Lattice polarons are deemed important especially in the paramagnetic but also to a degree in the ferromagnetic phase. Due to the energy scale mismatch, the Jahn-Teller effect is unlikely to be the cause for these lattice polarons, at least for the layered samples.

INTRODUCTION

The recent observation of "colossal" magnetoresistance (the CMR effect) in the doped manganese-oxide ceramics (manganites) has sparked a great amount of effort aimed at understanding the unusual electronic and magnetic properties of these materials. Before coming to this realization, one of the most important tasks will be to determine the relevant energy scales for the problem. This paper reviews some of our efforts to do this using high resolution X-ray Absorption Spectroscopy (XAS) to probe the unoccupied density of states and Angle Resolved Photoemission Spectroscopy (ARPES) to probe the k-dependent occupied density of states.

SAMPLES

The samples studied in this experiment were bulk single crystalline samples grown by the floating zone method [1], except for the $La_{.1}Sr_{.9}MnO_3$ sample of figure 1, which was a polycrystalline ceramic [2]. The samples can be described by the general formula $(A,B)_{n+1}Mn_nO_{3n+1}$, where n is the number of Mn-O planes per unit cell [3]. So $(A,B)MnO_3$ corresponds to the n=infinity case (3-dim), $(A,B)_3Mn_2O_7$ is the n=2 case, and $(A,B)_2MnO_4$ is the n=1 case. While the n=1 case does not show any Metal-Insulator transition, the n=2 case does show the transition with CMR effect for $d^{3.6}$ (x=.4) doping [1]. The single or double MnO_2 plane in the layered compounds are isolated by the two La(Sr)O planes, keeping the two dimensional (2d) character. This 2d nature proved very valuable for our experiments in that we were able to exploit a polarization dependence of the spectra to deconvolve the states of different symmetries.

EXPERIMENT

The oxygen 1s X-ray Absorption (XAS) experiments were performed at the SGM beamline 10-1 of the Stanford Synchrotron Radiation Lab (SSRL). The energy resolution was between 0.1 and 0.2 eV, and the absorption intensity was determined by measuring the current flow from the sample to ground, then normalized by the incoming photon flux. The photon energy of each spectrum was individually referenced (with an uncertainty of less than 0.02 eV) by the Cr 2p edge. All samples were cleaved in-situ, and kept in a vacuum better than 2×10^{-9} torr at room temperature. All the XAS data presented in this paper were normalized at 600eV photon energy (approximately 70 eV above the edge).

Mat. Res. Soc. Symp. Proc. Vol. 494 © 1998 Materials Research Society

The ARPES measurements were performed at the undulator beamline 5 of SSRL using a 50 mm hemispherical analyzer mounted on a 2-axis goniometer. The energy resolution was typically 40 meV FWHM and the angular resolution ±1 degree, giving a **k**-resolution (for hv=22.4 eV) better than 5% of the length of the first Brillouin zone edge. The chamber pressure was typically 4×10^{-11} torr, and the samples were cleaved and measured at 50K.

RESULTS
X-ray Absorption
Figure 1 Shows the O 1s XAS pre-edge region of two 3-dimensional manganese perovskites - the $d^{3.1}$ sample $La_{.1}Sr_{.9}MnO_3$ and the $d^{3.6}$ sample $La_{.6}Sr_{.4}MnO_3$. We focus on the first peak, which corresponds to the addition of an extra electron in to an e_g up-spin state. We see that there is a clear splitting of this peak for the x=.4 sample, but not for the x=.1 sample. We ascribe this splitting to the Jahn-Teller effect, i.e. the breaking of the degeneracy of the e_g levels, which gives an electronic energy gain to the e_g electron. The sample near d^3 doping does not initailly have an e_g electron and so does not have a driving force for the distortion and is found to be cubic. The intermediately doped samples such as the x=.4 one shown here show no long-range order to the distortion, and so the distortion can not be observed by a diffraction experiment. Local probes such as neutron PDF [4] and EXAFS [5] do show some distortions - here we also show the distortion and also determine the energy scale. The splitting of the levels is expected to be near $4E_{J-T}$, where E_{J-T} is the energy gain for the distortion [6]. Since the splitting is near 1 eV, we find that E_{J-T} is approximately .25 eV.

Figure 2 shows the O 1s XAS pre-edge region forthe d^3 end member of the single layer (n=1) compound Sr_2MnO_4. In this layered sample we can distinguish between the $d_{3z^2-r^2}$ (hereafter called d_{z^2}) out-of-plane and $d_{x^2-y^2}$ in-plane-states by the polarization effect. We highlight the polarization effects by performing the subtraction of the two spectra, as shown in the lowest curve (in-plane minus out-of-plane). This is also very helpful for determining the positions of the various states, although the uncertainty will grow as their widths increase. We also note that the t_{2g} symmetry states are expected to have a much weaker polarization effect and so their contribution to the subtracted curves should be minimal (the t_{2g} down-spin states will likely be centered near 529.5 eV).

From the subtracted curve in figure 2, we see that the d_{z^2} up state (~ 528 eV) has a lower energy than the $d_{x^2-y^2}$ up state (~ 529 eV). Furthermore, we observe another set of out-of-plane states (~ 530.7 eV) and in-plane states (~ 531.7 eV). This is expected since the e_g down-spin states should exist above the spin up states due to the additional energy cost to add a down spin electron compared to an up spin electron, which will be our definition for the exchange energy **J**.

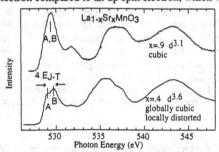

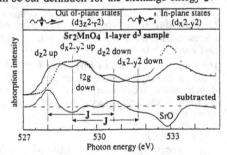

Figure 1: Oxygen 1s XAS showing the unoccupied densities of states from two 3-dimensional manganites. The lowest energy peak corresponds to the eg up spin states. They are doubly-degenerate for the $d^{3.1}$ sample, and due to the Jahn-Teller distortion, are split for the doped sample. The splitting $4 E_{J-T}$ is about 1 eV.

Figure 2: Oxygen 1s XAS data from Sr_2MnO_4 - a single layer d^3 sample. The solid curves were the measurements at grazing incidence (E field out-of-plane) and the dotted curves were measurements at normal incidence (E field in-plane). The Hund's rule energy J, the cost to add a down spin vs. up-spin electron, is found to be 2.7 eV.

From either the d_{z^2} or $d_{x^2-y^2}$ states we can determine that **J** is ~2.7 eV, which is in very good agreement with the theoretical values for the exchange integral (about 0.9 eV) [2,7] since there are 3 t_{2g} electrons to couple to. The similar exchange splitting (2.7 eV) observed for both symmetry states confirms the consistency of the data and the subtraction procedure.

Figure 3 shows polarization-dependent XAS data for x=.4 ($d^{3.6}$) samples as a function of layer number (the globally cubic sample (n=infinity) shows no polarization effects). The large polarization effect of the peak shown near 533 eV (also shown in figure 1) is due to states in the (La, Sr)-O plane because it varies with layer number and does not exist in the 3-dim case. The general trend of the data and the polarization effects in the near-edge region are similar to that for the d^3 sample - the lowest energy portion (528.5eV) has principally out-of-plane character (d_{z^2}) while the higher energy portion (529.5eV) has principally in-plane character ($d_{x^2-y^2}$). An important difference between the data from the doped samples and the data from the end members is that the peaks from the doped sample are in general broader and more washed out. We attribute this to an increase in the electron itinerancy energy for the doped samples.

From the similarity of the pre-edge structures of the layered and cubic samples in figure 3, it is natural to assume that the two bumps in the pre-edge of all samples are of the same origin, i.e. they are from e_g-symmetry states which have had their degeneracy broken by a distortion of the MnO_6 octahedra. For the layered samples this is a static elongation along the z-axis, while for the cubic sample it should be due to a Jahn-Teller distortion. This is the same interpretation we arrived at from the doping dependence of the compounds (figure 1). We note that different assignments for the XAS peaks in the cubic samples have been proposed by Park et al. and Pelligren et al.[8]. The extra information obtained from the comparison to the layered samples gives us confidence that our interpretation is the most reasonable.

Angle-resolved photoemission

Fig. 4 shows the valence bands (panel a) and near-E_F region (panel b) of an x=.4 bilayer manganite sample ($La_{1.2}Sr_{1.8}Mn_2O_7$) along the $(0,0)\rightarrow(\pi,0)$ high symmetry direction in the Brillouin zone. The valence band is composed of oxygen 2p states, t_{2g} symmetry up-spin states, and near E_F, e_g up-spin states which are found to disperse as a function of k and which are most repsonsible for the physical properties of the system. More details about the dispersion,

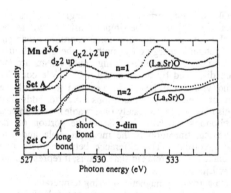

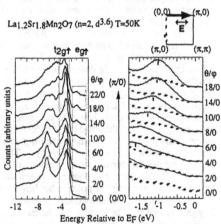

Figure 3. Oxygen 1s XAS data from a variety of samples with x=.4 ($d^{3.6}$), including 3-dimensional samples (set C) and layered samples with one (set A) and two (set B) MnO_2 layers per unit cell .

Figure 4. ARPES data from $La_{1.2}Sr_{1.8}Mn_2O_7$. Solid lines correspond to electron emission in the same direction as the photon E field, while the dotted curves correspond to electron emission 90° from the E field. The intensity modulation with polarization implies that the dispersive features have predominantly $d_{x^2-y^2}$ symmetry.

temperature dependence, and lineshape of these low-energy states can be found in the paper by Saitoh et al., in this same issue [9]. Here we concentrate on the polarization dependence of the data, shown in panel b. Two sets of data are shown - one taken in the direction of the exciting photons Electric field vector (solid lines) and one taken perpendicular to that direction (dotted lines), although both sets cover the same k-space locations. The large intensity modulation with the polarization can be analyzed in a standard way [10], with the result that the dispersive features have predominantly d_{x2-y2} symmetry. Finding d_{x2-y2} symmetry is not surprising, as the d_{x2-y2} states will arise from the in-plane sigma anti-bonding states.

Fig. 5 shows a combination of the XAS and ARPES data from the bi-layer .4 sample, taken from figures 3 and 4, respectively. The occupied part of the d_{x2-y2} band starts 1.5eV below the Fermi level and the unoccupied part of the d_{x2-y2} band can be seen at least 1.5eV above the Fermi level . The total bandwidth of the d_{x2-y2} state is therefore at least 3 eV. This is similar to the predictions of the LSDA band structure calculations [9], but is very different from a simple ionic picture in which the Fermi level should be on the d_{z2} state and no d_{x2-y2} states should be occupied. Because the bandwidth W is slightly larger than J (3 eV vs. 2.7 eV), we expect that the up and down spin d_{x2-y2} bands will not be completely separated but will overlap slightly. The three-dimensional compounds will be expected to have a similar J but greater bandwidth (up to 4.5 eV for a non-distorted compound) and so may have even greater overlap between the up and down spin bands. This overlap will lead to a slight decrease of the spin polarization through hybridization of the up and down spin bands and may affect many properties of the manganites, including the potential for high performance magnetic tunnel junctions (MTJ's). For these types of devices, it might be preferable to use the layered compounds or three dimensional materials with highly distorted Mn-O-Mn bond angles.

In the usual interpretation, we would argue that the large one-electron bandwidth (~ 3 eV or more) determined above implies that there is no mass enhancement to the carriers ($m*/M_{LDA} \sim 1$), i.e. there should be minimal polaronic effects. However, as is detailed more thoroughly in Saitoh's paper in this issue [9], the ARPES spectra of the manganites are very unusual, displaying a pseudogap (depression of spectral weight) centered at E_F and very broad ARPES peaks. These features imply that even the low temperature phase of the manganites should be considered a very strange metal, and requires a reinterpretation of the usual way to understand the ARPES spectra. This reinterpretation will tell us that polaronic effects are very important both in the low and high temperature regions of the manganites, and will also tell us the energy scale and hence coupling parameters for the polaronic distortions.

In this reinterpretation, the ARPES peaks do not correspond to a single quasiparticle excitation but instead to an envelope of many excitations. This can be simply illustrated within the Franck-Condon picture, as shown in figure 6. We consider a configurational coordinate diagram where there is a parabola describing the phonon potential energy curve both for the ground state and the final state of the photoemission aborption process. Due to the strong lattice effects around the ground state electron, the two parabolas are displaced by an amount Q_1-Q_0. We consider transitions between the ground state and any of the (vibrational) levels of the excited state, with the photoemission spectrum being a superposition of the possible transitions (due to additional broadening effects in the solid the individual levels may not be seperately observable). The lowest energy transition occurs without exciting any phonons and will give rise to the portion of the spectrum closest to the Fermi energy. The vertical transition is the most probable and will correspond to the most intense portion of the spectrum. The difference between these two represents the difference in energy between the relaxed and unrelaxed state, that is, it represents the lattice relaxation energy or polaron binding energy E_B.

Figure 7 shows temperature dependent data across the magnetic ordering temperature of $La_{1.2}Sr_{1.8}Mn_2O_7$. The k-value for the measurement was at the point of closest approach of the peak to E_F along the $(\pi,0) \rightarrow (\pi,\pi)$ cut (see reference [9]). From this data, we estimate the lattice relaxation or polaron binding energy to be .65 eV in the ferromagnetic case (50K) and .8 eV for the paramagnetic case (200K). We presume that this is NOT from the Jahn-Teller effect because the energy scale for the Jahn-Teller effect (.25 eV) is significantly smaller, so we argue that other phonons such as due to breathing modes may be relevant, or a completely different type of collective excitation may be considered.

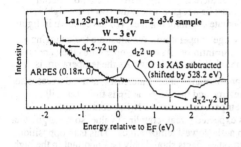

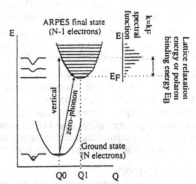

Figure 5. A combination of ARPES and XAS subtracted data from La1.2Sr1.8Mn2O7. The XAS data has been shifted by 528.2 eV so that the dz2 peak would be located just above the Fermi level. The total bandwidth of the dx2-y2 up-spin state is at least 3 eV.

Fig 6. A schematic of photoemission from a strongly coupled electron-phonon system, where Q is a generalized distortion. In the spirit of the Franck-Condon approximation, the vertical transition is the most probable and will correspond to the most intense portion of the spectrum.

This strength is important in comparison to the full bandwidth W, which we found to be 3 eV for the in-plane d_{x2-y2} states for this sample . Using the approximation $\lambda = 8E_B/W$ [11], we obtain a coupling parameter λ of 1.7 for the ferromagnetic case and near 2.5 for the paramagnetic case, indicating that these materials are in or near the strong coupling regime in both the ferromagnetic and paramagnetic states. Figure 8 summarizes how we expect the coupling parameter λ to affect the crucial property the density of states at E_F. As the coupling is increased, the quasiparticle weight is reduced by the factor $1/Z$ while the quasiparticle band should exhibit reduced dispersion, or an effective mass increase by the same factor Z. The net result of this is that in the weak coupling limit $N(E_F)$ is unchanged as a function of λ. In the very strong coupling limit, the system is composed of completely localized small polarons which can transport current

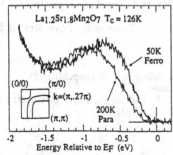

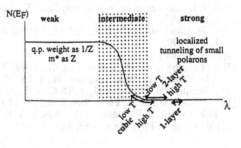

Figure 7 Temperature dependent photoemission data across the magnetic ordering temperature of La1.2Sr1.8Mn2O7. The k-value for the measurement was at the point of closest approach of the peak to E_F along the $(\pi,0) \rightarrow (\pi,\pi)$ cut. The high temperature data was taken immediately after the cleave, followed by the low temperature data.

Figure 8. Schematic diagram showing the spectral weight at E_F as a function of electron-phonon coupling parameter λ. Included are approximate placements of three-dimensional, bi-layer and single-layer manganites as a function of temperature.

only by tunneling from one lattice site to another, i.e. there are no free carriers and no spectral weight at E_F. Connecting these two limits there must be a region where the spectral weight is of an intermediate value, which is termed the intermediate coupling regime. We have shown in figure 7 how the coupling parameter λ changes as we change temperature. We illustrate the changing λ with temperature on this plot for the n=1, n=2, and n=infinity. At high temperatures they are all in the strong coupling regime where $N(E_F)$ is zero and they are insulating. As the temperature is lowered double-exchange takes effect, the electron itinerancy energy increases, and λ decreases so that the system enters the intermediate coupling regime and the weight at E_F is finite but still drastically reduced. Continued lowering of the temperature increases $N(E_F)$ continuously. However, $N(E_F)$ is always well reduced from its expected (weak-coupling) value, so we argue that the coupling to the lattice is very significant even at the lowest temperatures. This is in opposition with the usual arguments which state that the polaronic effects should only be important in the high temperature state, with the low temperature state consisting of essentially free electrons.

CONCLUSIONS
        Lattice polarons are deemed important especially in the paramagnetic phase but also to a degree in the ferromagnetic phase (see figure 8). Due to the energy scale mismatch, the Jahn-Teller effect is unlikely to be the cause for these lattice polarons, at least for the layered samples.

ACKNOWLEDGEMENTS
        We acknowledge helpful discussions with Noriaki Hamada, Andy Millis, and Atsushi Fujimori, as well as the use of Jo Stohr and Mahesh Samant's XAS chamber. The Colorado work was funded by an Office of Naval Research Young Investigator grant. The Stanford Synchrotron Radiation Laboratory is supported by the DOE office of Basic Energy Sciences, Division of Chemical Sciences. The Stanford work was supported by the Office's Division of Material Sciences. The work at JRCAT was supported by the New Energy and Industrial Technology Development Organization (NEDO).

* Currrent address: Center for Integrated Research in Science and Engineering (CIRSE), Nagoya University, Nagoya 464-01, Japan

REFERENCES
[1] Y. Moritomo, A. Asamitsu, H. Kuwahara, and Y. Tokura, Nature 380, 141 (1996)
[2] T. Saitoh et al., Phys. Rev. B 51, 13942 (1995)
[3] R. A. Mohan Ram et al., J. Solid St. Chem. 70, 82 (1987)
[4] S. J. L. Billinge et al., Phys. Rev. Lett. 77, 715 (1996); D. Louca et al., Phys. Rev B. **56**, R8475 (1997)
[5] C. H. Booth et al., Phys. Rev. B 54, R15606 (1996); T. A. Tyson et al., Phys. Rev. B 53, 13985 (1995)
[6] Explain $4E_{J-T}$
[7] S. Satpathy et al., Phys. Rev. Lett. 76, 910 (1996)
[8] J.-H. Park et al., Phys. Rev. Lett. 76, 4215 (1996), E. Pellegrin et al. (unpublished).
[9] T.Saitoh, D.S. Dessau et al., MRS Conf. Proc. (this issue)
[10] G.W. Gobeli et al., Phys. Rev. Lett., **12**, 94 (1964), E. Dietz et al., Phys. Rev. Lett., 36, 1397 (1976), E. R. Ratner et al., Phys. Rev. B 48, 10482 (1993)
[11] A. J. Millis, Phys. Rev. B 53, 8434 (1996)

SPIN TUNNELING IN CONDUCTING OXIDES

Alexander BRATKOVSKY
Hewlett-Packard Laboratories, 3500 Deer Creek Road, Palo Alto, CA 94304-1392,
alexb@hpl.hp.com

ABSTRACT

Different tunneling mechanisms in conventional and half-metallic ferromagnetic tunnel junctions are analyzed within the same general method. Direct tunneling is compared with impurity-assisted, surface state assisted, and inelastic contributions to a tunneling magnetoresistance (TMR). Theoretically calculated direct tunneling in iron group systems leads to about a 30% change in resistance, which is close to experimentally observed values. It is shown that the larger observed values of the TMR might be a result of tunneling involving surface polarized states. We find that tunneling via resonant defect states in the barrier radically decreases the TMR (down to 4% with Fe-based electrodes), and a resonant tunnel diode structure would give a TMR of about 8%. With regards to inelastic tunneling, magnons and phonons exhibit opposite effects: one-magnon emission generally results in spin mixing and, consequently, reduces the TMR, whereas phonons are shown to enhance the TMR. The inclusion of both magnons and phonons reasonably explains an unusual bias dependence of the TMR.

The model presented here is applied qualitatively to half-metallics with 100% spin polarization, where one-magnon processes are suppressed and the change in resistance in the absence of spin-mixing on impurities may be arbitrarily large. Even in the case of imperfect magnetic configurations, the resistance change can be a few 1000 percent. Examples of half-metallic systems are CrO_2/TiO_2 and CrO_2/RuO_2, and an account of their peculiar band structures is presented. The implications and relation of these systems to CMR materials, which are nearly half-metallic, are discussed.

INTRODUCTION

Tunnel magnetoresistance (TMR) in ferromagnetic junctions, first observed more than a decade ago,[1,2] is of fundamental interest and potentially applicable to magnetic sensors and memory devices.[3] This became particularly relevant after it was found that the TMR for $3d$ magnetic electrodes reached large values at room temperature[4,5], and junctions demonstrated a non-volatile memory effect. These observations have ignited a world-wide effort towards using this effect in various applications, with memories and sensors being the most natural choices.

A simple model for spin tunneling has been formulated by Julliere[1] and further developed in Refs. [6,7]. This model is expected to work rather well for iron, cobalt, and nickel based metals, according to theoretical analysis[6] and experiments.[4] However, it disregards important points such as impurity-assisted and inelastic scattering, tunneling into surface states, and the reduced effective mass of carriers inside the barrier. These effects are important for proper understanding of the behavior of actual devices, like peculiarities in their $I - V$ curves, as considered in Ref. [8] and the present paper. I shall also discuss a couple of *half-metallic* systems which should in principle achieve the ultimate magnetoresistance at room temperatures and low fields.

ELASTIC AND INELASTIC TUNNELING, MODEL

The model that we will consider below includes a Hamiltonian for non-interacting conducting spin-split electrons $\mathcal{H}_0$, electron-phonon interaction $\mathcal{H}_{ep}$, and exchange interaction with localized d_l electrons $\mathcal{H}_x$, the later giving rise to the electron-magnon interaction. Impurities will be described by a short-range confining potential V_i,

$$
\begin{aligned}
\mathcal{H} &= \mathcal{H}_0 + \mathcal{H}_{ep} + \mathcal{H}_x + \mathcal{H}_i, \\
\mathcal{H}_i &= \sum_{n_l} V_i(\mathbf{r} - \mathbf{n}_i)
\end{aligned}
\tag{1}
$$

where $\mathbf{r}$ stands for the coordinate of the electron and $\mathbf{n}_l$ denotes the impurity sites.

The non-interacting part of the Hamiltonian $\mathcal{H}$ describes electrons in the ferromagnetic electrodes and insulating barrier according to the Schrödinger equation[7]

$$
(\mathcal{H}_{00} - \mathbf{h} \cdot \hat{\sigma})\psi = E\psi,
\tag{2}
$$

where $\mathcal{H}_{00} = -(\hbar^2/2m_\alpha)\nabla^2 + U_\alpha$ is the single-particle Hamiltonian with $U(\mathbf{r})$ the potential energy, $\mathbf{h}(\mathbf{r})$ the exchange energy ($= 0$ inside the barrier), σ stands for the Pauli matrices; indices $\alpha=1$, 2, and 3 mark the quantities for left terminal, barrier, and right terminal, respectively ($\mathcal{H}_0$ is the expression in brackets). We shall also use the following notations to clearly distinguish between left and right terminal: $\mathbf{p} = \mathbf{k}_1$ and $\mathbf{k} = \mathbf{k}_3$. Solution to this problem in the limit of a thick barrier provides us with the basis functions for electrons in the terminals and barrier to be used in Bardeen's tunneling Hamiltonian approach.[9,10] We assume that all many-body interactions in the electrodes are included in the effective parameters of (2). To fully characterize tunneling we add to Bardeen's direct tunneling term $\mathcal{H}_T^0$ the contributions from $\mathcal{H}_x$ and $\mathcal{H}_{ep}$:

$$
\begin{aligned}
\mathcal{H}_T &= \mathcal{H}_T^0 + \mathcal{H}_T^x + \mathcal{H}_T^{ep}, \\
\mathcal{H}_T^0 &= \sum_{\mathbf{p},ka} T_{\mathbf{p}a,ka}^0 r_{ka}^\dagger l_{\mathbf{p}a} + h.c.,
\end{aligned}
\tag{3}
$$

$$
T_{\mathbf{p}a,ka}^0 = -\hbar^2/(2m_2) \int_\Sigma dA \left(\bar{\psi}_{ka} \nabla \psi_{\mathbf{p}a} - \nabla \bar{\psi}_{ka} \psi_{\mathbf{p}a} \right);
\tag{4}
$$

$$
\mathcal{H}_T^x = -\sum_{an,k,\mathbf{p}} T_{k,\mathbf{p}}^{J,\alpha}(\mathbf{n}) \left[(S_n^3 - \langle S_n^3 \rangle)(r_{k\uparrow}^\dagger l_{\mathbf{p}\uparrow} - r_{k\downarrow}^\dagger l_{\mathbf{p}\downarrow}) + S_n^+ r_{k\downarrow}^\dagger l_{\mathbf{p}\uparrow} + S_n^- r_{k\uparrow}^\dagger l_{\mathbf{p}\downarrow} \right] + h.c.,
$$

$$
\mathcal{H}_T^{ep} = \sum_{aan,k,\mathbf{p}} T_{k,\mathbf{p}}^{ep,\alpha}(\mathbf{q}) r_{ka}^\dagger l_{\mathbf{p}a}(b_{q\alpha} - b_{-q\alpha}^\dagger) + h.c.
\tag{5}
$$

Here the surface Σ lies somewhere in the barrier and separates the electrodes, we have subtracted an average spin $S_n^3 - \langle S_n^3 \rangle$ in each of electrodes as part of the exchange potential, the exchange vertex is $T^J \sim J_n \exp(-\kappa w)$, and the phonon vertex is related to the deformation potential D in the usual way $[T^{ep}(q) \sim \imath D q(\hbar/2M\omega_q)^{1/2} \exp(-\kappa w)]$, where M is the atomic mass, $\mathbf{q}$ is the phonon momentum, $\mathbf{n}$ marks the lattice sites, and the vertices contain the square root of the barrier transparency.[10,11] The operators l_a and r_a annihilate electrons with spin a on the left and right electrodes, respectively. Two more things to note: (i) the summations over $\mathbf{p}$ and $\mathbf{k}$ always include densities of initial g_{La} and final g_{Rb} states, that makes both exchange *and* phonon contribution spin-dependent, (ii) when the magnetic moments on the electrodes are at a mutual angle θ, one has to express the operator r w.r.t. the lab system and then use it in $\mathcal{H}_T$ (5).

The tunnel current will be calculated within the linear response formalism as[10]

$$I(V, t) = \frac{\imath e}{\hbar} \int_{-\infty}^{t} dt' \langle [dN_L(t)/dt, \mathcal{H}_T(t')] \rangle_0, \tag{6}$$

where $N_L(t) = \sum_{\mathbf{pa}} l_{\mathbf{pa}}^{\dagger}(t) l_{\mathbf{pa}}(t)$ is the operator of the number of electrons on the left terminal in the interaction representation, $\langle \ \rangle_0$ stands for the average over $\mathcal{H}_0$,

$$\mathcal{H}_T(t) = \exp(-\imath e V t/\hbar) A(t) + h.c., \qquad A(t) = \sum_{\mathbf{pa,kb}} T_{\mathbf{pa,kb}}(t) r_{\mathbf{kb}}^{\dagger}(t) l_{\mathbf{pa}}(t),$$

the tunnel vertex T is derived for each term in (5), and V is the bias. We shall later consider impurity-assisted tunneling within the same general approach.

Elastic tunneling

We are now in position to calculate all contributions to the tunneling current, the simplest being direct elastic tunneling due to $\mathcal{H}_T^0$. It is worth noting that it can also be calculated from the transmission probabilities of electrons with spin a, $T_a = \sum_b T_{ab}$, which have a particularly simple form for a square barrier and *collinear* [parallel (P) or antiparallel (AP)] moments on the electrodes.[8] We obtain the following expression for the direct tunneling conductance, assuming $m_1 = m_3$ (below the effective mass in the barrier will be measured in units of m_1):

$$\frac{G^0}{A} = \frac{1}{A} \left(\frac{I}{V} \right)_{V \to 0} = G_{\text{FBF}}^0 (1 + P_{\text{FB}}^2 \cos(\theta)), \tag{7}$$

$$G_{\text{FBF}}^0 = \frac{e^2}{\pi \hbar} \frac{\kappa_0}{\pi w} \left[\frac{m_2 \kappa_0 (k_{\uparrow} + k_{\downarrow})(\kappa_0^2 + m_2^2 k_{\uparrow} k_{\downarrow})}{(\kappa_0^2 + m_2^2 k_{\uparrow}^2)(\kappa_0^2 + m_2^2 k_{\downarrow}^2)} \right]^2 e^{-2\kappa_0 w}, \quad \text{and} \tag{8}$$

$$P_{\text{FB}} = \frac{k_{\uparrow} - k_{\downarrow}}{k_{\uparrow} + k_{\downarrow}} \frac{\kappa_0^2 - m_2^2 k_{\uparrow} k_{\downarrow}}{\kappa_0^2 + m_2^2 k_{\uparrow} k_{\downarrow}}, \tag{9}$$

where P_{FB} is the effective polarization of the ferromagnetic (F) electrode in the presence of the barrier (B), $\kappa_0 = [2m_2(U_0 - E)/\hbar^2]^{1/2}$, and U_0 is the top of the barrier. Eq. (7) corrects an expression derived earlier[7] for the effective mass of the carriers in the barrier. By taking a typical value of G/A =4-5 $\Omega^{-1} \text{cm}^{-2}$ (Ref. [4]), $k_{\uparrow} = 1.09 \text{Å}^{-1}$, $k_{\downarrow} = 0.42 \text{Å}^{-1}$, $m_1 \approx 1$ (for itinerant d electrons in Fe)[6] and a typical barrier height for Al_2O_3 (measured from the Fermi level μ) $\phi = U_0 - \mu = 3\text{eV}$, and the thickness $w \approx 20$ Å, one arrives at the following estimate for the effective mass in the barrier: $m_2 \approx 0.4$.[13] These values give the renormalized polarization $P_{\text{FeB}} = 0.28$, which is less than the bulk value for iron $P_{\text{Fe}} = 0.4$ (Ref. [3,4]). Note that the neglect[7] of the mass correction makes $P_{\text{FeB}} < 0$, a result which is not corroborated by experimental evidence where the polarization in all systems studied was found positive, $P > 0$. The majority spin electrons in all cases were predominant in the tunnel current (Ref. [3], p.204).

In the standard approximation of a rectangular shape the barrier height is $U_0 = \frac{1}{2}(\phi_L + \phi_R - eV)$ and this leads to a quick rise of the conductance with bias, $G^0(V) = G^0 + const \cdot V^2$ at small V (ϕ_L and ϕ_R are the work functions of the electrodes). In practice, the barrier parameters should be extracted from independent experiments, such as internal photoemission, etc., but here we are concerned with the generic behavior, where the present formalism is sufficient for qualitative and even semi-quantitative analysis. Since the barrier shape depends in a non-trivial manner on image forces, the calculations have been performed numerically with the actual barrier shape at finite temperatures (Fig. 1).

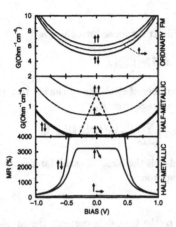

Figure 1: Conductance and magnetoresistance of tunnel junctions versus bias. Top panel: conventional (Fe-based) tunnel junction (for parameters see text). Middle panel: half-metallic electrodes. Bottom panel: magnetoresistance for the half-metallic electrodes. Dashed lines show schematically a region where a half-metallic gap in the minority spin states is controlling the transport. Even for imperfect antiparallel alignment ($\theta = 160°$, marked ↑↘), the magnetoresistance for half-metallics (bottom panel) exceeds 3000% at biases below the threshold V_c. All calculations have been performed at 300K with the inclusion of multiple image potential and exact transmission coefficients. Parameters are described in the text.

We note that the (undesirable) downward renormalization of the polarization rapidly goes with diminishing effective carrier mass in the barrier. The renormalization is completely absent in half-metallic ferromagnets with $\mathrm{Re}k_\downarrow = 0$, as we shall discuss below.

We define the magnetoresistance as the relative change in contact conductance with respect to the change of mutual orientation of spins from parallel (G^P for $\theta = 0$) to antiparallel (G^AP for $\theta = 180°$) as

$$MR = (G^\mathrm{P} - G^\mathrm{AP})/G^\mathrm{AP} = 2P_\mathrm{FB}P'_\mathrm{FB}/(1 - P_\mathrm{FB}P'_\mathrm{FB}). \qquad (10)$$

The most striking feature of Eqs. (3),(4) is that the MR tends to infinity for vanishing $\mathrm{Re}k_\downarrow$, i.e. when both electrodes are made of a 100% spin-polarized material ($P = P' = 1$), because of a gap in the density of states (DOS) for minority carriers up to their conduction band minimum $E_{CB\downarrow}$. Then G^AP vanishes together with the tunnel probability, since there is a zero DOS at $E = \mu$ for both spin directions.

Such half-metallic behavior is rare, but some materials possess this amazing property, most interestingly the oxides CrO_2 and Fe_3O_4.[14] These oxides have potential for future applications in combination with lattice-matching materials, as we shall illustrate below.

A more accurate analysis of the $I - V$ curve requires a numerical evaluation of the tunnel current for arbitrary biases and image forces, and the results are shown in Fig. 1. The top panel in Fig. 1 shows $I - V$ curves for an iron-based F-B-F junction with the above-mentioned parameters. The value of TMR is about 30% at low biases and steadily decreases with increased bias. In a half-metallic case ($\mathrm{Re}k_\downarrow = 0$, Fig. 1, middle panel, where a threshold $eV_c = E_{CB\downarrow} - \mu = 0.3$ eV has been assumed), we obtain *zero* conductance G^{AP} in the AP configuration at biases lower than V_c. It is easy to see that above this threshold $G^{AP} \propto (V - V_c)^{5/2}$ at temperatures much smaller than eV_c.[8] Thus, for $|V| < V_c$ in the AP geometry one has $MR = \infty$. In practice, there are several effects that reduce this

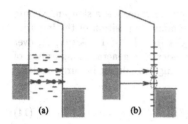

Figure 2: Schematic of tunneling via chains of the localized states in the barrier (a) and into the localized surface states (b).

MR to some finite value, notably an imperfect AP alignment of moments in the electrodes. However, from the middle and the bottom panels in Fig. 1 we see that even at 20° deviation from the AP configuration, the value of MR exceeds 3,000% within the half-metallic gap $|V| < V_c$, and this is indeed a very large value.

Impurity-assisted tunneling

An important aspect of spin-tunneling is the effect of tunneling through the defect states in the (amorphous) oxide barrier (Fig. 2). Since the devices under consideration are very thin, their $I - V$ curves and MR should be very sensitive to defect resonant states in the barrier with energies close to the chemical potential, forming "channels" with the nearly periodic positions of impurities (Fig. 2).[15] Generally, channels with one impurity (most likely to dominate in thin barriers) would result in a monotonous behavior of the $I - V$ curve, whereas channels with *two or more* impurities would produce intervals with negative differential conductance.[15]

Impurity-assisted spin tunneling at zero temperature (at non-zero T one should include an integration with the Fermi functions) has a resonant form[15,8]

$$G_a = \frac{2e^2}{\pi\hbar} \sum_i \frac{\Gamma_{La}\Gamma_{Ra}}{(E_i - \mu)^2 + \Gamma^2},$$ (11)

where $\Gamma = \Gamma_{La} + \Gamma_{Ra}$ is the total width of the resonance given by the sum of the partial widths Γ_L (Γ_R) corresponding to electron tunneling from the impurity state at the energy E_i to the left (right) terminal. For the tunnel width we have

$$\Gamma_{(L,R)a} = 2\pi^2 \kappa_0 (\hbar^2/m_2)^2 \sum_{\mathbf{k}_{(L,R)a}} |\psi_{\mathbf{k}_{(L,R)a}}(\mathbf{n}_i)|^2 \delta(E_\mathbf{k} - E_i),$$ (12)

where $\psi_{\mathbf{k}_{(L,R)a}}(\mathbf{n}_i)$ is the value of the electrode wave function, exponentially decaying into the barrier, at an impurity site $\mathbf{n}_i$. For a rectangular barrier we have[8]

$$\Gamma_{La} = \epsilon_i \frac{2m_2 k_a}{\kappa_0^2 + m_2^2 k_a^2} \frac{e^{-\kappa_0(w+2z_i)}}{\kappa_0(\frac{1}{2}w + z_i)},$$ (13)

where z_i is the coordinate of the impurity with respect to the center of the barrier, $\epsilon_i = \hbar^2\kappa_0^2/(2m_2)$. For e.g. P configuration and electrodes of the same material, the conductance would then be proportional to $\left[(E_i - \mu)^2 + 4\Gamma_{0a}^2 \cosh^2(2\kappa_0 z_i)\right]^{-1}$, where Γ_{0a} equals

191

(13) without the factor $\exp(-2\kappa_0 z_i)$ [c.f. Eq. (15)]. The conductance has a sharp maximum $(= e^2/(2\pi\hbar))$ when $\mu = E_i$ and $\Gamma_L = \Gamma_R$, i.e. for the symmetric position of the impurity in the barrier $|z_i| < 1/\kappa_0$ in a narrow interval of energies $|\mu - E_i| < \Gamma$. Averaging over energies and positions of impurities in Eq. (11), and considering a general configuration of the magnetic moments on the terminals, we get the following formula for impurity-assisted conductance in the leading order in $\exp(-\kappa w)$:

$$\frac{G^1}{A} = G^1_{\text{imp}}(1 + \Pi^2_{\text{FB}}\cos(\theta)), \tag{14}$$

where we have introduced the quantities

$$G^1_{\text{imp}} = \frac{e^2}{\pi\hbar}N_1, \quad N_1 = \pi^2\nu\Gamma_1/\kappa_0,$$

$$\Gamma_1 = \epsilon_i\frac{e^{-\kappa_0 w}}{\kappa_0 w}(r_\uparrow + r_\downarrow)^2, \quad \Pi_{\text{FB}} = (r_\uparrow - r_\downarrow)/(r_\uparrow + r_\downarrow), \quad \text{and}$$

$$r_a = [m_2\kappa_0 k_a/(\kappa_0^2 + m_2^2 k_a^2)]^{1/2}, \tag{15}$$

with N_1 being the effective number of one-impurity channels per unit area, and Π_{FB} is the 'polarization' of the impurity channels. When the total number of one-impurity channels $\mathcal{N}_1 = N_1 A \gg 1$, then the conductance will be a self-averaged quantity, otherwise it will depend on a specific arrangement of impurities (regime of mesoscopic fluctuations).

Comparing the direct and the impurity-assisted contributions to conductance, we see that the latter dominates when the impurity density of states $\nu \geq (\kappa_0/\pi)^3\epsilon_i^{-1}\exp(-\kappa_0 w)$, and in our example a crossover takes place at $\nu \geq 10^{17}\text{cm}^{-3}\text{eV}^{-1}$. When the resonant transmission dominates, the magnetoresistance is given by

$$MR_1 = 2\Pi\,\Pi'/(1 - \Pi\,\Pi'), \tag{16}$$

which is just 4% in the case of Fe. Thus, we have a drastic reduction of the TMR due to non-magnetic impurities in the tunnel barrier, and in the case of magnetic impurities the TMR will be even smaller.

With standard ferromagnetic electrodes, the conductance is exponentially enhanced $[G^1 \propto \exp(-\kappa_0 w)$, whereas $G^0 \propto \exp(-2\kappa_0 w)]$ but the magnetoresistance is reduced in comparison with the 'clean' case of a low concentration of defect levels. These predictions[8] have been confirmed by recent experiments.[12,16]

With further increase of the defect density and/or the barrier width, the channels with two- and more impurities will become more effective than one-impurity channels described above, as has been known for quite a while.[17,15] The contribution of the many-impurity channels, generally, will result in the appearance of irregular intervals with negative differential conductance on the $I - V$ curve.[15] Thus, the two-impurity channels define random fluctuations of current with bias. This is due to the fact that the energy of defect states depends on bias as $\epsilon_i = \epsilon_i^0 + eVz_i/w$. With increasing bias (i) the total number of two-impurity channels increases but (ii) some of these channel go off resonance and reduce their conductance. Accidentally, the number of two-impurity channels going off resonance may become larger than a number of new channels, leading to a suppressed overall conductance. If we denote by Γ_2 the width of the two-impurity channels, then the fluctuations would obviously occur on a scale $\Delta V < \Gamma_2/e$. Then, according to standard arguments, the change in current will be

$$\frac{\Delta I}{I} = \frac{\Delta V}{V} \pm \left(\frac{e\Delta V}{\Gamma_2}\right)\mathcal{N}^{-1/2}, \tag{17}$$

where $\mathcal{N} = eV\mathcal{N}_2/\Gamma_2$ is the number of the two-impurity channels contributing at the bias $V > \Gamma_2/e$, $\mathcal{N}_2$ is the total number of the two-impurity channels, $\mathcal{N}_2 = A\pi^3 w^3 \nu^2 \Gamma_2^2 \kappa_0^{-1}$, and $\Gamma_2 = (4\epsilon_i \Gamma_1/(\kappa_0 w))^{1/2}$.[15] When $eV/\Gamma_2 > \mathcal{N}_2(\kappa_0 w)^2$, then the second (random) term in (17) exceeds the first term, and this leads to random intervals with negative differential conductance. Obviously, with increasing temperature or/and bias in thick enough barriers longer and longer impurity channels will be 'turned on'. A corresponding microscopic model should include impurity states coupled to a phonon bath, and such a model has been solved in Ref. [18]. The authors found an average conductance due to an n-impurity chain in the limit $eV \ll T$, which gives for $n = 2$ $G_2(T) \propto T^{4/3}$. In the opposite limiting case of $eV \gg T$, the result is:[18] $G_2(V) \propto V^{4/3}$, and this crossover behavior is indeed in very good agreement with experiments on a-Si barriers.[19]

One may try to fabricate a resonant tunnel diode (RTD) structure to sharply increase the conductance of a system. We can imagine an RTD structure with an extra thin non-magnetic layer placed between two oxide barrier layers producing a resonant level at some energy E_r. The only difference from the previous discussion is the effectively $1D$ character of the transport in the RTD in comparison with $3D$ impurity-assisted transport. However, the transmittance will have the same resonant form as in (11) and the widths (13). The estimated magnetoresistance in the RTD geometry is, with the use of (11),

$$MR_{RTD} = \left[(r_\uparrow^2 - r_\downarrow^2)/(2r_\uparrow r_\downarrow)\right]^2, \tag{18}$$

which is about 8% for Fe electrodes. We see that the presence of random impurity levels or a single resonant level reduces the value of the magnetoresistance as compared with direct tunneling.

The general reason for the MR being reduced even by non-magnetic impurities is the downward renormalization of spin polarization of tunneling current by *non*-magnetic insulator (proximity effect). Since the exchange potential vanishes in non-magnetic insulator like Al_2O_3, matching of the corresponding wave functions results in a reduced difference in the density of majority and minority states, i.e. reduced polarization. Same, of course, is true of electron states in electrodes overlapping with a non-polarized defect state in the barrier. As we have seen, this proximity effect is enough to reduce MR down to small values, and if we were to include exchange effects (spin mixing) on impurities the MR will be even less. This prediction[8] has indeed been confirmed experimentally.[12,16]

Roughness
As we have seen, the conductance is dominated by the exponentially small barrier transparency, $\propto \exp(-2w(\kappa_0^2 + \mathbf{k}_\parallel^2)^{1/2})$, so that the contribution comes mainly from electrons tunneling perpendicular to the barrier, i.e. with small parallel momenta $|\mathbf{k}_\parallel| < (\kappa_0/w)^{1/2}$. For barriers with a rough interface $w = \overline{w} + h$, where h is the height of asperities and $\overline{w}$ is the average barrier thickness. Each asperity will contribute a factor of $\exp(2\kappa_0 h)$ to the conductance, which we have to average. We assume a normal distribution for roughness, $P(h) = (2\pi h_0^2)^{-1/2} \exp\left(-h^2/(2h_0^2)\right)$. Then, the average conductance $\overline{G}$ becomes

$$G = \overline{G} \int_{-\infty}^{\infty} dh \exp(2\kappa_0 h) P(h) = \overline{G} \exp(2\kappa_0^2 h_0^2) \propto \exp[-2\kappa_0(\overline{w} - \kappa h_0^2)]. \tag{19}$$

This result means that the effective thickness of the barrier is reduced by κh_0^2 in comparison with the observed average thickness $\overline{w}$. The generalization for the case of correlated roughness is straightforward and does not change this result.

Tunneling via Surface States

Direct tunneling, as we have seen, gives TMR of about 30%, whereas in recent experiments the TMR is approaching 40%.[12,23] As we shall see shortly, this difference is unlikely to come from the inelastic processes. Up to now we have disregarded the possibility of localized states at metal-oxide interfaces (Fig. 2). Keeping in mind that the usual barrier AlO_x is amorphous, the density of such states may well exceed that at typical semiconductor-oxide surfaces. If this is true, then we have to take into account tunneling into/from those states. If we assume that electrons at the surface are confined by a short-range potential then we can estimate the tunneling matrix elements as described above. The corresponding tunneling MR is given by[11]

$$
\frac{G_{bs}(\theta)}{A} = \frac{e^2}{\pi\hbar}B\overline{D}_s(1 + P_{FB}P_s\cos(\theta)),
$$

$$
P_s = \frac{D_{s\uparrow} - D_{s\downarrow}}{D_{s\uparrow} + D_{s\downarrow}},
$$

$$
\overline{D}_s = \frac{1}{2}(D_{s\uparrow} + D_{s\downarrow}),
$$

$$
B = 8\pi^2\frac{\epsilon_s}{\kappa_0 w}\frac{m_2\kappa_0(k_\uparrow + k_\downarrow)(\kappa_0^2 + m_2^2 k_\uparrow k_\downarrow)}{(\kappa_0^2 + m_2^2 k_\uparrow^2)(\kappa_0^2 + m_2^2 k_\downarrow^2)}\exp(-2\kappa_0 w), \tag{20}
$$

where P_s is the polarization and $\overline{D}_s$ is the average density of surface states, $\epsilon_s = \hbar^2\kappa_0^2/(2m_2)$. The corresponding magnetoresistance would be $MR_{bs} = 2P_{FB}P_s/(1 - P_{FB}P_s)$.

Comparing (20) with (7), we see that the bulk-to-surface conductance exceeds bulk-to-bulk tunneling at moderate densities of surface states $D_s > D_{sc} \sim 10^{13}\text{cm}^{-2}\text{eV}^{-1}$ per spin, comparable to those found in MOSFET structures.

If on both sides of the barrier the density of surface states is above critical value D_{sc}, the magnetoresistance will be due to surface-to-surface tunneling with a value given by

$$
MR_{ss} = 2P_{s1}P_{s2}/(1 - P_{s1}P_{s2}),
$$

and if the polarization of surface states is larger than in the bulk, as is often the case even for imperfect surfaces,[20] then it would result in enhanced TMR. This mechanism may be even more relevant for Fe/Si and other ferromagnet-semiconductor structures.[21]

INELASTIC TUNNELING, 'ZERO-BIAS' ANOMALY

So far we have disregarded all inelastic processes, such as phonon emission by the tunneling electrons. These processes were long thought to be responsible for a so-called 'zero-bias' anomaly observed in a variety of non-magnetic[22] and magnetic junctions.[12,23] Magnetism in electrodes introduces new peculiarities into the problem, which we will now discuss. The obvious one is related to emission of magnons. At temperatures well below the Curie temperature and not very large biases, one can describe spin excitations by introducing magnons. Then the calculations of exchange- and phonon-assisted currents become very similar. Thus, we obtain from (6) and (5) the following expression for magnon-assisted current in e.g. parallel configuration (corresponding expressions can be easily found for other configurations

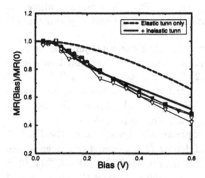

Figure 3: Fit to experimental data for the magnetoresistance of $Co/Al_2O_3/NiFe$ tunnel junctions [12] with inclusion of elastic and inelastic (magnons and phonons) tunneling. The fit gives for magnon DOS $\propto \omega^{0.65}$ which is close to the standard spectrum $\propto \omega^{1/2}$.

as well):

$$I_P^x(V,T) = \frac{2\pi e}{\hbar} \sum_{q\alpha} X^\alpha \left(g_\uparrow^L g_\downarrow^R \left(eV + \omega\right) \left[\frac{N_\omega}{1 - \exp(-\beta(eV + \omega))} + \frac{N_\omega + 1}{1 - \exp(\beta(eV + \omega))} \right] \right.$$
$$\left. + g_\downarrow^L g_\uparrow^R \left(eV - \omega\right) \left[\frac{N_\omega + 1}{1 - \exp(-\beta(eV - \omega))} + \frac{N_\omega}{1 - \exp(\beta(eV - \omega))} \right] \right), \qquad (21)$$

where $N_\omega = [\exp(\beta\omega) - 1]^{-1}$, $\beta = 1/T$ is the inverse temperature, $\omega = \omega_q^\alpha$ and X^α is the magnon incoherent vertex related to the $|T_{p,k}^{x,\alpha}(2S_n/N)^{1/2}|^2$ (5) with all momenta parallel to the barrier integrated out.[11] To get this expression, we have also assumed that the electron densities of states g in (21) vary on a larger scale than the bosonic contributions do, and, therefore, substituted them by representative values at the nominal Fermi levels. If there are some fine features in the electron DOS, then the integral over electron energies should remain, thus necessarily smoothing out any such fine features in the electron DOS.

For the limiting case of $T = 0$, we obtain for inelastic tunneling current:

$$I_P^x = \frac{2\pi e}{\hbar} \sum_\alpha X^\alpha g_\downarrow^L g_\uparrow^R \int d\omega \rho_\alpha^{mag}(\omega)(eV - \omega)\theta(eV - \omega),$$
$$I_{AP}^x = \frac{2\pi e}{\hbar} \left[X^R g_\uparrow^L g_\uparrow^R \int d\omega \rho_R^{mag}(\omega)(eV - \omega)\theta(eV - \omega) \right.$$
$$\left. + X^L g_\downarrow^L g_\downarrow^R \int d\omega \rho_L^{mag}(\omega)(eV - \omega)\theta(eV - \omega) \right]. \qquad (22)$$

where $\theta(x)$ is the step function, $\rho_\alpha^{mag}(\omega)$ is the magnon density of states that has a general form $\rho_\alpha^{mag}(\omega) = (\nu + 1)\omega^\nu/\omega_0^{\nu+1}$, ν can be used as a fitting parameter to define a dispersion of the relevant magnons, and ω_0 is the maximum magnon frequency. For phonon-assisted current at $T = 0$ we have

$$I_P^{ph} = \frac{2\pi e}{\hbar} \sum_{a\alpha} g_a^L g_a^R \int d\omega \rho_\alpha^{ph}(\omega) P^\alpha(\omega)(eV - \omega)\theta(eV - \omega), \qquad (23)$$
$$I_{AP}^{ph} = \frac{2\pi e}{\hbar} \sum_{a\alpha} g_a^L g_{-a}^R \int d\omega \rho_\alpha^{ph}(\omega) P^\alpha(\omega)(eV - \omega)\theta(eV - \omega). \qquad (24)$$

One can show that the ratio of phonon to exchange vertex is $P(\omega)/X = \gamma\omega/\omega_D$, where γ is a constant depending on the ratio between deformation potential and exchange constants,[11] and ω_D is the Debye frequency.

The elastic and inelastic contributions together will define the total junction conductance $G = G(V,T)$ as a function of the bias V and temperature T. We find that the inelastic contributions from magnons and phonons (22)-(24) grow as $G^x(V,0) \propto (|eV|/\omega_0)^{\nu+1}$ and $G^{ph}(V,0) \propto (|eV|/\omega_D)^4$ at low biases. These contributions saturate at higher biases: $G^x(V,0) \propto 1 - \frac{\nu+1}{\nu+2}\frac{\omega_0}{|eV|}$ at $|eV| > \omega_0$; $G^{ph}(V,0) \propto 1 - \frac{4}{5}\frac{\omega_D}{|eV|}$ at $|eV| > \omega_D$. This behavior would lead to sharp features in the $I - V$ curves on a scale of 30-100 mV (Fig. 3).

It is important to highlight the opposite effects of phonons and magnons on the TMR. If we take the case of the same electrode materials and denote $D = g_\uparrow$ and $d = g_\downarrow$ then we see that $G^x_P(V,0) - G^x_{AP}(V,0) \propto -(D-d)^2(|eV|/\omega_0)^{\nu+1} < 0$, whereas $G^{ph}_P(V,0) - G^{ph}_{AP}(V,0) \propto +(D-d)^2(|eV|/\omega_D)^4 > 0$, i.e. spin-mixing due to magnons *kills*, whereas the phonons tend to *enhance* the TMR.[24]

Finite temperature gives contributions of the same respective sign as written above. For magnons: $G^x_P(0,T) - G^x_{AP}(0,T) \propto -(D-d)^2(-TdM/dT) < 0$, where $M = M(T)$ is the magnetic moment of electrode at given temperature T. The phonon contribution is given by a standard Debye integral with the following results: $G^{ph}_P(0,T) - G^{ph}_{AP}(0,T) \propto +(D-d)^2(T/\omega_D)^4 > 0$ at $T \ll \omega_D$, and linear temperature dependence at high temperatures $G^{ph}_P(0,T) - G^{ph}_{AP}(0,T) \propto +(D-d)^2(T/\omega_D)$ at $T \gg \omega_D$.[11] We note again an opposite effect of magnons and phonons on the tunneling magnetoresistance.

It is worth mentioning that the magnon excitations are usually cut off by e.g. the anisotropy energy K_{an} at $\omega_c = 2g\mu_B K_{an}/M_s$, where M_s is the saturation magnetization. Therefore, at low temperatures $T \ll \omega_c$ the magnon finite temperature contribution to conductance will be exponentially small, $\propto \exp(-\omega_c/T)$, and the conductance at zero bias will be almost independent on temperature at $T \ll (\omega_c, \omega_D)$.

We have not included Kondo[25] and other correlation effects that might contribute at very low biases, since they usually do not help to quantitatively fit the data.[19]

The role of phonons is illustrated by my fit to recent experiments carried out at HPL:[12] it appears that only after including phonons is it possible to get a sensible fit to the magnon DOS with $\nu = 0.65$, which is close to the bulk value $\frac{1}{2}$ and $\gamma \approx 0.1$ (Fig. 3).

100% POLARIZATION

It is very important that *in the case of half-metallics* $r_\downarrow = 0$, $\Pi_{FB} = 1$, and even with an imperfect barrier magnetoresistance can, at least in principle, reach any value limited by only spin-flip processes in the barrier/interface and/or misalignment of moments in the half-metallic ferromagnetic electrodes.[8] We should note that the *one-magnon* excitations in half-metallics are suppressed by the half-metallic gap, as immediately follows from our discussion in the previous section. Spin-mixing can only occur on magnetic impurities in the barrier or interface, because the allowed *two*-magnon excitations in the electrodes do not result in spin-mixing.

Therefore, these materials should combine the best of both worlds: very large magnetoresistance with enhanced conductance in tunnel MR junctions. One should be aware, however, that defects in the barrier (like unpaired electrons) will induce the spin flips, so the magnetoresistance could vanish with an increasing concentration of defects. In the case of conventional systems (e.g. NiFe electrodes) we have seen, however, that resonant tunneling

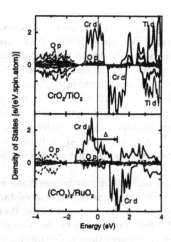

Figure 4: Density of states of CrO_2/TiO_2 (top panel) and $(CrO_2)_2/RuO_2$ (bottom panel) half-metallic layered structures calculated with the use of the LMTO method.

significantly reduces the tunnel MR by itself, so the possibility of improving the conductance and still having a very large magnetoresistance resides primarily with half-metallics. I shall finish with a couple of examples of systems with half-metallic behavior, CrO_2/TiO_2 and CrO_2/RuO_2[8] (Fig. 4). They are based on half-metallic CrO_2, and all materials have the rutile structure with almost perfect lattice matching, which should yield a good interface and should help in keeping the system at the desired stoichiometry. TiO_2 and RuO_2 are used as the barrier/spacer oxides. The half-metallic behavior of the corresponding multilayer systems is demonstrated by the band structures calculated within the linear muffin-tin orbitals method (LMTO) in a supercell geometry with [001] growth direction and periodic boundary conditions. The calculations show that CrO_2/TiO_2 is a perfect half-metallic, whereas $(CrO_2)_2/RuO_2$ is a weak half-metallic, since there is some small DOS around E_F, and an exact gap opens up at about 0.58 eV above the Fermi level (Fig. 4). In comparison, there are only states in the majority spin band at the Fermi level in CrO_2/TiO_2. An immediate consequence of the fact that minority spin bands are fully occupied is an exact *integer* value of the magnetic moment in the unit cell ($=2\mu_B$/Cr in CrO_2/TiO_2), and this property is a simple check for possible *new* half-metallics.

The electronic structure of CrO_2/TiO_2 is very interesting in that it has a half-metallic gap which is 2.6 eV wide and extends on both sides of the Fermi level, where there is a gap either in the minority *or* majority spin band. Thus, an huge magnetoresistance should in principle be seen not only for electrons at the Fermi level biased up to 0.5 eV, but also for *hot* electrons starting at about 0.5 eV above the Fermi level. We note that states at the Fermi level are a mixture of Cr(d) and O($2p$) states, so that $p - d$ interaction within the first coordination shell produces a strong hybridization gap, and the Stoner spin-splitting moves the Fermi level right into the gap for minority carriers (Fig. 4).

An important difference between the two spacer oxides is that TiO_2 is an insulator whereas RuO_2 is a good metallic conductor. Thus, the former system can be used in a tunnel junction, whereas the latter will form a metallic multilayer. In the latter case the physics of conduction is different from tunneling but the effect of vanishing phase volume for transmitted states still works when current is passed through such a system *perpendicular to planes*. For the P orientation of moments on the electrodes, CrO_2/RuO_2 would have a normal metallic conduction, whereas in the AP one we expect it to have a semiconducting type of transport, with a crossover between the two regimes. One interesting possibility is to form three-terminal devices with these systems, like a spin-valve transistor,[26] and check the effect in the hot-electron region. CrO_2/TiO_2 seems to a be a natural candidate to check the present predictions about half-metallic behavior and for a possible large tunnel magnetoresistance. An important advantage of these systems is almost perfect lattice matching at the oxide interfaces. The absence of such a match of the conventional Al_2O_3 barrier with Heusler half-metallics (NiMnSb and PtMnSb) may have been among other reasons for their moderate performance.[27]

By using all-oxide half-metallic systems, as described herein, one may bypass many materials issues. Then, the main concerns for achieving a very large value of magnetoresistance will be spin-flip centers and imperfect alignment of moments. As for conventional tunnel junctions, the present results show that the presence of defect states in the barrier, or a resonant state like in a resonant tunnel diode type of structure, reduces their magnetoresistance by several times but may dramatically increase the current through the structure.

Finally, we can mention the CMR materials. Experiment[28] and LDA calculations[29] indicate that manganites are close to half-metallic behavior as a result of a significant spin-splitting presumably due to strong Hund's rule coupling on Mn. Manganites are strongly correlated materials, likely with electronic phase separation,[30] which makes their study a real challenge. There are a number of studies of systems, where transport is going across grain boundaries or between MnO_2 layers in tailored derivatives of the perovskite phase.[31] A hope is that some of these structures with manganites might operate at low fields and reasonably high temperatures.[32] The low field (below 1000 Oe) TMR in polycrystalline $La_{2/3}Sr_{1/3}MnO_3$ perovskite and $Tl_2Mn_2O_7$ pyrochlore is about 30% and is likely due to intergrain carrier transport. It would be interesting to apply the results of the present work to tunneling phenomena in the CMR-based layered/inhomogeneous structures. For instance, CrO_2 junctions would help to check on the relevance of the half-metallic behavior to conduction in the CMR materials. In particular, it should be signaled by a plateau in the tunneling magnetoresistance as a function of bias within the half-metallic band gap (Fig. 1).

I am grateful to J. Nickel, T. Anthony, J. Brug, and J. Moodera for sharing their data, and to G.A.D. Briggs, N. Moll, and R.S. Williams for useful discussions.

REFERENCES

1. M. Julliere, Phys. Lett. **54A**, 225 (1975).
2. S. Maekawa and U. Gäfvert, IEEE Trans. Magn. **18**, 707 (1982).
3. R. Meservey and P.M. Tedrow, Phys. Reports **238**, 173 (1994).
4. J.S. Moodera *et al.*, Phys. Rev. Lett. **74**, 3273 (1995); J. Appl. Phys. **79**, 4724 (1996).
5. T. Miyazaki and N. Tezuka, J. Magn. Magn. Mater. **139**, L231 (1995).
6. M.B. Stearns, J. Magn. Magn. Mater. **5**, 167 (1977); Phys. Rev. B **8**, 4383 (1973).
7. J.C. Slonczewski, Phys. Rev. B **39**, 6995 (1989).
8. A.M. Bratkovsky, Phys. Rev. B **56**, 2344 (1997); JETP Lett. **65**, 452 (1997).
9. J. Bardeen, Phys. Rev. Lett., **6**, 57 (1961).
10. G.D. Mahan, Many-Particle Physics, 2nd ed., Plenum, New York, 1990, Ch. 9; C.B. Duke, Tunneling in Solids, Academic Press, New York, 1969, Ch. 7.
11. A.M. Bratkovsky (to be published).

12. J.Nickel, T.Anthony, and J. Brug, private communication.
13. Q.Q. Shu and W.G. Ma, Appl. Phys. Lett. **61**, 2542 (1992) give even smaller $m_2 = 0.2$ for Al-Al$_2$O$_3$-metal junctions.
14. V.Y. Irkhin and M.I. Katsnelson, Sov. Phys. - Uspekhi **164**, 705 (1994).
15. A.I. Larkin and K.A. Matveev, Zh. Eksp. Teor. Fiz. **93**, 1030 (1987); I.M.Lifschitz and V.Ya. Kirpichenkov, Zh. Eksp. Teor. Fiz. **77**, 989 (1979).
16. R. Jansen and J.S. Moodera (1997), to be published.
17. M. Pollak and J.J. Hauser, Phys. Rev. Lett. **31**, 1304 (1973); A.V. Tartakovskii *et al.*, Sov. Phys. Semicond. **21**, 370 (1987); E.I. Levin *et al.*, Sov. Phys. Semicond. **22**, 401 (1988); J.B. Pendry, J. Phys. C **20**, 733 (1987).
18. L.I. Glazman and K.A. Matveev, Sov. Phys. JETP **67**, 1276 (1988).
19. Y. Xu, D. Ephron, and M. Beasley, Phys. Rev. B **52**, 2843 (1995).
20. A.V. Smirnov and A.M. Bratkovsky, Phys. Rev. B **54**, R17371 (1996); *ibid.*, **55**, 14434 (1997).
21. A. Chaiken, R.P. Michel, and M.A. Wall, Phys. Rev. B **53**, 5518 (1996).
22. C.B. Duke, S.D. Silverstein, and A.J.Bennett, Phys. Rev. Lett. **19**, 315 (1967).
23. J. Moodera, private communication.
24. In a recent attempt to explain the $I-V$ curves of ferromagnetic junctions S. Zhang, P.M. Levy, A.C. Marley, and S.S.P. Parkin [Phys. Rev. Lett. **79**, 3744 (1997)] have apparently neglected a strong bias dependence of the direct tunneling and did not consider the effect of phonons.
25. J. Appelbaum, Phys. Rev. Lett. **17**, 91 (1966).
26. D.J. Monsma *et al.*, Phys. Rev. Lett. **74**, 5260 (1995).
27. C.T. Tanaka and J.S. Moodera, J. Appl. Phys. **79**, 6265 (1996).
28. Y. Okimoto *et al.*, Phys. Rev. B **55**, 4206 (1997); Phys. Rev. Lett. **75**, 109 (1995).
29. W.E. Pickett and D.J. Singh, Phys. Rev. B **53**, 1146 (1996); D.J. Singh, Phys. Rev. B **5**, 313 (1997).
30. G. Allodi *et al.* Phys. Rev. B **56**, 6036 (1997); E.L. Nagaev, JETP Lett. **6**, 484 (1967); Phys. Rev. B **54**, 16608 (1996); *ibid.* **56**, 14583 (1997).
31. M.K. Gubkin *et al.* Phys. Sol. State **35**, 728 (1993); H.Y. Hwang *et al.*, Phys. Rev. Lett. **77**, 2041 (1996); J.Z. Sun *et al.*, Appl. Phys. Lett. **69**, 3266 (1996); T. Kimura *et al.*, Science **274**, 1698 (1996).
32. H.Y. Hwang and S.W. Cheong, Nature **389**, 942 (1997).

FORMATION OF FERROMAGNETIC/FERROELECTRIC SUPERLATTICES BY A LASER MBE AND THEIR ELECTRIC & MAGNETIC PROPERTIES

Hitoshi TABATA, Kenji UEDA and Tomoji KAWAI
ISIR-Sanken, Osaka University, 8-1 Mihogaoka, Ibaraki, Osaka 567 Japan
tabata@sanken.osaka-u.ac.jp

ABSTRACT

We have constructed artificial superlattices with a combination of magnetic/magnetic and ferroelectric/ferromagnetic materials using a laser ablation technique.   An ideal hetero-epitaxy can be obtained due to the similar crystal structure   of the perovskite type di/ferroelectric $BaTiO_3$, $Pb(Zr,Ti)O_3$ (so-called PZT), $SrTiO_3$ and ferro/antiferromagnetic $LaFeO_3$, $LaCrO_3$, $(La,Sr)MnO_3$.   First of all, we have controlled ferromagnetic order on $LaFeO_3/LaCrO_3$ superlattices formed on $SrTiO_3(111)$ substrate.   Such a spin structure(ferromagnetic order) can't be got in bulk condition.   In   the   heterostructured   ferromagnetic   /ferroelectric   devices, $(La,Sr)MnO_3/PZT$ there are remarkable and interesting phenomena.   The electric properties of the ferromagnetic material can be controlled by the piezoelectric effect via distortion of the crystal structure.

INTRODUCTION

Transition metal oxides have an interesting potential for use in functional electric devices.   Early transition metal oxides (titanium oxide, niobium oxide; for example)   show ferroelectric and dielectric properties corresponding to their band insulative characteristics.   The $PbTiO_3$ and $BaTiO_3$ are famous ferroelectric materials which are expected to be used in DRAM and FRAM devices.[1,2]   Much research is being   done on ferroelectric devices.   Superconducting properties can be observed on copper oxides (late transition metal oxide).[3]   The magnetic characteristics are interesting for the middle transition metal oxides groups (Mn, Fe, Co etc.) with increasing   number of spins per unit.[4,5]   Until now, such material research has been independently performed.

From a crystal chemistry point of view, these transition metal oxides have the same crystal structure so called "perovskite; ABO_3".   They almost have a similar lattice constant, thermal expand coefficiency,   and they have common alkaline earth group ions (A-site ions). Therefore, ideal hetero-epitaxy can be expected with these transition metal oxides.[6,7] Furthermore, the electric and magnetic characteristics can be dramatically changed by controlling only the B-site ions.

In this paper, we report the fabrication of a new type spin ordered magnetic superlattices and a new concept field effect type device consisting of a ferromagnetic /ferroelectric heterostructure formed by a laser MBE (Fig.1). And we describe the interesting electro-magnetic properties of the ferromagnetic layer by changing the lattice distortion via changing the applied voltage of the ferroelectric layer.

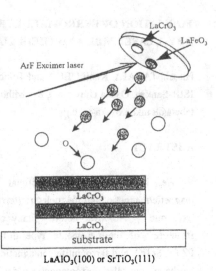

Fig. 1   Schematic diagram of constructing a superlattice by a Laser MBE

Creation of a new type spin order on LaCrO₃/LaFeO₃ magnetic superlattices

According to the Goodenogh-Kanamori-Anderson (GKA) theory [8,9], ferromagnetic interaction can be expected in the d3-O-d5 order (for example, such as Fe(3+)-O-Cr(3+) order). This is a so-called 180° superexchange interaction. Not only GKA theory but also quantum calculation support the ferromagnetic 180° superexchange interaction in the metal dimer which have d3-d5 electron state. Figure 2 shows a schematic model used in our theoretical calculations using Gaussian 94 (UHF, B3LYP). A spin-spin interaction between the two metals can be calculated with changing the bond angle of $\angle$M-O-M (M=Fe^{3+}, Cr^{3+}). In our calculation, the distance between Metal and Oxygen is fixed as 2.0Å. The calculated effective exchange integrals (J_{ij}) of Fe-O-Cr and Fe-O-Fe cluster models are shown in Fig.3 (a) and (b), respectively. The value of J_{ij} is positive and negative in the Fe-O-Cr and Fe-O-Fe model, corresponding to ferromagnetic and antiferromagnetic spin order of them.

From these results, it is believed that if Fe^{3+} and Cr^{3+} ions can be ordered alternately at the B site of perovskite-type transition metal oxides(ABO_3), ferromagnetic property can be occurred.

Some attempts to synthesize the material by sintering method were made, but this method separate it into two phase - Fe and Cr phase. So, the magnetic ordered phase can't be obtained and the materials became to be antiferromagnetic [10,11]. It is because, the Fe-O and Cr-O phase separate each other due to their thermodynamic limitation.

By using a superlattice technique, on the other hand, we can control the stacking order of atoms(spins) as we desire. In this study, we succeed in synthesizing new ferromagnetic artificial superlattice for the first time made by layering one unit cell (4 Å) of $LaCrO_3$ and $LaFeO_3$ alternately using laser MBE method. Such materials can not be gotten in bulk condition since they aren't stable in the viewpoint of thermodynamics. But the method for making artificial superlattices make us synthesize these new materials. There is possibility to make new region for research in the field of oxide electronics as well as semiconductor by using the method.

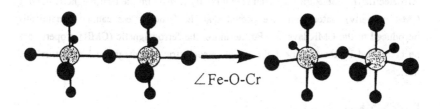

Fig.2   Calculation model of $M_2O_{11}^{16-}$ cluster ($M=Fe^{3+}$, Cr^{3+})

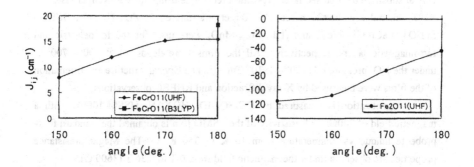

Fig.3 J_{ij} vs bond angle on Fe-O-Cr(left) and Fe-O-Fe(right) clusters.

Characteristics of ferroelectric/ferromagnetic heterostructure

"Colossal Magnetoresistant" (CMR) materials were discovered in manganite-perovskites such as (La,Sr)MnO$_3$.[12-14] The resistance in the CMR materials was reduced by 99.9% (normalized to the low field resistance) with the application of a several tesla (T) field.   In general, CMR materials have been observed in perovskites of the general formula A$_{1-x}$B$_x$MnO$_3$ where A=La, Nd, Pr and B=Ba, Sr, Ca, Pb. At high temperatures, they are paramagnetic insulators, undergoing a magnetic transition to a ferromagnetic metal as the material is cooled through its Curie temperature (Tc).   Also, a large magnetoresistant property occurs around the Tc. The Tc can be controlled by valence control or applying a lattice stress with isostatic pressure.

The ferroelectric PZT, on the other hand, shows a large piezo effect.   A lattice stress can be introduced by applying a voltage to the PZT.   When the ferroelectric (piezoelectric) material is epitaxially formed on the ferromagnetic (having CMR property) material, it is expected that the lattice stress can be dynamically introduced to the CMR layer.   Furthermore, the ferromagnetic (CMR) property can be controlled by the electric field in the ferroelectric/ferromagnetic heterostructured devices.

EXPERIMENT

Ferroelectric and ferromagnetic thin films and ferroelectric/ferromagnetic heterostructures were constructed on a SrTiO$_3$ (100), (111) and LaAlO$_3$ (100) single crystal substrate by a pulsed laser deposition technique using an ArF excimer laser (λ =193 nm, pulse width=20 n sec).   Targets of sintered Pb(Zr$_{0.52}$Ti$_{0.48}$)O$_3$, BaTiO$_3$, SrTiO3, LaCrO$_3$, LaFeO$_3$ and (La$_{0.82}$Sr$_{0.18}$)MnO$_3$ were used for the ferroelectric and ferromagnetic layers, respectively.   All the films were deposited at 500 - 700 ℃ under the NO$_2$ pressure of 1x10^{-6} - 1x10^{-5} Torr.   The Crystal structure and orientation of the films were measured by X-ray diffraction and RHEED observations.

Magnetization is measured by DC-SQUID (HOXAN; HSSM-1000) with a magnetic field of 200G.   Resistivity of the LSMO film is obtained the standard four-probe technique vs. temperature from 10 K to 350 K.   The Magnetoresistance properties are observed under the magnetic field from 0 T to 0.36 T (3600 G).

The electric properties of the PZT films, such as its ferroelectric and dielectric properties, loss tangent and leakage current are measured using impedance analyzer (Hewlett-Packard; 4194A) and an ultrahigh resistivity meter (ADVANTEST; R8340A).   The ferroelectrical D-E hysteresis loops are observed using a Sawyer-Tower circuit at a voltage from 0 to 20V (0 to 800 kV/cm).

Figure 4 shows a cross-sectional view of a ferromagnetic/ferroelectric heterostructured FET device.   A ferroelectric (piezoelectric) PZT layer is formed on the ferromagnetic (La,Sr)MnO$_3$ layer.   The LSMO layer is used as the source and drain electrode.   Thickness of the PZT (piezoelectric) layer and LSMO (ferromagnetic) layer are 4000 and 1000 Å, respectively.   An aluminum metal layer is used as the top electrode with a size of 200 x 200 μ m.   A gate voltage is applied to this electrode.   Resistivity of the LSMO layer is measured by the standard four-probe technique by applying a gate voltage at the Al electrode in a magnetic field of 0 and 0.36 T (3600 G).

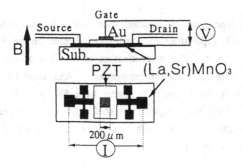

Fig.4   Cross-sectional view of a ferromagnetic/ferroelectric heterostructure FET device.
Ferroelectric (Piezoelectric) PZT layer is sandwiched by ferromagnetic (Ls,Sr)MnO$_3$ layer (bottom
electrode; source and drain electrode) and Au metal (top electrode; gate electrode).

RESULTS AND DISCUSSION

All the ferroelectric and ferromagnetic compound films could be epitaxially formed on SrTiO$_3$ single crystal substrates at a temperature of 700 °C under an NO$_2$ pressure of 1x10^{-5} Torr.   During the superlattice formation, streaked RHEED pattern can be observed (see Fig.5).   Figure 5 (a) and (b) show RHEED oscillation of   1/1 and 3/3 superlatices.   By monitoring the oscillation of the RHEED intensity, we can control the film growth with atomic order accuracy.

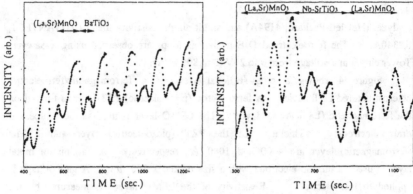

Fig.5 RHEED oscillation pattern of (a)1/1 and (b)3/3 superlattices.

Artificial spin order on LaFeO₃/LaCrO₃ suprlattices

Temperature dependence of magnetization and hysteresis curve (M-H curve) of LaCrO₃/LaFeO₃ superlattice (1/1 sequence) formed on the SrTiO₃(111) substrate is shown in Fig. 6. The magnetization of the film ascends drastically at 375K and raised with lowering the temperature. And clear hysteresis is observed in M-H curves in the temperature region from 6K to 350K. These results are typical feature of the ferromagnetic material. The ferromagnetic spin order is realized in the artificial lattice with a stacking combination of one by one layer for the first time (Fig. 7).

(a)

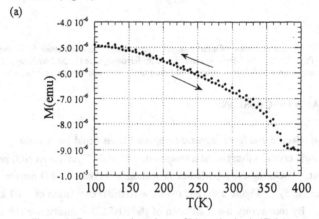

Fig. 6   (a)Temperature dependence of Magnetization in 1T field

(b)

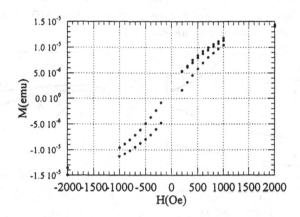

Fig. 6   (b) Hysteresis curve at 6K for LaCrO₃/LaFeO₃/SrTiO₃ (111).

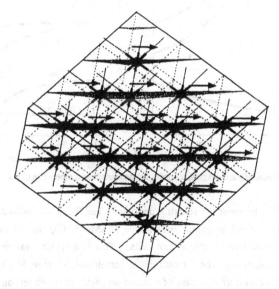

Fig.7   Spin order of   LaCrO₃/LaFeO₃ (1/1) superlattice on SrTiO₃ (111).

Ferroelectric/Ferromagnetic Superlattices

The ferroelectric/ferromagnetic superlattices have been formed with a combination of LSMO and STO or BTO. In these superlattices, the magnetic layer of LSMO is isolated by inserting the non-magnetic layers such as STO and BTO. The resistance and magnetization vs. temperature are shown in Fig. 8 and 9.

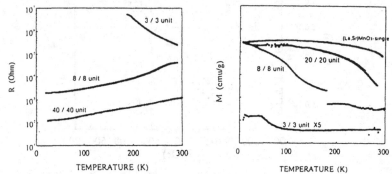

Fig.8 Resistance(left) and magnetization(right) dependent of temperature of $(La,Sr)MnO_3/SrTiO_3$.

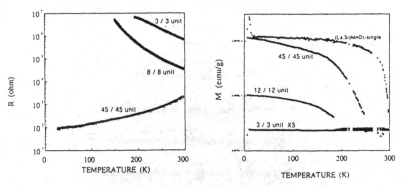

Fig.9 Resistance(left) and magnetization(right) dependent of temperature of $(La,Sr)MnO_3/BaTiO_3$.

The relation between the stacking periodicity and Tc is re-arranged in Fig.10. In case of no-strained magnetic superlattice (LSMO/STO), the Tc rapidly decrease below the periodicity of several unit cells. This is a typical feature of short range magnetic correlation. The ferromagnetic correlation between Mn-Mn disappeared due to the 2D character. In case of strained magnetic superlattices, on the other hand, the Tc decrease more rapidly than that of no-strained.

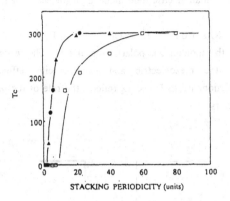

Fig.10 Tc vs. stacking periodicity of LSMO/STO(▲,●) and LSMO/BTO(□).

Dynamic properties of Ferroelectric/Ferromagnetic Heterostructure

For the ferroelectric/ferromagnetic structure, a ferromagnetic-paramagnetic transition temperature, the Curie temperature (Tc), is shifted to a higher value (about 8 K) by applying a gate voltage of 3 V. By doing uniaxial compressive pressure effect experiments [10], the Tc is raised at the rate of

$$d \ln (Tc) /dP = +4 \quad [10^{-2} \text{ GPa}^{-1}] \qquad \cdots \quad (1)$$

To explain an upshift of 8 K, it would require a compressive stress of about 0.6 GPa. Also the displacive value (U) and pressure effect of the PZT are shown as follows,

$$U = n \cdot d_{31} \cdot V \qquad \cdots \quad (2)$$
$$dP/dU = 0.16 \text{ [GPa/ \%]} \qquad \cdots \quad (3)$$

where n, d_{31}, V are the number of unit cells, piezoelectric coefficiency (94 x 10^{-12} [C/N]) and applied voltage, respectively. To introduce the lattice stress of 0.6 GPa, about 30 V is needed for the 2000 Å thick PZT layer. This is too large to be realistic for a piezoelectric effect. The piezoelectric coefficiency used above is an estimate for the bulk PZT sample. It may be different from that of the PZT film. Anyway, the magnetic properties can be controlled by the electric field, which is originally not dependent on the magnetization, via lattice distortion.

Next, the resistance of the LSMO layer is measured by applying a gate bias voltage from -3V to +3V. In Fig. 11(a), it is shown as the dependence of the modulation, -dR/R, on the value and direction of the applied field at a typical temperature of 300 K is shown. The dependence has an asymmetric parabolic nature. This asymmetry can be attributed to the properties of the gate PZT layer.

The strain (X) occurs from an electric field-induced piezoelectric and electrostrictive effect.

$$X = QPs^2 + 2Q\varepsilon\varepsilon_0 Ps \cdot E + Q(\varepsilon\varepsilon_0)^2 \cdot E_2 \qquad \cdots (4)$$

The first term shows the spontaneous polarization effect. The second term and the third term indicate the piezoelectric and electrostricive effect, respectively. Therefore, the proportionality to E and E_2 reflects the role of the piezoelectric and electrostrictive characters.

(a) (b)

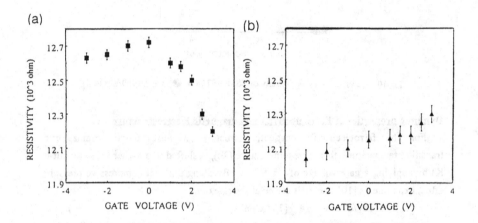

Fig.11 Relation between the resistivity of ferroelectric LSMO layer and the gate voltage applied at a gate electrode (a) without magnetic filed and (b) with magnetic field of 0.3 T (3000 G).

When the magnetic filed is induced for the hetero device, on the other hand, the modulation, dR/R, has an apparently different behavior (shown in Fig. 11 (b)). The resistance linearly increases with increasing gate voltage. The dR/R value is about 1.5 % and -1.2 % at the gate voltage of +3V and -3V, respectively. In this case, the conductivity depends on the carrier density of the conducting electrons, that is, the hole concentration in the LSMO layer. The induced hole concentration (Δn) increased due to the applied charge at the gate electrode that can be estimated as follows, [15,16]

$$\Delta n = \varepsilon\varepsilon rVG / etd \qquad \cdots (5)$$

where ε_0, ε_n, VG, e and td are the dielectric constant of vacuum (8.85×10^{-12} F/m) and PZT (~ 500), gate voltage, elementary electric charge (1.6×10^{-19} C) and thickness of the LSMO layer(2000 Å), respectively.

By using the above equation, the areal density at the 3 V gate voltage can be estimated as follows,

$$\Delta n = 1.8 \times 10^{13} \; [cm^{-2}] \qquad \cdots (6)$$

The mobile hole concentration in the (La,Sr)MnO$_3$ layer is about $\sim 10^{21}$ [cm^{-3}]. Therefore, the areal hole density is about $\sim 10^{15}$ [cm^{-2}] at a film thickness of 2000 Å. The resistant change with a percent order can be explained by changing of the induced electric field.

Two types of conducting systems consisting of a ferroelectric/ferromagnetic heterostructure are expected. Without a magnetic field or above the Tc, the transport is dominated by the semiconducting matrix in which the localized spin fluctuated due to the thermal vibration. Therefore, the conducting spin(electron) can not be easily transport by scattering these localized spins. Therefore, the resistivities are dramatically influenced by the slight lattice distortion caused by the piezoelectric and electroelastic effects. When the magnetic field is applied to the device, on the other hand, the localized spins are ordered in a ferromagnetic direction. Therefore, the conducting spin can be easily transported. In this case, the resistance of the ferromagnetic layer depends on the carrier (hole) concentration.

CONCLUSION

We have constructed a new concept device with a combination of ferroelectric and ferromagnetic materials using a laser ablation technique. The superlattice of LaFeO$_3$/LaCrO$_3$ with a stacking periodicity of 1/1 on SrTiO$_3$(111) shows ferromagnetic character. It was proved that new materials that has structure Fe and Cr ions arranged alternately were formed by constructing of artificial lattices. We can also easily introduce the lattice stress by applying a voltage to the piezoelectric compounds. In the heterostructured ferromagnetic/ferroelectric devices, there are remarkable and interesting phenomena. The electric properties of the ferromagnetic material can be controlled by the piezoelectric effect via distortion of the crystal structure. Furthermore, two types of conducting systems consisting of a ferroelectric/ferromagnetic heterostructure exist. Without a magnetic field, the lattice distortion is important for the resistance of the LS MO layer. Under a magnetic field, on the other hand, hole concentration plays an important role in the conductivity. These results will allow the development of new types of devices in the future.

ACKNOWLEDGEMENT

This research was supported in part by the Grant-in-Aid for Scientific Research on Priority Areas "Physics and Chemistry of Functionally Graded Materials" from the Ministry of Education, Science Sports and Culture and TOKUYAMA Scholarship Foundation. We wish to thank them for their generous financial assistance.

REFERENCES

1. Ahn C.H., Triscone J.M., Archibald N., Decroux M., Hammond R.H., Geballe T.H. Fisher O. and Beaseley M.R., Science 269, 373 (1995).

2. Erbil A., Kim Y. and Gerhardt A., Phys.Rev.Lett. 77, 1628 (1996).

3. RAmesh R., Inam A., Wilkens B., Chan W.K., Sands T., Tarascon J.M., Fork D.K., Geballe T.H. Evans J. and Bullinton J., Appl.Phys.Lett. 59, 1782 (1991).

4. Ogale S.B., Talyansky V., Chen C.H., Ramesh R., Greene R.L. and Venkatesan T., Phys.Rev.Lett. 77, 1159 (1996).

5. Yoshiasa A, Inoue Y, Kanamaru F and Koto K, J.Solid State Chem. 86, 75 (1990)

6. Tabata H., Tanaka H. and Kawai T., Appl.Phys.Lett. 65, 1970 (1994).

7. Tabata H., Tanaka H., Kawai T. and Okuyama M., Jpn.J.Appl.Phys., 34 544 (1995).

8. J. Kanamori : J. Phys. Chem. Solids, 10 (1958) 87

9. J.B.Goodenough : Phys. Rev., 100 (1955) 564

10. A. Wold and W. Croft, : J. Phys. Chem., 63 (1959) 447

11. A. Belayachi, M. Nogues, J. -L. Dormann, and M.Taibi : Eur. J. Solid. State. Inorg. Chem., t.33 (1996) 1039

12. Zhao G., Conder K., Keller H. and Muller K.A., Nature 381, 676 (1996).

13. Helmolt R., Wecker J., Holzapfel B., Schultz L. and Samwer K., Phys.Rev.LEtt. 71, 2331 (1993).

14. Tokura Y., Urushibara A., Moritomo Y., Arima T., Asamitsu A., Kido G. and Furukawa N., J.Phys.Soc.Jpn. 63, p.3931 (1994).

15. J.Manhart J. Schlom D.G., Bednorz J.G. and Muller K.A., Phys.Rev.Lett. 67, 2099 (1991).2099

16. Walkenhorst A., Doughty C., Xi X.X., Li Q., Lobb C.J., Mao S.N. and Venkatesan T., Phys.Rev.Lett. 69, p.2709 (1992).

LOW ENERGY k-DEPENDENT ELECTRONIC STRUCTURE OF THE LAYERED MAGNETORESISTIVE OXIDE La$_{1.2}$Sr$_{1.8}$Mn$_2$O$_7$

T. SAITOH[1], D. S. DESSAU[1], C.-H. PARK[2], Z.-X. SHEN[2], P. VILLELLA[1], N. HAMADA[3]*, Y. MORITOMO[3]**, Y. TOKURA[3,4]

[1]Department of Physics, University of Colorado, Boulder, CO 80303-0390
[2]Department of Applied Physics, Stanford University, Stanford, CA 94305
[3]Joint Research Center for Atom Technology, Tsukuba 305, JAPAN
[4]Department of Applied Physics, University of Tokyo, Bunkyo-ku, Tokyo 113, JAPAN

ABSTRACT

We have studied the k-dependent electronic structure of the layered colossal magnetoresistive manganite La$_{1.2}$Sr$_{1.8}$Mn$_2$O$_7$ using high-resolution angle-resolved photoemission spectroscopy. We found dispersive energy bands as a function of the crystal momentum k near the Fermi level (E_F). We have also performed local spin density approximation (LSDA)+U band-structure calculations on the current system. The overall experimental dispersion relation is basically in agreement with the band-structure calculations yet close to E_F there is a significant deviation from the predicted dispersions. Instead of clear Fermi-surface (FS) crossings, we observe a depression of the features as the FS is approached as if there is a "pseudo" gap in the excitation spectrum. The pseudogap continuously opens with temperature and does not show further significant opening above T_C, corresponding to the metal-insulator transition. Those unusual aspects of the spectra has been discussed from the viewpoint of the strong electron-lattice coupling model.

INTRODUCTION

Colossal magnetoresistance (the CMR effect) has recently been observed [1] in doped manganese oxides (manganites), sparking a great amount of effort aimed at understanding the electronic and magnetic properties of these materials. At low temperatures, properly doped manganites exhibit ferromagnetic metallic or nearly metallic behavior, while at high temperatures they exhibit a paramagnetic insulating behavior. This generic behavior, as well as the MR effect which occurs near the transition, is understood to first order within the framework of double exchange theory, as developed in the 1950's and 60's by Zener, de Gennes, and Anderson and Hasegawa [2]. Recently there has been an increasing realization that although double-exchange is clearly important for understanding the behavior of the manganites, it is not enough. For instance, it has been shown that the insulating behavior above T_C cannot be understood solely with the spin-disorder scattering inherent in the double-exchange model [3]. It appears that an additional effect such as the creation of polarons due to strong electron-lattice coupling [3] is necessary to explain the observed behavior.

In fact, there is a growing body of evidence for lattice polaron effects in the high temperature paramagnetic state of the manganites [4]. In the low temperature ferromagnetic state however, theories have typically assumed that the manganites are standard metals, where polaronic or other effects (such as correlations) are not dominant. To gain insight into this problem, we have performed high energy-resolution angle-resolved photoemission (ARPES) measurements on the critical near-Fermi level (E_F) electronic structure of the manganites by determining the E vs. k relationship and interaction effects in these materials. We have also performed band-structure calculations on the layered magnetoresistive oxide [5], and have compared and contrasted the experimental and theoretical data. Our results indicate that the low temperature ferromagnetic state of the manganite should instead be considered a very strange metal. Our analysis indicates that the very strange behavior is most likely due to significant electron-lattice coupling even in the ferromagnetic state, the possibility of which had been mostly ignored previously.

The structure for these in comparison to the cubic manganites is shown in Ref. 6. One of the key features of these layered samples for our experiments is that they cleave easily between the two ionically-bonded (La,Sr)O planes, yielding a mirror-like surface that should be representative of

213

the bulk properties. The data shown in this paper were obtained from the bilayer material $La_{1.2}Sr_{1.8}Mn_2O_7$. This material has a nominal doping level of 0.4 holes per Mn site ($d^{3.6}$) and more than two orders of magnitude decrease in resistivity at the ferromagnetic T_C of 126K [6].

EXPERIMENT

Single crystals of $La_{1.2}Sr_{1.8}Mn_2O_7$ were grown by the floating-zone method [6]. Good LEED patterns absent of extra superlattice spots are easily obtainable, confirming the high quality of the surfaces. The photoemission experiments were performed at the undulator beamline 5 of the Stanford Synchrotron Radiation Laboratory (SSRL) using a 50 mm hemispherical analyzer mounted on a 2-axis goniometer. Spectra shown in this paper were all taken with a photon energy of 22.4 eV, an energy resolution of 40 meV FWHM, and an angular resolution of ±1 degree (giving a k-resolution better than 5% of the length of the first Brillouin zone edge). The chamber pressure was typically 4×10^{-11} torr, and the samples were cleaved and measured *in-situ* at temperatures ranging from 10 to 200 K. At these temperatures, the cleaved sample surfaces showed minimal changes within the time frame of a few hours to half a day, enabling a few cuts in k-space to be completed on a freshly cleaved surface.

RESULTS

Figure 1 a) shows the full valence band spectra from $La_{1.2}Sr_{1.8}Mn_2O_7$ taken along the (0,0)–(π,0) high symmetry direction. k-space markings refer only to the near-E_F portion of the spectra. We observe a high density-of-states region between 1.5 eV and 7 eV, which is primarily composed of O $2p$ and Mn $3d$ t_{2g} (d_{xy}, d_{yz} and d_{zx}) up spin states. The low density-of-states region near E_F is expected to be composed of the e_g–symmetry ($d_{x^2-y^2}$ and $d_{3z^2-r^2}$) states. Figures 1 b), c), d) and e) show these near-E_F states along various k-space directions, measured at 10K (ferromagnetic phase). Directions of these cuts in the two-dimensional Brillouin zone are indicated in Fig. 1 f). Concentrating first on the spectra along the (0,0) $\rightarrow$ (π,0) line (panel b)), we observe a strong feature first visible at (0.27π,0) which disperses towards E_F as we progress towards (π,0). In addition, there is a weak and broad feature at about -0.6 eV which is strongest near the (0,0) point (at higher photon energies this feature evolves into a clearly resolvable peak, so we are confident that it is real). Panel d) shows a continuation of the dispersion along the (π,0) $\rightarrow$ (π,π) direction. In the first part of this cut the peak continues to disperse towards E_F, but very surprisingly never reaches E_F. Instead, it attains its minimum energy near the angle (π,0.27π), at which point it very rapidly loses intensity as if weight was transferred above E_F. Beyond this point, the spectra in addition exhibit some evidence of bending back away from E_F in a similar way from what would be expected for the opening of an excitation gap centered at E_F. A very similar result is seen for the cut shown in panel c), with the minimum energy at the k-position (0.63π,0.27π).

Figure 1 g) shows a plot the peak positions (indicated by the tick marks in panels b) and d)) vs. crystal momentum along the (0,0) $\rightarrow$ (π,0) $\rightarrow$ (π,π) direction compared to the up-spin dispersion predicted by our local spin density approximation (LSDA+U) band structure calculations [5]. The calculation predicts a set of dispersive bands of e_g symmetry mostly separated from a region of higher binding-energy bands. The band crossing E_F near the (0,0) point is predicted to have principally $d_{3z^2-r^2}$ out-of-plane character, while the two bands crossing between (π,0) and (π,π) are predicted to have primarily $d_{x^2-y^2}$ in-plane character. We find that there is a correspondence between many aspects of the experimental and theoretical data. First, the agreement in both energy position and dispersion rate between the experiment and theory along the (0,0)–(π,0) line is surprisingly good, especially considering that we have not rescaled or shifted the energy scales to account for the often observed renormalization effects. Second, by taking advantage of the polarization of the incident photons we have performed a symmetry analysis on the main dispersive features (the ones predicted to cross E_F along (π,0)–(π,π)) and found them to have primarily $d_{x^2-y^2}$ character [7], in agreement with the band theory prediction. Third, the location of the experimental minimum in binding energy as well as the location where the spectral weight is rapidly being depleted agree well with the predicted Fermi surface (FS) crossing.

Despite these agreements there are clear deviations between the experiment and theory along the $(\pi,0)$–(π,π) line, signaling additional physics not contained in the calculation. These deviations can tell us many of the details of the interactions responsible for the very unusual properties of the manganites. In particular, the experimental peaks never approach closer than approximately 0.6–0.7 eV to E_F, while theoretically they are expected to reach E_F, and there is a vanishingly small spectral weight at E_F, even though the measurements were made in the ferromagnetic "metallic" state of the compound. To make sure that we simply did not miss a FS crossing, we have made measurements along all the high symmetry directions as well as along many off-symmetry points (not shown), have used a variety of photon polarizations and photon energies, and have repeated the measurements on more than 10 cleaves.

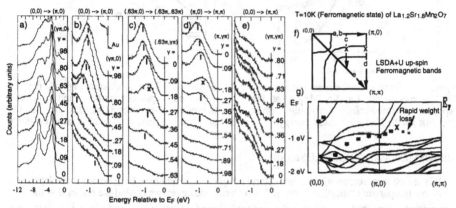

Figure 1. a): Valence band spectra of $La_{1.2}Sr_{1.8}Mn_2O_7$ along the $(0,0) \rightarrow (\pi,0)$ direction. b)–e): Low temperature high resolution ARPES spectra along various high symmetry directions, as indicated at the top of each panel and by the arrows along the two-dimensional Brillouin zone of panel f). The three curved lines in panel f) are the Fermi surfaces for the up-spin bands in LSDA+U band theory calculations. The two x's are the experimental locations of closest approach to E_F. g) : The up-spin bands in an LSDA+U band theory calculation vs. experimentally determined peak centroids from panels b) and d) (tick marks).

We term the above depression of spectral weight a pseudogap. We find that this pseudogap affects *both* the ferromagnetic and paramagnetic states, although it affects the paramagnetic states most severely, i.e. the pseudogap is larger above T_C. The temperature dependence of near-E_F ARPES spectra at the expected FS crossing point $(\pi,0.28\pi)$ is shown indicated in Fig. 2. From 175 K to 125 K, one observes no appreciable change in the lower set of spectra. However, from 125 K to 100 K, there can be seen a jump of the leading edge located around -0.5 eV and the peak of the spectra as well as a considerable change of the lineshape in the upper set. Below 100 K, the edge and the peak keeps approaching E_F upon cooling down to 50 K. Nevertheless, the spectral weight at E_F always remains very small, resulting in the lineshape change or the spectral-weight transfer from the higher binding-energy (> 0.7–0.8 eV below E_F) towards the near-E_F region (from E_F to 0.7–0.8 eV below E_F). The tick marks in the figure show the peak positions at the highest and the lowest temperatures, which clearly represents the large temperature-dependent change. We note that the ferromagnetic T_C is 126 K [6], and hence all the changes that we observed should closely be related to the paramagnetic-to-ferromagnetic and the insulator-to-"metal" transitions. In addition, the fact that the temperature-dependent changes begin at T_C gives much added confidence that the photoemission spectra we present are representative of the bulk properties of the samples, and also indicates that the pseudogap should be responsible for the drastic changes in the conductivity across the magnetic ordering temperature.

215

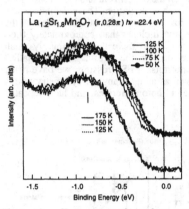

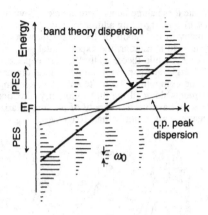

Figure 2. Temperature-dependent near-E_F ARPES spectra of $La_{1.2}Sr_{1.8}Mn_2O_7$ at the expected Fermi-Surface crossing point $(\pi, 0.28\pi)$ with 22.4 eV photon energy.

Figure 3. Schematic drawing of the k-dependent photoemission and inverse photoemission spectra of $La_{1.2}Sr_{1.8}Mn_2O_7$ near an expected Fermi-Surface crossing point.

DISCUSSIONS AND CONCLUSIONS

It is at first very surprising that the spectral weight at E_F could have been so severely depressed in the low temperature "metallic" state of this material. However, we remind the reader that the low temperature resistivity for this material is unusually high ($> 3\times10^{-3}$ Ωcm) and that it is in fact even increasing at the lowest temperatures [6]. In addition, the low frequency optical conductivity of the doped bilayer manganites has recently been measured, and it has been found that the Drude peak is essentially completely absent for these materials, even at the lowest temperatures [8]. We believe that the pseudogap we have observed is responsible for all of this very unusual behavior. On the other hand, the perovskite-type manganites such as $La_{1-x}Sr_xMnO_3$ measured at low temperature do show finite photoemission spectral weight at E_F and do show a clear Drude peak in the optical measurements [9,10]. In both cases however, the E_F weight or the Drude weight are reduced by at least a factor of 10 from expectations [10,11]. We believe that this reduction of spectral weight is also due to the same reason as the pseudogap opening in the present system, but with reduced strength compared to it.

We have found that the pseudogap affects the entire Fermi surface to a similar degree (including the data shown here and unpublished data along other k-space cuts). While we can not yet say that there is no anisotropy to the gap, the general lack of k-space dependence most likely signals that a local effect, such as a local lattice distortion, is responsible for the pseudogap behavior (note that in general there is a large amount of k-space dependence in the data - however, the pseudogap measured along the expected FS does not have k-dependence). This contrasts with the recently observed pseudogap effects in the high-T_C superconductors which have a pronounced k-space dependence [12]. In that case, the maximal effect is near the $(\pi,0)$ point of the Brillouin zone, signaling a possible origin from the antiferromagnetic fluctuations with wave-vector near (π,π).

The ARPES data presented above is very unusual and indicates that the low temperature state of the bi-layer manganites should be considered to be a very strange metal. From this point on our paper will get a little more speculative, as we try to give a first-order explanation for many of the unusual spectral and physical properties. Because of the lack of k-dependence of the pseudogap mentioned above, we will focus on strong (hence essentially local) electron-lattice coupling. Accurate calculations of strong electron-lattice coupling effects in a solid with real k-dependences have not been performed yet. However, we believe that we can extrapolate known results to explain our measured spectra, with important implications both for the physics of manganites and for the more general problem of polarons in solids. An exactly solvable model that we believe contains much of the relevant physics is the classic problem of the coupling of a single electron to a bath of Einstein phonons with frequency ω_0 [13]. The electron spectral function for this problem is an envelope of many individual peaks separated by ω_0, as illustrated in Fig. 3. The multiple peaks indicate that a single electron is not an eigenstate of the system - therefore the removal of an electron from the system occurs with a probability of shaking off a certain number of phonons. The quasiparticle (QP) peak or "coherent" part of the spectrum is the one with zero phonons shaken off and is the peak closest to E_F. In the strong coupling case the Poisson distribution function is broad and the QP peak will have very little spectral weight. An important point about this result is that irrespective of the strength of the coupling, the centroid of the distribution $<\omega>$ is equal to the energy of the electron in the absence of the coupling [13]. A known example of this type of distribution is the measured gas-phase photoemission spectrum of H_2, which shows a clear progression of many peaks, corresponding to the many different vibrational levels [14].

In contrast with the usual interpretation, we argue that the dispersive peaks we have measured should not be considered to be a single QP peak but should be considered to be an envelope of many individual peaks, in the spirit of the strong coupling arguments above. Second, in analogy to the single-electron calculation above, we argue that the centroid of the ARPES spectrum should have an energy equal or similar to the energy in the absence of the coupling, which in this case is the LSDA band energy. This explains the good agreement of the experiment and theory through much of the zone (see Fig. 1 g)). The QP peak corresponds to the portion of the spectrum nearest E_F, and is found to have an almost vanishingly small weight for this material, indicating that the coupling is strong (this is also seen by the large width of the peaks). When the QP peak crosses E_F, the entire photoemission peak must rapidly lose weight because the hole excitation can no longer be created. The centroid of the envelope therefore always stays well below E_F with a minimum binding energy equal to the distance between the centroid of the distribution and the QP peak. Since weight is transferred above E_F, the centroid of the spectrum must include both the occupied and unoccupied states, and so should still agree with the band theory calculation, even though the centroid of the *photoemission* spectrum need not match the band theory result.

The above scenario is in strong contrast with our usual findings and expectations (for instance in the high T_C superconductors [15]) where the centroid of the photoemission peak essentially reaches E_F. The key difference here is that the photoemission peak in the high T_C's sharpens significantly as it reaches E_F, indicating there is a large QP content to it. In the underdoped cuprates the situation may be more similar to the manganites, as portions of the Brillouin zone show pseudogap effects and significantly broadened photoemission peaks.

In conclusion, we have investigated both experimentally and theoretically the electronic structure of $La_{1.2}Sr_{1.8}Mn_2O_7$ using high-resolution ARPES and LSDA+U band-structure calculations. The overall experimental dispersion relation near E_F is essentially in agreement with the band-structure calculations yet closer to E_F there is a significant deviation from the predicted dispersions. Instead of clear FS crossings, we observe a depression of the features as the FS is approached as if there is a pseudogap in the excitation spectrum. The pseudogap continuously opens with temperature and does not show further significant opening above T_C, corresponding to the metal-insulator transition. The many unusual aspects of the spectra has been discussed from the viewpoint of the strong electron-lattice coupling model.

ACKNOWLEDGMENTS

We acknowledge helpful discussions with A. Bishop, S. Doniach, A Fujimori, T. Katsufuji, A. Millis, T. Mizokawa, L. Radzihovsky, and G. Sawatzky. This work was funded by an Office of Naval Research Young Investigator grant through University of Colorado. SSRL is supported by the Department of Energy. The Stanford work was supported by the Office's Division of Materials Science. The work at JRCAT was supported by the New Energy and Industrial Technology Development Organization (NEDO).

REFERENCES

* Current address: Department of Physics, Science University of Tokyo, Noda 278, JAPAN.
**Current address: Center for Integrated Research in Science and Engineering, Nagoya University, Nagoya 464-01, JAPAN.

1. R. M. Kusters, J. Singleton, d. A. Keen, R. McGreevy, and W. Hayes, Physica B **155**, 362 (1989); Y. Tokura, A. Urushibara, Y. Moritomo, T. Arima, A.Asamitsu, G. Kido, and N. Furukawa, J. Phys. Soc. Jpn. **63**, 3931 (1994); S. Jin, T. H. Tiefel, M. McCormack, R. Fastnacht, R. Ramesh, and L. H. Chen, Science **264**, 413 (1994).
2. C. Zener, Phys. Rev. B **82**, 403 (1951); P.-G. de Gennes, Phys. Rev. **118**, 141 (1960); P. W. Anderson and H. Hasegawa, Phys. Rev. **100**, 675 (1955).
3. A. J. Millis, B. I. Shraiman, and R. Mueller, Phys. Rev. Lett. **77**, 175 (1996); A. J. Millis, R. Mueller, and B. I. Shraiman, Phys. Rev. B **54**, 5389 (1996); ibid. **54**, 5405 (1996).
4. S. J. L. Billinge, R. G. DiFrancesco, G. H. Kwei, J. J. Neumeier, and J. D. Thompson, Phys. Rev. Lett. **77**, 715 (1996); C. H. Booth, F. Bridges, G. J. Synder, and T. H. Geballe, Phys. Rev. B **54**, R15606 (1996); G. Zhao, K. Conder, H. Keller, and K. A. Müller, Nature **381**, 676 (1996); D. Louca, T. Egami, E. L. Brosha, H. Röder, and A. R. Bishop, Phys. Rev. B **56**, R8475 (1997).
5. N. Hamada *et al.* (unpublished). A U of 2 eV was used for this calculation. Compared to a U of 0, a gap at E_F is opened in the down-spin bands, and only very small effects are found in the up-spin bands.
6. Y. Moritomo, A. Asamitsu, H. Kuwahara, and Y. Tokura, Nature **380**, 141 (1996).
7. C.-H. Park, D. S. Dessau, T. Saitoh, Z.-X. Shen, Y. Moritomo, and Y. Tokura (preprint); D. S. Dessau, T. Saitoh, C.-H. Park, Z.-X. Shen, Y. Moritomo, and Y. Tokura, appear in Science and Technology of Magnetic Oxides, edited by M. Hundley, J. Nickel, R. Ramesh, and Y. Tokura (Mater. Res. Soc. Proc.).
8. T. Ishikawa, T. Kimura, T. Katsufuji, and Y. Tokura (preprint).
9. D. D. Sarma, N. Shanthi, S. R. Krishnakumar, T. Saitoh, T. Mizokawa, A. Sekiyama, K. Kobayashi, A. Fujimori, E. Weschke, R. Meier, G. Kaindl, Y. Takeda, and M. Takano, Phys. Rev. B, **53**, 6874 (1996); J.-H. Park, C. T. Chen, S.-W. Cheong, W. Bao, G. Meigs, V. Chakarian, and Y. U. Idzerda, Phys. Rev. Lett. **76**, 4215 (1996).
10. Y. Okimoto, T. Katsufuji, T. Ishikawa, T. Arima, and Y. Tokura, Phys. Rev. B **55**, 4206 (1997).
11. T. Saitoh, D. S. Dessau, C.-H. Park, Z.-X. Shen, P. Villella, T. Kimura, Y. Moritomo, and Y. Tokura (unpublished).
12. D. S. Marshall, D. S. Dessau, A. G. Loeser, C.-H. Park, A. Y. Matsuura, J. N. Eckstein, I. Bozovic, P. Fournier, A. Kapitulnik, W. E. Spicer, and Z.-X. Shen, Phys. Rev. Lett. **76**, 4841 (1996); A. G. Loeser Z.-X. Shen, D. S. Dessau, D. S. Marchall, C.-H. Park, P. Fournier, and A. Kapitulnik, Science **273**, 325 (1996); H. Ding, T. Yokoya, J. C. Campuzano, T. Takahashi, M. Randeria, M. R. Norman, T. Mochiku, K. Kadowaki, J. Giapintzakis, Nature **382**, 51 (1996); Z.-X. Shen and J. R. Schrieffer, Phys. Rev. Lett. **78**, 1771 (1997).
13. G. D. Mahan, Many Particle Physics, 2nd ed. (Plenum Press, New York, 1990), p.293.
14. S. Hüfner, Photoelectron Spectroscopy, (Springer-Verlag, Berlin, 1995), p.150.
15. Z.-X. Shen and D. S. Dessau, Physics Reports **253**, 1 (1995).

Part IV

Metallic Magnetic Oxide Devices and Multilayers

SUB-200 Oe GIANT MAGNETORESISTANCE IN MANGANITE TUNNEL JUNCTIONS

GANG XIAO*, A. GUPTA**, X. W. LI*, G. Q. GONG*, J. Z. SUN**
*Department of Physics, Brown University, Providence, RI 02912
** IBM Research Division, T. J. Watson Research Center, Yorktown Heights, New York, 10598

ABSTRACT

Metallic manganite oxides, $La_{1-x}D_xMnO_3$ (D=Sr, Ca, etc.), display "colossal" magnetoresistance (CMR) near their magnetic phase transition temperatures (T_c) when subject to a Tesla-scale magnetic field. This phenomenal effect is the result of the strong interplay inherent in this class of materials among electronic structure, magnetic ordering, and lattice dynamics. Though fundamentally interesting, the CMR effect achieved only at large fields poses severe technological challenges to potential applications in magnetoelectronic devices, where low field sensitivity is crucial. Among the objectives of our research effort involving manganite materials is to reduce the field scale of MR by designing and fabricating tunnel junctions and other structures rich in magnetic domain walls. The junction electrodes were made of doped manganite epitaxial films, and the insulating barrier of $SrTiO_3$. The interfacial expitaxy has been imaged by using high-resolution transmission electron microscopy (TEM). We have used self-aligned lithographic process to pattern the junctions to micron scale in size. Large MR values close to 250% at low fields of a few tens of Oe have been observed. The mechanism of the spin-dependent transport is due to the spin-polarized tunneling between the half-metallic electrodes, in which the spins of the conduction electrons are nearly fully polarized. We will present results of field and temperature dependence of MR in these structures and discuss the electronic structure of the manganite inferred from tunneling measurement. Results of large MR at low fields due to the grain-boundary effect will also be presented.

INTRODUCTION

Magnetotransport properties in magnetic solids and structures have been the subject of intense research in the last ten years [1-12]. The investigation provides better understanding on the physics of spin-dependent transport in heterogeneous systems. Perhaps more significantly, such studies have contributed to new generations of magnetic devices for information storage and magnetic sensors. There exist a diverse group of magnetoresistive materials. Many layered structures exhibit the so called giant magnetoresistance (GMR) effect, due to the spin-dependent scattering off changing magnetization configurations in various layers [1,2]. Resistance, typically, changes tens of percent in a small magnetic field of tens of Oe. More recently, researchers have focused on a different magnetotransport mechanism, known as the spin-polarized tunneling[9-12]. Magnetic tunnel junctions (MTJ), such as $Ni_{79}Fe_{21}/Al_2O_3/Co$, have been shown to demonstrate superior properties[11,12]. Not only do they possess large MR values similar to GMR systems and at small fields (tens of Oe), their tunneling resistance can also span more than five orders of magnitude upon varying barrier and area parameters, while keeping the large MR primarily intact.

Metallic manganite oxides, $La_{1-x}D_xMnO_3$ (D=Sr, Ca, etc.), are yet another important class of magnetoresistive materials[5-8]. They display "colossal" magnetoresistance (CMR), of the order of hundreds to thousands of percent, near their magnetic phase transition temperatures (T_c) when subject to a Tesla-scale field. Though the CMR materials are fundamentally interesting, the large field required for observing CMR might prevent them from being feasible for low-field applications.

The origin of CMR stems from the strong interplay among the electronic structure, magnetism, and lattice dynamics in manganites. Doping of divalent Ca or Sr impurities into the trivalent La sites create mixed valence states of Mn^{3+} (fraction: 1-x) and Mn^{4+} (fraction: x). Mn^{4+} ($3d^3$) has a localized spin of $S=3/2$ from the low-lying t_{2g}^3 orbitals, whereas the e_g obitals are empty. Mn^{3+} ($3d^4$) has an extra electron in the e_g orbitals, which can hop into the neighboring Mn^{4+} sites. The spin of this conducting electron is aligned with the local spin ($S=3/2$) in the t_{2g}^3 orbitals of Mn^{3+} due to a strong Hund's coupling. When the manganite becomes ferromagnetic, the electrons in the e_g orbitals are fully spin-polarized. The band structure is such that all the conduction electrons are in the majority band. This kind of metal with empty minority band is generally called a half-metal. Because of the rare occurrence of half metals, manganites naturally become a good candidate for the study of spin-polarized transport.

We have taken advantage of the manganites' half-metallic nature and fabricated MTJ using manganites as electrodes. Very large MR has been obtained in sub-200 Oe magnetic field and at low temperatures. We have confirmed the spin-polarized tunneling as the active mechanism. The low saturation field comes from the fact that manganites are magnetically soft, having a coercive field as small as 10 Oe. We have also made epitaxial manganite films with a large number of the grain boundaries and observed large MR at low fields. Here the mechanism has been attributed to the spin-dependent scattering across the grain boundaries that serve to pin the magnetic domain walls. In this paper, we will describe the fabrication and investigation of the above-mentioned manganite structures.

EXPERIMENT

Our MTJ structures consist of top and bottom electrodes made of the same manganite $La_{0.67}Sr_{0.33}MnO_3$ (LSMO). The tunnel barrier is $SrTiO_3$ (STO) that share the same perovskite structure and similar lattice constant of LSMO. The substrate used is either STO or $LaAlO_3$ (LAO). We have used a multitarget pulsed laser deposition system to grow the LSMO/STO/LSMO trilayers in situ on (100)-oriented STO substratres. The thickness of both LSMO electrodes is about 50 nm and the STO barrier thickness is in the range 2-6 nm, based on rate calibrations. The substrate temperature was kept at about 700°C during deposition, and the oxygen pressure maintained at 400 mTorr in the vacuum chamber. The films were cooled at a rate of 15°C/min to room temperature in 700 Torr oxygen, with no subsequent thermal treatment. The cation stoichiometry of the LSMO films determined from Rutherford backscattering spectroscopy was within 5% of the nominal target composition. Fig. 1 shows a high-resolution cross-sectional TEM lattice image of the LSMO/STO/LSMO junction region. One observes that heteroepitaxial growth and the interfaces are of good quality for this junction.

We have used a self-aligned photolithographic process to pattern the MTJ structures. A schematic diagram of the finished junction is shown in Fig. 1. The bottom electrode is just the bulk LSMO film, whereas the top LSMO electrode is patterned into a rectangular shape with dimensions in the micron scale. Au contact terminals are used for current and voltage probes as shown in Fig. 1.

The actual lithographic process is illustrated in Fig. 2. First, we used ion-milling to pattern the trilayer film into the shape of the bottom electrode (Fig. 2(b)). Then the top electrode was defined by ion-milling the unwanted areas of the top layers down to the surface of the bottom LSMO layer (Fig. 2 (c)). Next we coated the junctions with a 150 nm layer of SiO_2 by sputtering (Fig. 2(d)). The photoresist left after the previous ion-milling step was used as the lift-

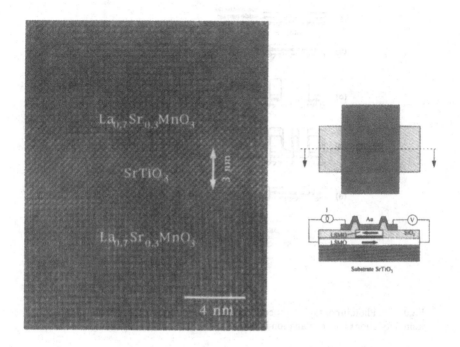

Figure 1. (Left): High resolution TEM micrograph of a cross-sectional lattice image of a magnetic tunnel junction $La_{0.67}Sr_{0.33}MnO_3/SrTiO_3/$ $La_{0.67}Sr_{0.33}MnO_3$. (Right): Schematic diagram of a tunnel junction. The bottom electrode is a bulk film of $La_{0.67}Sr_{0.33}MnO_3$ and the top electrode is a patterned rectangle made of the same $La_{0.67}Sr_{0.33}MnO_3$. Also shown are current and voltage contacts.

off stencil to open self-aligned contact holes to the top electrodes (Fig. 2(e)). Finally, we deposited and patterned a 300-nm Au layer for contact to the top electrode (Fig. 2(f)).

We have characterized the magnetic properties of our films using a Quantum Design SQUID magnetometer. A four-probe technique was used to measure the I-V curves and the tunneling resistance of junctions in a magnetic field.

In addition to junction samples, we have also grown polycrystalline LSMO and LCMO ($La_{0.67}Ca_{0.33}MnO_3$) thin films. The polycrystalline STO substrates were obtained by cutting and optically polishing sintered pellets (97% density). The time and temperature of sintering were varied to obtain samples with three different average grain sizes (3, 14, and 24 μm). Scanning and transmission electron microscopy (SEM and TEM) analysis has confirmed a well-defined grain morphology reminiscent of the underlying polycrystalline structure of the substrates. The grain boundary region has an average width of about 1 nm.

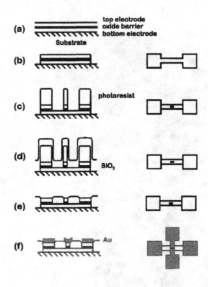

Figure 2. Photolithographic process for the patterning of micron-scale magnetic tunnel junctions (see explanation in text).

MAGNETIC PROPERTIES AND TUNNELING CHARACTERISTICS

To assess the basic magnetic properties of the (100) LSMO film, we have measured the spontaneous magnetization (M_s) and coercivity (H_c) in the temperature range between 5 and 380 K. The results are shown in Fig. 3. The value of M_s at each temperature was extrapolated from an M-H curve between 0.5 and 5 T. We have also measured the magnetic hysteresis loops, from that we obtained H_c. At T = 5K, M_s is 622 emu/cm^3. The ferromagnetic phase transition is rather sharp at T_c =347 K, reflecting a good phase homogeneity. The film is magnetically soft, having a small H_c of 49 Oe at 5 K and 9.5 Oe at 290 K. The two representative hysteresis loops at 5 and 300 K, shown in Fig. 3, are square-like. The low H_c values are particularly noteworthy for potential low-field applications.

The measurement of vertical transport in layer structures is challenging. The thin insulating barriers may contain pinholes and other defects detrimental to tunneling. The lithographic process may cause poor junction-edge definition that leads to junction shorts. In the junctions that we studied, the lead resistance is about three orders of magnitude smaller than the junction resistance. Therefore a uniform current density in the junctions is assured. To evaluate the transport mechanism, we have measured I-V curves of junctions at various temperatures. Fig. 4 shows such results for one of the junctions (area 1×16 µm^2) between T = 4K and 300 K. At T <

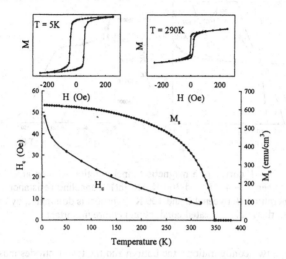

Figure 3. Temperature dependence of the spontaneous magnetization, M_s, and coercivity, H_c, of a bulk $La_{0.67}Sr_{0.33}MnO_3$ epitaxial film. Also shown are two magnetic hysteresis loops at T = 5 and 290 K.

140 K, the *I-V* curves are non-linear, and non-Ohmic. They are highly characteristic of electron tunneling. At T > 200 K, the *I-V* curves become gradually more linear, indicating tunneling is no longer the dominant transport mechanism at the high end of the temperature range.

Fig. 4 (Right) shows the zero-bias resistance of the same junction as a function of temperature. The barrier resistance is found to be thermally activated with a gap of 0.11 eV between 120 and 300 K. Therefore the thin STO barrier is an extrinsic semiconductor, which may be due to impurities or the extreme thin-film nature of the barrier.

SPIN-POLARIZED TUNNELING

In an MTJ, the resistance of the junction depends on the relative orientation of the magnetization vectors in the two electrodes. When the vectors are parallel to each other, tunneling probability is maximized because electrons from those states with a large density of states (DOE) can tunnel into the same states in the other electrode and vise versa. On the other hand, when the magnetization vectors are antiparallel, there will be a mismatch between the tunneling states on each side of the junction. This leads to a diminished tunneling probability, hence, a larger tunneling resistance. Assuming no spin-flipping events, the MR ratio between the parallel and antiparallel configurations can be shown to give the following relation[13],

$$(\Delta R / R_p) = (R_{\uparrow\downarrow} - R_{\uparrow\uparrow}) / R_{\uparrow\uparrow} = 2P^2 /(1 - P^2), \qquad (1)$$

where R_p is the resistance for the parallel configuration, P is the spin-polarization parameter for a magnetic solid. A half-metal corresponds to $P = 1$. It is noted that relation (1) should be modified if the electron energy is away from Fermi surface, or if there exist spin-flipping scatterings.

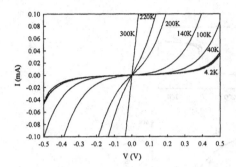
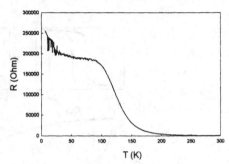

Figure 4. (Left): *I-V* curves of a magnetic tunnel junction with an area 1×16 μm²
measured between T = 4.2 K and 300 K. (Right): Tunneling resistance *R* at zero
bias vs. temperature. Between 4.2 and 100 K, transport is dominated by tunneling.
Above 100 K, thermally activated conduction is more important.

To achieve the two configurations, the bottom and the top electrodes must have different
switching fields. If the electrodes are made of different materials, the switching fields naturally
differ from one another. In our LSMO/STO/LSMO junctions, however, since both electrodes are
made from the same material, we have to use other means to differentiate the switching fields.
Fortunately, since the bulk coercivity is rather small for LSMO, we can take advantage of the
magnetic shape anisotropy to increase the switching field. By making the top electrode into a
rectangular shape, we were able to increase the switching field to 100-200 Oe. The bottom
electrode, being a part of the bulk film, still attains a smaller switching field close to the bulk
coercivity (<60 Oe).

We have measured the zero-bias junction resistance as a function of magnetic field at
various temperatures. One of the examples is shown in Fig. 5. The top electrode has a
rectangular shape with dimensions of 2.5×12.5 μm². The field direction is along the longer axis,
or the magnetic easy axis, of the electrode. Using the film thickness of 50 nm, we have estimated
that the switching field, H_{c2}, due to the shape anisotropy should be about 160 Oe for the top
electrode. The switching field, H_{c1}, for the bottom electrode should be close to the bulk
coercivity of 50 Oe at T = 5K.

As seen in Fig. 5, when the magnitude of *H* exceeds 200 Oe, i.e., *H* > 200 Oe or *H* < 200
Oe, magnetization vectors in both electrodes become parallel. The tunneling resistance (*R*) is
minimized. In the range of 60 < |*H*| < 160 Oe, the magnetizations are antiparallel, leading to a
maximum *R*, as expected from spin-polarized tunneling. The MR ratio, Δ*R*/*R*, is 83% at T = 4.2
K for this sample. This is a rather large value considering the fact that the field scale is sub-200
Oe, which is substantially lower than the Tesla-scale field required for observing the CMR
effect.

We have also measured the thermal effect on magnetic tunneling. As we increase
temperature, the field scale in the MR remains in sub-200 Oe range. However, the MR value
decreases with increasing T. To find out the cause, we have measured Δ*R*/*R* and *R* as functions of
temperature. Fig. 6 shows the results as well as the dynamic conductance vs. bias voltage taken
at T = 4.2 K. The bias dependence of conductance is indicative of tunneling transport at low
temperatures. Between 4.2 and 90 K, Δ*R*/*R* remains quite large (> 60%). However, beyond 100K,

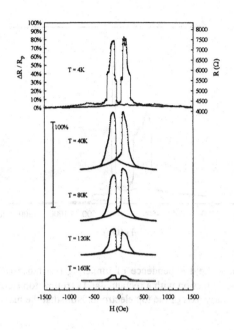

Figure 5. Tunneling resistance (R) and magnetoresistance ratio ($\Delta R/R$) vs. magnetic field for a tunnel junction with a rectangular 2.5×12.5 μm^2 top electrode. Field was applied along the easy axis of the top electrode.

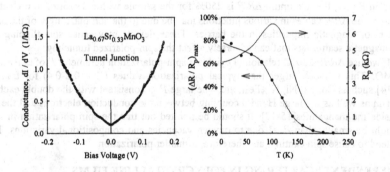

Figure 6. (Left) Dynamic conductance vs. bias voltage and (Right) temperature dependence of $\Delta R/R$ and tunneling resistance R_p (parallel magnetization) for the tunnel junction used in Figure 5.

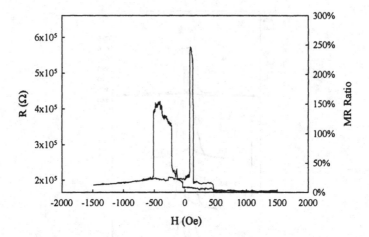

Figure 7. Magnetic field dependence of tunneling resistance (R) and MR ratio ($\Delta R/R$) for a tunnel junction with a rectangular 1×16 μm^2 top electrode. Field was applied along the easy axis of the top electrode. Note that the maximum MR value is 250%.

$\Delta R/R$ drops rapidly with increasing T, and so does the zero-bias tunneling resistance R. This behavior can be attributed to the increasing role of the thermally activated conduction at high temperatures. In order to observe the intrinsic temperature dependence of $\Delta R/R$ resulting from spin-polarized tunneling, a better insulating barrier would be highly desirable.

In some of our samples, the $\Delta R/R$ values at low temperatures exceed 80%. For example, as shown in Fig. 7, the maximum $\Delta R/R$ is 250% for the sample with a $1 \times 16 \mu m^2$ top electrode. The non-unique $\Delta R/R$ values in various junctions may be due to the non-uniformity of the barrier parameters or magnetic impurities in the barrier. These defects may cause spin-flipping spin-orbit or magnetic scatterings that can adversely affect the spin-polarized tunneling.

If we use $\Delta R/R=250\%$, relation (1) yields a spin-polarization parameter P of at least 0.75 for LSMO. This is much larger than typical polarization values ($P = 0.2$-0.4) for transition metals[14] such as Co, Fe, Ni, $Ni_{79}Fe_{21}$, etc. The large P is consistent with the double exchange mechanism as well as the large Hund's coupling between the conduction electrons and the local spins inside the manganites[15-17]. It should be pointed out that the spin-polarization in actual films might be smaller than $P =1$ due to oxygen vacancies and compositional variations. These defects tend to cause spin canting, and therefore, a smaller polarization.

SPIN-DEPENDENT SCATTERING IN POLYCRYSTALLINE FILMS

Because of its large spin polarization in manganites, one anticipates that spin-dependent scatterings across magnetic boundaries might be enhanced. A magnetic heterogeneous layered structure is often used to create such magnetic scattering centers, as in GMR multilayer or spin-valve systems [1,2]. We have used a different route to achieve inhomogeneous structures by creating polycrystalline films of manganites. Magnetic Kerr microscope revealed that the grain

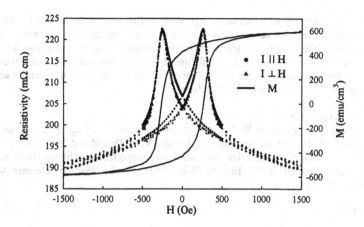

Figure 8. Resistivity and magnetization as functions of magnetic field, measured at $T = 10$ K, for a polycrystalline LCMO film with 3-μm average grain size. Two current directions are used, parallel or perpendicular to magnetic field. The MR is nearly isotropic.

boundaries are pinning centers for magnetic domain walls. The coercivity is about 250 Oe at $T = 10$ K for a polycrystalline LCMO film with 3-μm average grain size. As we sweep field across H_c, the magnetic configuration evolves from a multidomain structure to magnetic saturation, as observed in the Kerr microscope.

Fig. 8 shows the magnetic hysteresis loop and the corresponding field dependence of resistivity of the 3-μm polycrystalline LCMO film measured at T =10 K. The MR is negative and almost isotropic for both current directions, parallel and perpendicular to the applied field. This is one signature of spin-dependent scattering. Under the same condition of measurement, the epitaxial manganite film shows an anisotropic MR and a magnitude of only 0.25% up to 1500 Oe. The cause for the anisotropic MR is likely the spin-orbit scattering in single crystalline film. The polycrystalline sample, as shown in Fig. 9, develops a MR of 17.5% at 1500 Oe. Though smaller comparing with tunneling junctions, it is enhanced by 70 fold over the MR value of the epitaxial counterpart. We believe that the enhanced MR is due to the magnetic scattering centers near the grain boundaries [18, 19]. The scattering cross-section varies with the changing field strength. A large field reduces the scattering because of the alignment of the magnetic domains associated with the grains.

CONCLUSIONS

We have obtained large magnetoresistance at low magnetic fields in two manganite artificial structures, tunneling junctions and polycrystalline films. Rather than directly using the CMR mechanism associated with the magnetic phase transition and large magnetic field, we take advantage of the fact that manganite is nearly a half-metallic system. Its large spin-polarization leads to a MR of 250% at T = 4.2 K in sub-200 Oe field. This corresponds to a spin-polarization parameter of $P = 0.75$, which is much larger than that of magnetic transition metals. In the temperature region (4.2 – 100 K) where tunneling is the dominant transport channel, the MR value decreases moderately with increasing temperature. However, it diminished rapidly beyond 100 K as thermally activated conduction become important, possibly duo to the less than ideal insulating nature of the barrier. The magnetotransport in polycrystalline manganite films is also substantially enhanced over that in epitaxial films by nearly two orders of magnitude at low fields and low temperatures. We attribute this to the strong spin-dependent scattering across the magnetic domains induced by the grain boundaries.

ACKNOWLEDGEMENTS

We wish to thank T. R. McGuire, Yu Lu, J. Slonczeski, and W. J. Gallagher for discussions. We are grateful to P. R. Duncombe for preparing targets, V. P. Dravid and Y. Y. Wang for TEM measurements on some of our samples. This work was supported partially by National Science Foundation Grants Nos. DMR 9414160 and DMR 9701578 and partially by Defense Advanced Research Projects Agency.

REFERENCES

1. M. N. Babich, J. M Broto, A. Fert, F. Nguyen Van Dau, F. Petroff, P. Eitenne, G. Creuzet, A. Friederich, and J. Chazelas, Phys. Rev. Lett. **61**, 2472 (1988).
2. S. S. P. Parkin, R. Bhadra, and K. P. Roche, Phys. Rev. Lett. **66**, 2152 (1991).
3. A. E. Berkowitz, J. R. Mitchell, M. J. Carey, A. P. Young, S. Zhang, F. E. Spada, F. T. Parker, A. Hutten, and G. Thomas, Phys. Rev. Lett. **68**, 3745 (1992).
4. J. Q. Xiao, J. S. Jiang, and C. L. Chien, Phys. Rev. Lett. **68**, 3749 (1992)
5. R. von Helmolt, J. Wecker, B. Holzapfel, L. Shultz, and K. Samwer, Phys. Rev. Lett. **71**, 2331 (1993).
6. S. Jin, T. H. Tiefel, M. McCormack, R. A. Fastnacht, R. Ramesh, and L. H. Chen, Science, **264**, 413 (1994).
7. G. Q. Gong, C. L. Canedy, Gang Xiao, J. Z. Sun, A. Gupta, and W. J. Gallagher, Appl. Phys. Lett. **67**, 1783 (1995).
8. Gang Xiao, G. Q. Gong, C. L. Canedy, E. J. McNiff, Jr, and A. Gupta, J. Appl. Phys. **81**, 5324 (1997).
9. T. Miyazaki and N. Tezuka, J. Magn. Magn. Mater. **139** ,L231 (1995).
10. J. S. Moddera, L. R. Kinder, T. M. Wong, and R. Meservey, Phys. Rev. Lett, **74**, 3272 (1995).
11. W. J. Gallagher, S. S. P. Parkin, Yu Lu, X. P. Bian, A. Marley, K. P. Roche, R. A. Altman, S. A. Rishton, C. Jahnes, T. M. Shaw, and Gang Xiao, J. Appl. Phys. **81**, 3741 (1997); Yu Lu, X. W. Li, G. Q. Gong, Gang Xiao, A. Gupta, P. Lecoeur, J. Z. Sun, Y. Y. Wang, and V. P. Dravid, Phys. Rev. B **54**, R8357 (1996); J. Z. Sun, W. J. Gallagher, P. R. Duncombe, L. Krusin-Elbaum, R. A. Altman, A. Gupta, Yu Lu, G. Q. Gong, and Gang Xiao, Appl. Phys. Lett. **69**, 3266 (1996).
12. Yu Lu, R. A. Altman, A. Marley, S. A. Rishton, P. L. Trouilloud, Gang Xiao, W. J. Gallagher, and S. S. P. Parkin, Appl. Phys. Lett. **70**, 2610 (1997).
13. M. Julliere, Phys. Lett. A **54A**, 225 (1975).
14. R. Meservey and P. M. Tedrow, Phys. Rep. **239**, 174 (1994).
15. C. Zener, Phys. Rev. **82**, 403 (1951).
16. P. W. Anderson and H. Hasegawa, Phys. Rev. **100**, 675 (1955).
17. P.-G. de Gennes, Phys. Rev. **118**, 141 (196).
18. A. Gupta, G. Q. Gong, Gang Xiao, P. R. Duncombe, P. Lecoeur, P. Trouilloud, Y. Y. Wang, V. P. Dravid, and J. Z. Sun, Phys. Rev. **54**, R15629 (1996).
19. X. W. Li, A. Gupta, Gang Xiao, and G. Q. Gong, Appl. Phys. Lett. **71**, 1124 (1997).

LOW-FIELD COLOSSAL MAGNETORESISTANCE
IN MANGANITE TUNNEL JUNCTIONS

J. NASSAR*, M. VIRET**, M. DROUET*, J.P. CONTOUR*, C. FERMON* *, A. FERT*
*UMR CNRS-Thomson CSF, Domaine de Corbeville, F-91404 Orsay
**CEA Saclay, Service de l'Etat Condensé, F-91191 Gif sur Yvette

ABSTRACT

Large magnetoresistance values are obtained on tunnel junctions epitaxially deposited by pulsed-laser deposition and consisting of ferromagnetic manganite $La_{0.67}Sr_{0.33}MnO_3$ electrodes separated by various tunnel barriers: $SrTiO_3$, $PrBaCu_{2.8}Ga_{0.2}O_7$, and CeO_2. The magnetoresistance can be decomposed into a low-field and a high-field contribution. The latter is attributed to the presence of canted interfacial manganite phases, as confirmed by the temperature behaviour of the resistance. A low-field magnetoresistance ratio of 450% below 100 Oe is obtained on a sample with a $SrTiO_3$ barrier, indicating a spin polarization value in excess of 0.83 for the manganite.

INTRODUCTION

Mixed-valent manganese perovskites with general formula $A_{1-x}B_xMn^{3+}_{1-x}Mn^{4+}_x O_3^{2-}$ (where A is a trivalent ion like La or Pr, and B a divalent ion like Ca or Sr) have been increasingly studied in the last few years. Their main characteristic in the range $0.1<x<0.5$ is a very close correlation between ferromagnetic order and electrical conductivity, whose most dramatic evidence is the « colossal magnetoresistance » (CMR) observed in the vicinity of the Curie temperature T_c. In the Zener « double exchange » mechanism [1] the conductivity is due to the transfer of e_g electrons between neighbouring Mn^{3+} and Mn^{4+} cations through filled O^{2-} shells which is favoured when the spins of the two Mn ions are parallel. Spin ordering (obtained by applying high fields or lowering temperature) allows delocalization of 3d e_g electrons and broadening of the discrete 3d e_g levels of the manganese ions into a partially filled band of width W (Figure 1). This band is fully spin-polarized (the compound is then called « half-metallic ») if the Hund splitting J between the spin up and spin down states is larger than W. However the extent to which hybridization of the Mn 3d e_g levels with spin-unpolarized oxygen 2p levels decreases the polarization of the manganite at the Fermi level is still a subject of controversy.

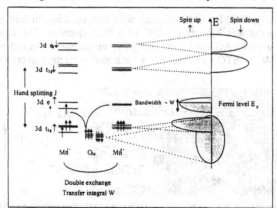

Figure 1 : schematic band structure of half-metallic manganite

Most attention has been focused on the CMR behaviour near T_c. However, the high field (in the Tesla range) needed to induce magnetic order makes the use of manganite compounds for applications difficult, indeed the resulting low-field sensitivities 1/R.dR/dH do not exceed those of conventional metallic GMR multilayers or magnetoresistive tunnel junctions. The latter are layered FM1/B/FM2 systems, where FM1 and FM2 are ferromagnetic transition metals, and B a thin insulating barrier through which electrons can tunnel. If the magnetizations of the two ferromagnets can be switched between parallel and antiparallel configurations, the relative difference of resistance TMR = $(R_{antipar}-R_{par})/R_{par}$ between both configurations is given by the Jullière [2] formula :

$$TMR = 2P_1P_2/(1-P_1P_2) \qquad (1)$$

where P_1 and P_2 are the spin polarizations of the carriers in the two ferromagnets, as defined by Tedrow and Meservey [3]. Such tunnel junctions show low-field TMR up to 34% at room temperature [4], because the spin polarization of the transition metals and alloys commonly used (Co, NiFe, CoFe) does not exceed 40%.

It is possible to take advantage of the 100% spin polarization of the manganites in the half-metallic temperature range (far below T_c) by making them into FM/B/FM structures with adjustable relative magnetic configuration. In the most ideal description, full spin polarization should lead to zero conductance in the antiparallel configuration, and hence to an infinite TMR. Artificial structures to investigate spin-polarized tunneling between manganites in a current in plane (CIP) geometry were successfully built by Mathur et al. [5] and Steenbeck et al. [6]. They used grain boundaries as a barrier between ferromagnetic regions which were synthetized by epitaxial growth of $LaCaMnO_3$ (LCMO) or $LaSrMnO_3$ (LSMO) films on $SrTiO_3$ (STO) bicrystals. Sun et al. obtained low-field MR values near 120% in layered LSMO/STO/LSMO tunnel devices structured by conventional lithography and etching techniques for current perpendicular to plane (CPP) measurements[7].

We report here on large magnetoresistance results obtained on planar tunnel junctions consisting of pulsed-laser-deposited epitaxial LSMO/B/LSMO trilayers structured by optical lithography for CPP measurements. The compound $La_{1-x}Sr_xMnO_3$ with x=0.3 was chosen because it is the manganite compound with the highest Curie temperature (T_c=370K) [8] ; this should allow half-metallic behaviour near room temperature. Three insulators epitaxially compatible with LSMO (lattice constant : a=0.380nm) were used as a barrier, on account of their promising dielectric properties[9,10] : STO (a=0.390nm), $PrBa_2Cu_{2.8}Ga_{0.2}O_7$ (PBCGO, a=0.391nm) and CeO_2 ($a/\sqrt{2}$=0.382nm).

Simple LSMO layers and LSMO/B/LSMO trilayers were deposited in situ from stoichiometric targets on STO [100] substrates heated at 700°C in a PLD chamber. The laser used was a frequency doubled and tripled YAG and the operating O_2 pressure during deposition was 0.3 torr. Epitaxial growth of LSMO on STO was checked on transmission electron microscopy transverse views (Figure 2).

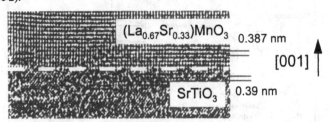

Figure 2 : transverse electron microscopy view of a LSMO film epitaxied on STO

Different thicknesses were used for the bottom and top LSMO layers (25 and 33 nm respectively) to induce a coercivity difference. Magnetic characterization of the samples was performed using a commercial SQUID magnetometer. A thermoremanence measurement at a field of 50 Oe showed that the Curie temperature of the simple manganite thin films is above room temperature and that some magnetic order persists above 350K (Figure 3). Magnetic hysteresis loops of the as-grown trilayers show very little decoupling of the two ferromagnetic layers. This coupling is probably induced by the large-scale structural defects (e.g. droplets) generated by laser ablation. However hysteresis loops measured after ion etching the top manganite layer into 6x6 μm² squares evidences a good magnetic decoupling in most of these squares (Figure 4).

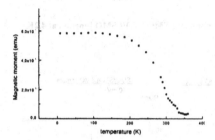

Figure 3 : Thermoremanence in a 50 Oe external field of a 40nm thick LSMO layer on a STO substrate, measured by SQUID

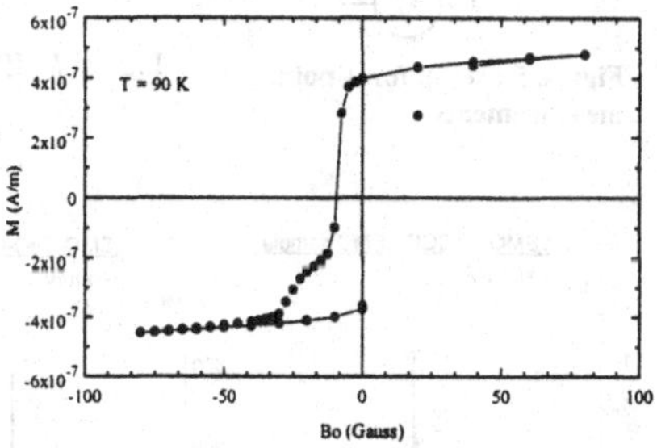

Figure 4 : magnetic hysteresis cycle of a LSMO/PBCGO/LSMO trilayer measured by SQUID at 90K after ion etching the top LSMO layer into 6x6 μm² squares

This prompted us to etch the samples for CPP transport measurements into series of 6x6 μm² square junctions, in the hope of finding a reasonable amount of them magnetically decoupled. The etching was carried out using conventional UV lithography techniques in the following steps: the bottom electrode was first defined by lithography and ion milling down to the substrate. The sample was then coated by sputtered SiO₂ before the remaining photoresist was lifted off. Copper was then evaporated on the whole sample and the top electrode was defined by UV lithography followed by ion etching timed to stop after removal of the top ferromagnetic layer. Aluminium wires were finally bonded onto the electrodes.

Four-point transport measurements were carried out using a retroacted Keithley 236 DC voltage source (Figure 5). At low temperature all samples show non-linear I(V) curves typical for tunnel conduction [11]. (Figure 7), and large magnetoresistance ratios (400 to 450%) are obtained for the three kinds of barriers. The magnetoresistance curves (Figure 6) are characterized by plateaus with steep edges at low-fields and a continuous decrease of resistance at high fields.

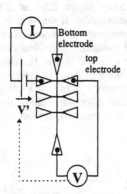

Figure 5 : setup for 4-point measurements

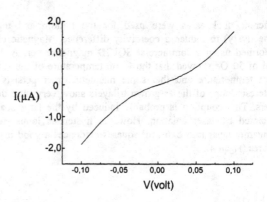

Figure 7 : I(V) curve of LSMO/STO/LSMO junction at 4.2K

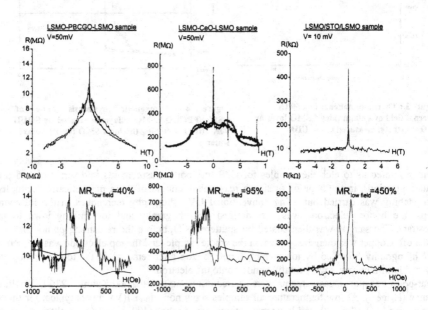

Figure 6 : junction magnetoresistance curves at 4.2K

The low-field MR effects are large, especially for the LSMO/STO/LSMO sample where the magnetoresistance reaches 450% at fields below 100 Oe. The associated field sensitivity reaches 9%/Oe. The spin polarization that can be deduced from this value using eq. (1) is 83%. This suggests that the hybridization of the Mn 3d e_g with the O 2p levels is weak.

We attribute the high-field magnetoresistance of the junctions with PBCGO and CeO$_2$ to the presence of a canted LSMO phase at he interface with the barrier (Figure 8). Canted phases are expected in oxygen-deficient manganites [12.] In our systems it is likely that the insulators act as oxygen getters and pump out some oxygen from a few interfacial manganite monolayers during growth. This interfacial layer can be viewed as an insulating layer (or a layer with a low density of carriers). Very high fields are needed to reduce the interface spin disorder of canted layers and enhance the carrier density. This accounts for a variation of resistance of more than 200% for the CeO$_2$ and PBCGO junctions at high fields. As the tunnel resistance is more than 1000 times higher than the electrode resistance, these high-field effects cannot be due to the MR of the electrode. On the other hand, the very weak decrease of the high field resistance of the STO junction indicates that the interface is much less canted.

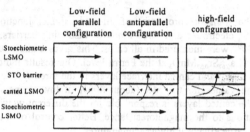

Figure 8 : interpretation for the low-field and high-field contributions to the junction magnetoresistance

This is confirmed by the analysis of the temperature dependence of the resistance for the three kinds of samples (Figure 9) where the junction resistance shows a broad maximum in every case. We believe that the maximum occurs at the mean ordering temperature of the canted interfacial layer. The increase of the tunnel resistance between 4.2K and the maximum could be due to the

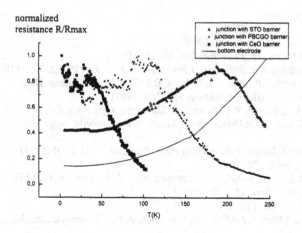

Figure 9 temperature dependence of the resistance for the tunnel junctions and for the bottom electrode

reduction of carrier concentration as the spin disorder increases. This temperature is lower as the manganite is more canted, and the maximum resistivity point for the STO junction (obtained at about 190K) is indeed at a higher temperature than the maximum for the PBCGO and CeO_2 samples. When the resistance of the bottom electrode alone is measured in plane, the canted layer is shunted and the resistance exhibits the typical metallic-like temperature dependence of a bulk manganite with a Curie point above room temperature. The spin disorder of the oxygen-deficient layer probably also accounts for the rapid decrease of the magnetoresistance ratio with temperature [13]. Optimization of the oxydation conditions during and after growth is under way and should lead to large effects at higher temperatures.

CONCLUSION

We have compared the transport properties of all-oxide tunnel junctions with nominally identical $La_{0.67}Sr_{0.33}MnO_3$ electrodes and three kinds of insulating barriers. Magnetoresistance values of more than 400% were measured in all cases. This gives a minimum value of 83% for the spin polarization of $La_{0.67}Sr_{0.33}MnO_3$ at the Fermi level. Our results indicate the presence of more or less oxygen deficient and canted interface layers in the three systems. All three systems lead to similar absolute values for the MR (about 400% at low temperature). However the different character of the canted layers is responsible for the different weight of the high-field and low-field contributions to the magnetoresistance. Better control of the oxidization of the interface should lead to large effects at higher temperatures.

ACKNOWLEDGEMENTS

We thank J.L. Maurice for TEM characterizations and P. Seneor for reviewing the manuscript.

REFERENCES

1. C.Zener, *Phys. Rev.* 82,3 403 (1951)
2. M.Jullière, *Physics Letters*, 54, 3, 225 (1975)
3. P.M. Tedrow and R.Meservey, *Phys. Rev.Lett.* 26, 192 (1971)
4. J.S.Moodera, E.F.Gallagher, K.Robinson, J.Nowak, *Appl.Phys.Lett* 70, 22, 3050 (1997)
5. N.D. Mathur, G. Burnell, S.P.Isaac, T.J. Jackson, B.S. Teo, J.L.MacManus-Driscoll, L.F.Cohen, J.E. Evetts, M.G. Blamire, *Nature*, 387, 266 (1997)
6. K.Steenbeck, T.Eick, K.Kirsch, K O'Donnel, E. Steinbess, *Appl. Phys. Lett.* 71, 7, 968 (1997)
7. J.Z.Sun, L.Krusin-Elbaum, P.R.Duncombe, A.Gupta, R.B.Laibowitz, *Appl. Phys. Lett.* 70, 13, 1769 (1997)
8. J.M.D Coey, M.Viret, L.Ranno, K.Ounadjela *Phys. Rev. Lett.* 75,21,3910 (1995)
9. Y.Xu, W.Guan, *Physica C* 206, 59 (1993)
10. A.Walkenhorst, M.Schmitt, H.Adrian, K.Petersen, *Appl. Phys. Lett.* 64, 14, 1871 (1994)
11. J.G. Simmons, *J. Appl. Phys*, 43, 1793 (1963)
12. P.G. de Gennes, *Phys. Rev.*118, 1, 141 (1959)
13. M.Viret, M.Drouet, J.Nassar, J.P.Contour, C.Fermon, A.Fert, *Europhys. Lett.* 39, 5, 545 (1997)

Observation of Large Low Field Magnetoresistance in Ramp-Edge Tunneling Junctions Based on Doped Manganite Ferromagnetic Electrodes and A SrTiO$_3$ Insulator

C. Kwon, Q. X. Jia, Y. Fan, M. F. Hundley, D. W. Reagor, M. E. Hawley,
and D. E. Peterson
Materials Science and Technology Division, Mail Stop K763
Los Alamos National Laboratory, Los Alamos, NM 87545

ABSTRACT

We report the fabrication of ferromagnet-insulator-ferromagnet junction devices using a ramp-edge geometry based on (La$_{0.7}$Sr$_{0.3}$)MnO$_3$ ferromagnetic electrodes and a SrTiO$_3$ insulator. The multilayer thin films were deposited using pulsed laser deposition and the devices were patterned using photolithography and ion milling. As expected from the spin-dependent tunneling, the junction magnetoresistance depends on the relative orientation of the magnetization in the electrodes. The maximum junction magnetoresistance (JMR) of 30 % is observed below 300 Oe at low temperatures (T < 100 K).

INTRODUCTION

There has been a great deal of interest in spintronic devices, in which the spin-dependent transports are utilized to generated a large magnetoresistance in conventional ferromagnetic metal thin films and multilayers. For example, the large junction magnetoresistance (JMR) in magnetic tunnel junctions is observed at room temperature [1]. Due to the uneven spin distribution of conduction electrons at the Fermi level in the ferromagnets, one can expect tunneling probability to be dependent on the relative magnetization orientation of the ferromagnet electrodes. By assuming that spin is conserved in the tunneling process and tunneling current is dependent on the density of states at Fermi level of two electrodes, Jullier showed the maximum change in the tunneling resistance (ΔR) as [2]

$$\Delta R/R_A = (R_A - R_p)/R_A = 2P_1P_2/(1 + P_1P_2),$$

where R_A and R_p are the junction resistances when the magnetizations are antiparallel and parallel, respectively, and P_1 and P_2 are the spin polarizations of the two electrodes.

Doped manganites, (R$_x$M$_{1-x}$)MnO$_3$ where R is a rare earth element such as La, Pr, and Nd and M is believed to be a half-metallic material due to the strong Hund's coupling and a relatively narrow conduction band [3]. The half-metallic characteristics in La$_{0.7}$Sr$_{0.3}$MnO$_3$ have been observed in spin-resolved photoemission measurements well below the Curie temperature, T$_c$ [4]. Half-metallic systems are characterized by the coexistence of metallic behavior for one electron spin and insulating behavior for the other. Hence, the density of states has 100 % spin polarization at the Fermi level and the conductivity is completely dominated by the metallic single-spin charge carriers. Since a half-metallic ferromagnet has 100% spin polarization for conduction electrons, it offers potential as a ferromagnetic metal electrode in devices based on spin-dependent transport effects. Compared to the tunneling junctions based on the conventional ferromagnetic metal electrodes, MR in the tunneling junctions made of the manganites is expected to be larger. Sun et al. have demonstrated the existence of large magnetoresistance at low fields and low temperatures in the trilayer sandwich junctions using doped manganites [5].

In this paper, we report on the fabrication of ramp-edge ferromagnet-insulator-ferromagnet junction devices using (La$_{0.7}$Sr$_{0.3}$)MnO$_3$ (LSMO) and SrTiO$_3$ (STO) as the ferromagnet and insulator layers, respectively. As we have learned from the study of thin-film high temperature superconductor applications that the control of interfaces in metal-oxide heterostructures is quite difficult. A ramp-edge structure has technical advantages in metal-oxide based junction devices due to a small junction area by nature of the design. Spin dependent tunneling transport characteristics are observed. A large junction magnetoresistance of 30 % (JMR = [(R_j(H) - R_j(1000 Oe)/R_j(1000 Oe)]) in fields less than 300 Oe is obtained at low temperatures.

EXPERIMENT

A schematic diagram of a ramp-edge junction fabrication process is shown in Fig. 1. At first, a bottom electrode (LSMO) and a thick insulation layer (STO) were deposited on a LaAlO$_3$ substrate using pulsed laser deposition. A 308 nm XeCl excimer laser was used with an energy density of 2 J/cm^2 and a repetition rate of 10 Hz. The oxygen background pressure was 400 mTorr and the heater block temperature was 700 °C. Using conventional photolithography and ion milling with Ar ions, a bottom electrode was defined (Fig. 1(b)) and the ramp-edge was created (Fig. 1(c)). After removing the photoresist, a thin insulating barrier of STO and a top electrode of LSMO were deposited using *in situ* pulsed laser deposition under 400 mTorr oxygen and at 700 °C (Fig. 1(d)). The insulating barrier and the top electrode were patterned to form a junction (Fig. 1(e) and (f)). The contact pads were defined by photolithography and gold was sputtered (Fig. 1(g)). After the lift-off process, gold formed the contact pads (Fig. 1(f)). For this work, the thicknesses of a bottom electrode of LSMO and a top electrode of LSMO were 1100 and 900 Å, respectively. The junction area is determined by the length, about 0.4 μm as estimated from the ramp-angle and the thickness of the bottom electrode, and the width of the junction.

Figure 2 shows an atomic force microscope image of a ramp-edge after the Ar ion milling of the bottom electrode. It shows the ramp-angle is $(13 \pm 1)°$.

As-grown LSMO films under the above growth conditions have a Curie temperature, T$_C$, of 350 K and a sharp decrease of resistance at the same temperature (LSMO has paramagnetic metal-to-ferromagnetic metal transition near T$_C$ [6]). The resistive transitions of both top and bottom electrodes were measured after the device fabrication process and showed the same temperature dependence as the as-grown films.

The JMR was measured with a four-terminal ac resistance bridge in fields up to ±1 T in the temperature range of 16 - 300 K. The measurements performed using a dc current source gave the same results. The current was flowing across the insulating barrier layer similar to the current-perpendicular-to-plane (CPP) geometry in a trilayer sandwich junction. The magnetic field was applied in the plane either parallel or perpendicular to the current. No significant difference was observed in the field dependence of JMR.

RESULTS

The field dependent junction resistance, R$_j$(H), and JMR, ΔR$_j$/R$_j$(1000 Oe) of a device at 15.7 K is shown in Fig. 3. JMR values as large as 30 % are observed at low fields (between 180 Oe and 220 Oe). The shape of JMR versus field is similar to that of metal-electrode-based tunnel junctions [1]. At high fields, the junction resistance is low because the magnetization in both electrodes is parallel. When H is between the coercive fields of the top and bottom electrodes (i.e., the electrode magnetization vectors are antiparallel), the junction resistance reaches a maximum value. Generally the JMR increases nearly linearly with decreasing temperature and saturates at values as large as 30 % below 100 K.

In order to study the effect of the insulating barrier, we have made two devices using the same fabrication process with and without the STO insulating barrier layer. Figure 4 shows the temperature dependent junction resistance, R$_j$(T), of 5 μm wide ramp-edge junction devices. The device with a STO barrier has much larger junction resistance indicating the junction resistance is dominated by the barrier layer not by the interfaces. The device without the STO barrier layer exhibits similar R(T) as an as-grown LSMO film. The junction resistance for the device with a barrier layer decreases with temperature, however, the change is much smaller than that in the device without a barrier. Unlike the trilayer sandwich junction devices, the ramp-edge junction devices does not show any sign of variable range hopping behavior.

The high field JMR, ΔR$_j$(H)/R$_j$(1T), is presented in Fig. 5 for both devices. The device without a barrier shows small JMR (< 0.5%) while the device with a barrier exhibits large JMR

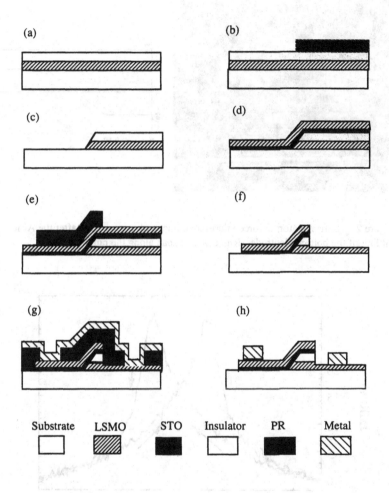

(a)

(b)

(c)

(d)

(e)

(f)

(g)

(h)

Substrate LSMO STO Insulator PR Metal

Figure 1. A schematic diagram of a ramp-dege junction fabrication process.

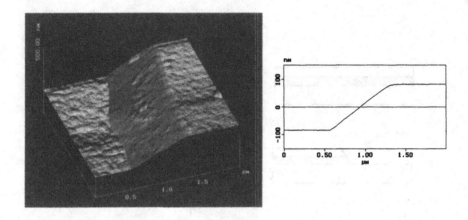

Figure 2. (Left) An atomic force microscope image of a ramp-edge after the Ar ion milling of the bottom electrode. (Right) A line scan along the ramp-edge.

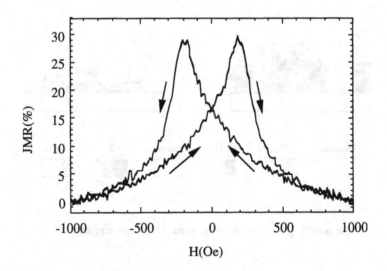

Figure 3. The field dependent junction magnetoresistance, JMR = $\Delta R_j/R(1000$ Oe$)$, of a device at 15.7 K.

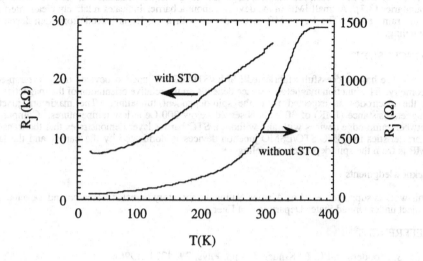

Figure 4. The temperature dependent junction resistance, R_j, of two ramp-dege devices with and without the STO barrier layer.

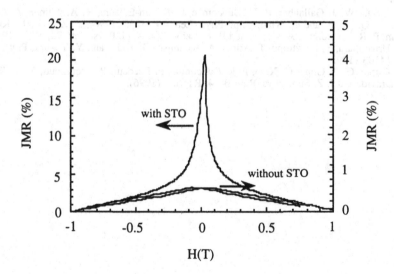

Figure 5. The high field JMR, $\Delta R_j(H)/R_j(1T)$, for two ramp-edge devices with and without the STO barrier layer.

(21 %).   There have been reports of a large low field magnetoresistance in samples with grain boundaries [3,7]. A small JMR in the device without a barrier indicates relatively clean interfaces in the ramp-edge.   The large JMR in the device with a barrier is indeed from spin-dependent tunneling.

CONCLUSIONS

We have successfully fabricated LSMO/STO/LSMO junction devices using a ramp-edge geometry. The junction magnetoresistance depends on the relative orientation of the magnetization in the electrodes as expected from the spin-dependent tunneling.   The maximum junction magnetoresistance (JMR) of 30 % is observed below 300 Oe at low temperatures. Comparison between ramp-edge devices with and without a STO barrier layer demonstrates that the transport characteristics in LSMO/STO/LSMO junction devices is dominated by the barrier and the large JMR is from the spin-dependent tunneling.

Acknowledgments

This work is supported as a Los Alamos National Laboratory Directed Research and Development Project under United States Department of Energy.

REFERENCES

[1] J. S. Moodera and L. R. Kinder, J. App. Phys. **79**, 4724 (1996).
[2] M. Juliere, Phys. Lett. **54A**, 225 (1975).
[3] H. Y. Hwang, S.-W. Cheong, N. P. Ong, and B. Batlogg, Phys. Rev. Lett. **77**, 2041 (1996).
[4] J. H. Park, E. Vescovo, H.-J. Kim, C. Kwon, R. Ramesh, and T. Venkatesan, Nature (1998).
[5] J. Z. Sun, W. J. Gallagher, P. R. Duncombe, L. Krusin-Elbaum, R. A. Altman, A. Gupta, Y. Lu, G. Q. Gong, and G. Xiao, Appl. Phys. Lett. **69**, 3266 (1996); J. Z. Sun, L. Krusin-Elbaum, P. R. Duncombe, A. Gupta, and R. B. Laibowitz, Appl. Phys. Lett. **70**, 1769 (1997).
[6] A. Urushibara, Y. Moritomo, T. Arima, A. Asamitsu, G. Kido, and Y. Tokura, Phys. Rev. B **51**, 14103 (1995).
[7] A. Gupta, G. Q. Gong, G. Xiao, P. R. Duncombe, P. Lecoeur, P. Trouilloud, Y. Y. Wang, V. P. Dravid, and J. Z. Sun, Phys. Rev. B **54**, R15629 (1996).

FABRICATION OF $La_{0.7}Sr_{0.3}MnO_3/La_{0.5}Sr_{0.5}CoO_3/La_{0.7}Sr_{0.3}MnO_3$ HETEROSTRUCTURES FOR SPIN VALVE APPLICATIONS

M. C. Robson[*], S. B. Ogale[*], R. Godfrey[*], T. Venkatesan[*,**], M. Johnson[***], and R. Ramesh[*,****]

[*] Center for Superconductivity Research, Department of Physics, University of Maryland, College Park, MD 20742

[**] also Department of Electrical Engineering, University of Maryland, College Park, MD 20742

[***] Naval Research Laboratory, Washington, D.C. 20375

[****] also Department of Nuclear and Materials Engineering, University of Maryland, College Park, MD 20742

ABSTRACT

Epitaxial growth of oxide heterostructures, which may be utilized in spin valve applications, has been demonstrated. The heterostructures consist of two ferromagnetic layers separated by a non-magnetic metallic interlayer. The ferromagnetic material used is the manganese perovskite oxide, $La_{0.7}Sr_{0.3}MnO_3$, while the metallic oxide interlayer is $La_{0.5}Sr_{0.5}CoO_3$. X-ray diffraction spectra demonstrate the high structural quality of the heterostructures. The magnetization of the heterostructure as a function of magnetic field measured at room temperature yields a double hysteresis loop that is characteristic of this type of spin valve structure. The behavior of this double hysteresis loop is also examined as a function of the metallic interlayer thickness.

INTRODUCTION

During the last fifty years, oxide materials have been extensively studied due to their exhibition of superconductivity, ferroelectricity, and colossal magnetoresistance. These oxides offer future advances in random access memory [1], hard disk-drive read heads [2-3], infrared detectors [4], and active microwave components. The oxides that exhibit colossal magnetoresistance are of the form $A_{1-x}B_xMnO_3$, where A is a trivalent ion (La, Nd, Pr) and B is a divalent ion (Ba, Ca, Sr). The doping of the trivalent ion site with a divalent ion causes the Mn to shift its valence state from Mn^{3+} to Mn^{4+}. This mixed valence state of the Mn ions allows for electron transfer through double exchange [5-6]. It has been suggested that colossal magnetoresistance occurs as a result of this double exchange mechanism coupled with Jahn-Teller distortions [7]. However, large magnetoresistance values have only been reported for low temperatures and high magnetic fields [8-10]. Room temperature magnetoresistance is only about 2% at 1000 Oe for single layer films [11]. Artificially layered structures based on conventional ferromagnetic metals, such as the GMR spin valves, have shown 20 % magnetoresistance at 50 Oe and room temperature [12]. However, the spin dependent scattering in these structures is relativley low due to the relatively small net spin polarization of the ferromagnetic metals. In the colossal magnetoresistance materials, the itinerant electrons are expected to be nearly 100 % polarized at low temperatures [13], so that the spin dependent scattering, and therefore the magnetoresistance, are expected to be higher. Since the growth of heterostuctures using conventional ferromagnetic metals has been shown to be useful, the growth of artificial multilayers utilizing the colossal magnetoresistance materials may provide interesting results with respect to the issues of magnetization coupling and transport.

In this paper, we report the successful epitaxial growth of an oxide trilayer heterostructure (an oxide spin valve) consisting of two ferromagnetic layers separated by a non-magnetic metallic interlayer. The ferromagnetic layers are from the family of manganese-based oxides, $La_{0.7}Sr_{0.3}MnO_3$ (LSMO), and the metallic interlayer is $La_{0.5}Sr_{0.5}CoO_3$ (LSCO). X-ray diffraction spectra are presented in order to prove the high structural quality of the heterostructure. Furthermore, magnetic hysteresis loops (M vs. H) display the double loop which is essential to the operation of a spin valve device. The dependence of this double hysteresis loop on the interlayer thickness is also examined.

Mat. Res. Soc. Symp. Proc. Vol. 494 © 1998 Materials Research Society

EXPERIMENT

The heterostructures have been fabricated using pulsed laser deposition. They consist of LSMO / LSCO / LSMO grown on (00l) oriented single crystal LaAlO$_3$ (LAO) substrate. The laser energy density on the target was ~2 J/cm^2. The first LSMO layer was grown at 800 °C after a 20 °C/min ramp-up from room temperature with an oxygen pressure of 400 mTorr. Then, the temperature was ramped down to 670 °C at 10 °C/min with the pressure maintained at 400 mTorr. Heterostructures were also made with the pressure maintained at 760 Torr during this cool down, but the properties reported in this paper did not change with this variation in pressure. At 670 °C LSCO and the top LSMO layer were grown at 100 mTorr and 400 mTorr, respectively. The pressure was increased to 760 Torr for the ramp down (at 10 °C/min) to room temperature. No post annealing was performed on the heterostructures. The thickness of the top and bottom LSMO layers was 750 Å, while the thickness of the LSCO interlayer was varied from 30 Å to 1100 Å. The crystal structure and orientation of the films were analyzed by means of a Siemens four circle x-ray diffractometer. The resistivity was measured as a function of temperature using a standard four probe configuration. Hysteresis loops were measured at room temperature using a vibrating sample magnetometer.

RESULTS

Spin valve structures require the presence of two independent ferromagnetic layers, which display different coercive fields, separated by a non-magnetic metallic layer. These different coercive fields give rise to a range of applied magnetic field (between the two coercive fields) where the two ferromagnetic layers are aligned anti-parallel. Outside of this unique magnetic field range, the two ferromagnetic layers are aligned parallel. The relative orientation of these two layers enhances or inhibits, respectively, the vertical transport of electrons. This leads to a magnetic field dependent resistance of the heterostructure, which can be termed a magnetoresistance. Our strategy for realizing two different coercive fields in the oxide spin valve is to grow the two ferromagnetic layers with different substrate temperatures during deposition, as explained above. The change in substrate temperature causes a change in the coercive field for two reasons. First, the two films will have different defect microstructures, which affects the ability of the spins in the ferromagnet to switch. Second, the proximity of room temperature to the Curie temperature will determine the coercive field observed at room temperature. The coercive field typically decreases when the temperature is increased towards the Curie temperature. Therefore, films with different Curie temperatures will display different coercive fields at room temperature. Fig. 1 shows room temperature hysteresis loops for single layer LSMO films grown at two different substrate

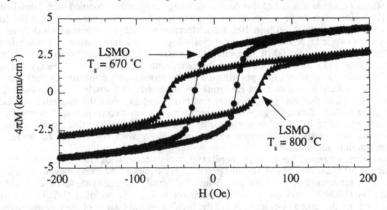

Figure 1. Hysteresis loops for single layer LSMO films grown at 670 °C (●) and 800 °C (▲).

temperatures, 670 °C and 800 °C. The coercive fields of their respective hysteresis loops are 25 Oe and 55 Oe. Therefore, we would expect that these layers would exhibit similar coercive fields when grown in the heterostructure.

A typical x-ray diffraction θ–2θ spectrum for the heterostructure is shown in Fig. 2. The data displayed is for the heterostructure with an interlayer thickness (t_{LSCO}) of 1100 Å. Each layer in the heterostructure is oriented in the (00l) direction. There is no peak for any other phase in the spectrum, which indicates that the heterostructure grows purely single phase. The bulk lattice constants for LSMO and LSCO are 3.889 Å and 3.844 Å, respectively, while LAO has a bulk lattice constant of 3.793 Å. Therefore, the first layer of LSMO experiences a compressive stress in the a- and b- axes, since thin films prefer to grow pseudomorphically if the lattice mismatch is not too large (< 9 %) [14]. The lattice mismatch between LSMO and LAO is only 2.5 %. This coherent growth causes the c-axis to lengthen according to Poison's relation (the volume of the unit cell must be maintained despite any forces that act on the unit cell). The c-axis lattice constant of LSMO in our heterostructure is 3.923 Å. The following LSCO layer has a lattice constant of 3.820 Å, which is smaller than the bulk value. This occurs according to Poison's relation, since in this case there are tensile stresses in the a- and b- axes which cause the c-axis to shorten. These tensile stresses in the a- and b- axes are again caused by the coherent growth of the LSCO with the underlying LSMO. The lattice mismatch between LSMO and LSCO is only 1.1 %. The top LSMO layer again grows pseudomorphically which causes compressive stresses in the a- and b- axes and lengthening of the c-axis to 3.923 Å. The top and bottom LSMO layers may not have exactly identical lattice constants, but they are close enough that within the sensitivity of our x-ray diffractometer we cannot distinguish two separate peaks.

The inset of Fig. 2 shows the rocking curves that were measured around the (002) θ–2θ peaks of the LSMO and LSCO layers of the heterostructure with t_{LSCO} = 1100 Å. The LSMO rocking curve shows a full width half maximum (FWHM) of 0.49°. Typical single layers of LSMO show rocking curves between 0.2 and 0.3°. Our increased FWHM in the heterostructure could be explained by the presence of two slightly different LSMO peaks in the θ–2θ scan (one from the top layer and the other from the bottom layer) which appear merged as one due to the sensitivity limit of the x-ray diffractometer. This would cause the rocking curve to appear unnaturally wide. The LSCO rocking curve shows a FWHM of 0.42°. However, the rocking curve of LSCO is complicated by the proximity of its θ–2θ peak to the substrate θ–2θ peak. The FWHM of the LSMO and LSCO layers are still relatively narrow and display the good crystalline quality of the heterostructures.

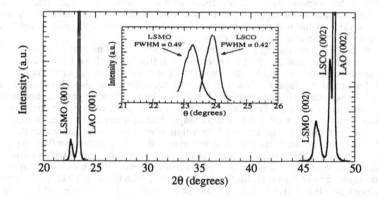

Figure 2.   The θ-2θ x-ray diffraction spectrum from a heterostructure with t_{LSCO} = 1100 Å. (Inset) Rocking curves of the LSCO and LSMO layers from a heterostructure with t_{LSCO} = 1100 Å. The FWHM for these layers is 0.42° and 0.49°, respectively.

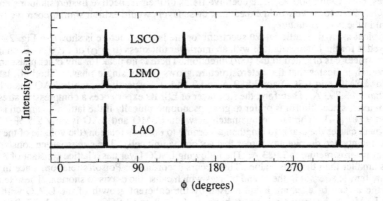

Figure 3. X-ray diffraction φ-scans of the LAO, LSMO, and LSCO layers from a heterostructure with $t_{LSCO} = 1100$ Å.

Figure 3 shows the φ-scans around the (202) peaks of each layer in the heterostructure with $t_{LSCO} = 1100$ Å. The LSCO and LSMO curves have been offset in order to ease viewing of thedata. Each layer shows only four peaks that are separated by 90° from each other. There are no anomalous peaks in the spectrum. This indicates that each layer has in-plane locking of the a- and b- axes. In addition, the layers grow cube on cube. This cube-on-cube growth minimizes thelattice mismatch between the layers. Therefore, our heterostructures show c-axis growth with in-plane locking of the a- and b- axes.

LSMO, which has a bulk resistive transition at T = 380 K, is ferromagnetic at room temperature. Independent resistivity measurements of the top and bottom LSMO layers in the heterostructure show that the magnetic transition for these layers occurs at a temperature greater than 350 K (the temperature limit of our measurements). Furthermore, the LSMO layer grown at 800 °C has a resistivity of 82 μΩ cm at room temperature and 5.2 μΩ cm at 10 K, while the layer grown at 670 °C has a resistivity of 415 μΩ cm at room temperature and 35 μΩ cm at 10 K. The resistivity measurement of the LSCO layer shows a resistive transition at 250 K, which coincides with the bulk value of the resistive transition. At room temperature, there is no net magnetic moment for the LSCO layer and the resistivity is 180 μΩ cm.

Therefore, when measuring the magnetization versus applied magnetic field (M vs. H) for the heterostructure, one would expect a hysteresis loop from each of the LSMO layers. These loops would either be independently visible or merged into one, depending on the coercive fields of the two ferromagnetic layers. However, the interlayer thickness also plays a role, since it determines whether the exchange coupling, J, between the two ferromagnets is dominating the behavior of the hysteresis loop. Theoretically, J is a monolayer interaction, but due to defects in the materials there is the possibility for a longer range residual interaction. If the two layers have different coercive fields, but are separated by a thin interlayer, it is possible that the residual coupling between the ferromagnets will cause the layer with the larger coercive field to pin part or all of the other layer. In this case, only one hysteresis loop would be visible with a coercive field equal to the larger coercive field of the two ferromagnetic layers. Figure 4 shows magnetization vs. applied magnetic field (M vs. H) for the heterostructure with $t_{LSCO} = 1100$ Å. Two magnetic hysteresis loops are clearly visible with coercive fields equal to $H_{c1} = 7.8$ Oe and $H_{c2} = 66.3$ Oe. The coercive fields are marked by arrows in the figure. H_{c1} comes from the top LSMO layer and H_{c2} comes from the bottom LSMO layer. This was determined from the hysteresis loops of single layer LSMO films as shown in Fig. 1. Therefore, our technique of growing the two ferromagnetic layers at different substrate temperatures successfully yielded a double hysteresis loop which is necessary for spin valve operation.

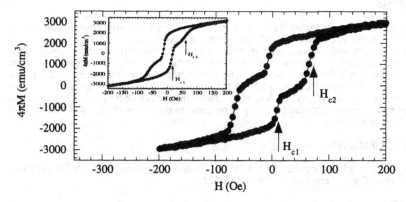

Figure 4. Hysteresis loop for a heterostructure with $t_{LSCO} = 1100$ Å. H_{c1} and H_{c2} are marked by arrows. (Inset) Hysteresis loop for a heterostructure with $t_{LSCO} = 190$ Å. H_{c1} and H_{c2} are marked by arrows.

The examination of the hysteresis loops as a function of the interlayer thickness serves two purposes. The first is to determine the extent of the residual coupling of the two ferromagnetic layers. Secondly and more importantly, is to approximate the spin diffusion length. We speculate that the mean free path in the oxides is ~ 100 Å and that the spin diffusion length can be up to ten times longer than the mean free path. Therefore, we grew films as thick as 1100 Å so that future magnetoresistance measurements on these heterostructures may give us an idea about the length scale of spin diffusion. Fig. 5 shows the LSCO thickness dependence of the coercive fields of the two ferromagnetic layers. H_{c1} and H_{c2} show coercive fields of 10 Oe and 65 Oe, respectively, over the entire range of LSCO thicknesses studied, 30 to 1100 Å. Between 200 and 1100 Å, the two hysteresis loops are distinct, as shown in Fig. 4. However, below 200 Å, the two hysteresis loops are barely identifiable as the transitions become skewed, as shown in the inset of Fig. 4. The reason behind this skewed behavior is still under investigation. However, several possibilities include residual coupling between the two ferromagnetic layers, decreased interfacial quality, and interdiffusion between the various layers.

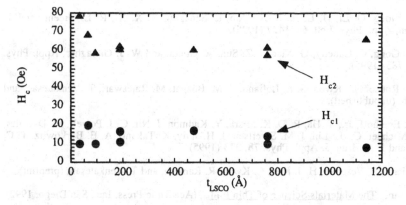

Figure 5. Coercive fields of the LSMO layers (H_{c1} (●) and H_{c2} (▲)) as a function of the LSCO interlayer thickness.

247

CONCLUSIONS

In summary, we have shown the epitaxial growth of oxide heterostructures that display the magnetic behavior that is characteristic of spin valves. X-ray diffraction spectra, including θ-2θ scans, rocking curves, and ϕ-scans, indicate the high crystalline quality of the heterostructures. Also, the magnetization as a function of the magnetic field (M vs. H) displays two sharp independent transitions in the hysteresis loop that originate from the top and bottom LSMO layers. Sharp transitions are evident down to LSCO thicknesses of 200 Å, although two loops are identifiable all the way down from 200 Å to 30 Å. This structure can be further explored as a spin valve device and measurements are presently in progress to determine the magnetoresistance in these oxide heterostructures.

ACKNOWLEDGEMENTS

The present work is supported by DARPA under contract no. N000149610770 and partly by the NSF-MRSEC under contract no. DMR-9632521.

REFERENCES

1. J. F. Scott and C. A. Paz De Araujo, Science **246**, 1400 (1989).

2. K. Derbyshire and E. Korczynski, Solid State Technology, Sept., 57 (1995).

3. J. A. Brug, T. C. Anthony, and J. H. Nickel, MRS Bulletin, Sept., 23 (1996).

4. M. Rajeswari, C. H. Chen, A. Goyal, C. Kwon, M. C. Robson, R. Ramesh, T. Venkatesan, and S. Lakeou, Appl. Phys. Lett. **68**, 3555 (1996).

5. C. Zener, Phys. Rev. **82**, 403 (1951).

6. P. -G. deGennes, Phys. Rev. **118**, 141 (1960).

7. A. J. Millis, Boris I. Shraiman, and R. Mueller, Phys. Rev. Lett. **77**, 175 (1996).

8. S. Jin, T. H. Tiefel, M. McCormack, R. A. Fastnacht, R. Ramesh, and L. H. Chen, Science **264**, 413 (1994).

9. G. C. Xiong, Q. Li, H. L. Ju, S. N. Mao, L. Senapati, X. X. Xi, R. L. Greene, and T. Venkatesan, Appl. Phys. Lett. **66**, 1427 (1995).

10. G.-Q. Gong, C. Canedy, G. Xiao, J. Z. Sun, A. Gupta, and W. J. Gallagher, Appl. Phys. Lett. **67**, 1783 (1995).

11. M. C. Robson, C. Kwon, S. E. Lofland, S. M. Bhagat, M. Rajeswari, T. Venkatesan, and R. Ramesh, (unpublished).

12. W. F. Egelhoff, Jr., T. Ha, R. D. K. Misra, Y. Kadmon, J. Nir, C. J. Powell, M. D. Stiles, R. D. McMichael, C. -L. Lin, J. M. Sivertsen, J. H. Judy, K. Takano, A. E. Berkowitz, T. C. Anthony, and J. A. Brug, J. Appl. Phys. **78**, 273 (1995).

13. J. H. Park, E. Vescovo, H.-J. Kim, C. Kwon, R. Ramesh, and T. Venkatesan, (preprint).

14. M. Ohring, The Materials Science of Thin Films, (Academic Press, Inc., San Diego, 1992), Ch. 7.

FABRICATION OF HIGH TEMPERATURE SUPERCONDUCTOR-COLOSSAL MAGNETORESISTOR SPIN INJECTION DEVICES

J. Kim, R. M. Stroud, R. C Y. Auyeung*, C. R. Eddy, D. Koller, M. S. Osofsky, R. J. Soulen Jr., J. S. Horwitz, and D. B. Chrisey

Naval Research Laboratory, Washington, DC 20375

* SFA Inc., Largo, MD 20774

ABSTRACT

Trilayer $YBa_2Cu_3O_{7-\delta}/(SrTiO_3, CeO_2)/La_{0.67}Sr_{0.33}MnO_{3-\delta}$ devices have been fabricated for the study of supercurrent suppression due to the injection of spin-polarized quasiparticle current. The critical current for a $YBa_2Cu_3O_{7-\delta}/100$ Å $SrTiO_3/La_{0.67}Sr_{0.33}MnO_{3-\delta}$ device was found to decrease from 118 mA to 12.6 mA, for an injection current of 60 mA. The effect of film microstructure on the critical current suppression was investigated. Defects in the $SrTiO_3$ and CeO_2 layers were found to control the device properties.

INTRODUCTION

Superconducting transistors have many potential advantages over traditional semiconducting transistors. The primary advantages are virtually loss-free operation, capacity for high current densities and easy integration with existing superconducting electronic components.

To achieve these advantages, a variety of approaches to fabricating superconducting transistors have been taken, including: the superconducting base transistor, dielectric base transistor, vortex flow transistors, electric field effects devices and quasiparticle injection devices. To-date, each approach has met with an insurmountable obstacle to wide-spread application, such as slow speed, or difficulties in reproducibility. The advantages and disadvantages of these superconducting transistor are discussed in a recent review. [1]

The most recent approach to the fabrication of a superconducting transistor is the spin injection device. These devices consist of a high temperature superconductor-chemical barrier-ferromagnet junction. They operate by the suppression of Cooper pair formation in the superconductor by the injection of spin-polarized current from the ferromagnet. The viability of this approach to fabricating a superconducting transistor was suggested by earlier work on the injection of spin-polarized current into normal metals [2], and low temperature superconductors [3]. The suppression of supercurrent due to spin-polarized current injection was first demonstrated for Permalloy/Au/ $YBa_2Cu_3O_{7-\delta}$ junctions [4]. In the most recent studies, all-oxide devices have been investigated: $La_{0.67}Sr_{0.33}MnO_{3-\delta}/La_2CuO_4/DyBa_2Cu_3O_{7-\delta}$ [5], $YBa_2Cu_3O_{7-\delta}/SrTiO_3/La_{0.67}Sr_{0.33}MnO_{3-\delta}$ [6,7] and $Nd_{0.67}Sr_{0.33}MnO_3/LaAlO_3/YBa_2Cu_3O_{7-\delta}$ [8]. The all-oxide devices use ferromagnetic materials, $La_{0.67}Sr_{0.33}MnO_{3-\delta}$ and $Nd_{0.67}Sr_{0.33}MnO_3$, which belong to the class of colossal magnetoresistive (CMR) manganites, which exhibit up to 100% spin-polarized transport current.

A comprehensive understanding of non-equilibrium superconductivity and the operation of spin injection devices is needed before there can be commercially useful spin injection transistors. In this paper, the suppression of supercurrent in $YBa_2Cu_3O_{7-\delta}/(SrTiO_3,CeO_2)/La_{0.67}Sr_{0.33}MnO_{3-\delta}$

trilayer junctions is demonstrated. The fabrication of these devices and the importance of barrier microstructure and injection geometry for optimum supercurrent suppression are discussed.

PROCEDURE

Trilayer $YBa_2Cu_3O_{7-\delta}$ /(SrTiO$_3$,CeO$_2$)/ La$_{0.7}$Sr$_{0.3}$MnO$_{3-\delta}$ junctions (Fig. 1) were fabricated by pulsed laser deposition (PLD) and wet-etch photolithography. La$_{0.67}$Sr$_{0.33}$MnO$_{3-\delta}$ (LSMO) films were deposited through a rectangular shadow mask onto (100) SrTiO$_3$ single-crystal substrates. The deposition conditions were: 2.8 J/cm^2 laser fluence, 750°C substrate temperature and 100 mTorr oxygen atmosphere. After deposition the films were quenched to room temperature in 1 atmosphere of oxygen. The resultant films were 1000Å thick x 2mm x 12.7mm. The insulating barrier and YBa$_2$Cu$_3$O$_7$ (YBCO) films were deposited sequentially through a rectangular shadow mask. The barrier layer was either SrTiO$_3$ (STO) or CeO$_2$ (CO) with a thickness of 50Å or 100Å. The STO barriers were deposited at a substrate temperature of 750°C in a 350 mTorr O$_2$ atmosphere at a laser fluence of 1.5 J/cm^2. The CO barriers were deposited at 850°C, 400mTorr O$_2$ and 3.05 J/cm^2. The 1000Å thick YBCO films 1000Å were deposited at 750°C in 320 mTorr O$_2$ at 1.4 J/cm^2. The YBCO films were patterned into 100μm wide microbridges by wet-etch photolithography with a 1% HCl etch solution, producing a YBCO-LSMO cross with a junction area of 2mm x 100μm.

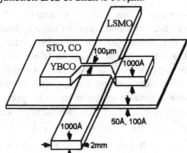

Figure 1. Device Schematic. The spin injection devices are trilayer
YBa$_2$Cu$_3$O$_7$/(CeO$_2$,SrTiO$_3$)/La$_{0.67}$Sr$_{0.33}$MnO$_{3-\delta}$ junctions.

X-ray diffraction (XRD) and room temperature resistivity (ρ_{RT}) were used to characterize each film layer. The superconducting critical temperature (T$_c$), the width of critical temperature (ΔT$_c$), and critical current density (J$_c$) were used to assess the quality of the YBCO film. The transport critical current at 78K as a function of current through the junction was measured to determine the suppression of YBCO supercurrent.

RESULTS

The XRD patterns for a YBCO/50Å STO/LSMO and a YBCO/100Å CO/LSMO device are shown in Fig. 2. The LSMO film shows (100) epitaxial growth on the STO substrate for both devices. The lattice parameter of the LSMO films is 3.85Å. The peak width of an ω scan about the (200) LSMO peak was resolution limited (< 0.16°). For the YBCO/100Å CO/LSMO device, the CO film shows (110) orientation, with the lattice parameter a = 5.41 Å. The CO peak widths indicate that the CO films are not well-ordered. The YBCO films in this device is c-axis oriented, but shows an expanded lattice parameter of 11.70Å. The XRD peaks of the STO film in the YBCO/100Å STO/LSMO device overlap the STO substrate peaks and cannot be resolved. This indicates that the STO film lattice parameters match that of the substrate, a = 3.91Å. The YBCO film in this device shows c-axis growth with a lattice parameter c = 11.68Å.

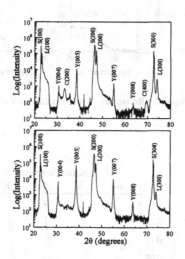

Figure 2. X-ray diffraction patterns for (top) YBCO/100Å CO/LSMO and (bottom) YBCO/100Å STO/LSMO.

Resistivity versus temperature graphs for the two devices are presented in Fig. 3. The resistivity of LSMO in the 100Å CO device was 5.10×10^{-4} Ω-cm at 89.2K. The superconductor critical temperature in this device was 89.2K, with a transition width of $\Delta T_c = 4K$. For the 100Å STO device the LSMO resistivity was 3.75×10^{-4} Ω-cm at 89.9K, T_c = 89.9K, ΔT_c = 1.8K. The resistivity of the CO and STO films in the film plane was found to exceed 1MΩ–cm at room temperature. The resistivity of the STO barrier perpendicular to the film plane was estimated from the current-voltage measurements of the barrier. A maximum resistance was calculated using R= dV/dI (V=0). The resistivity was given by $\rho = R*t*w/l \sim 6$ kΩ-cm.

The current-voltage characteristics of the YBCO films were measured as function of injection current. The suppression of critical current due to injection current YBCO/100 Å STO/LSMO device is shown in Fig. 4. For an injection (I_{inj}) of 60 mAat 78K, the critical current (I_c) is reduced from 118 mA to 12.6 mA. The current gain, $\Delta I_c / \Delta I_{inj}$, was -1.81.

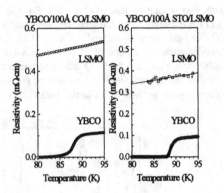

Figure 3. Resistivity vs. temperature of YBCO and LSMO.

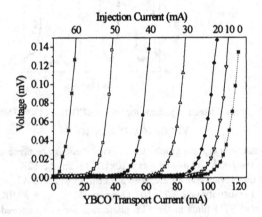

Figure 4 I-V characteristics of YBCO as a function of injection current for the
YBCO\100Å STO\LSMO device at 78K.

DISCUSSION

The operating principle of the spin injection devices is that a non-equilibrium distribution of quasiparticles with a net spin polarization injected into a superconductor can suppress the superconducting order parameter, by inhibiting the dynamic formation of Cooper pairs. The injection current is used to suppress supercurrent, resulting a negative current gain. This is similar to the use of gate current to control source-drain current in a transistor. In principle, the switching time in these devices is the spin-flip scattering time, ~ 10^{-9} sec, which would make them fast enough to be useful for many logic applications. However, the time response of the devices has not yet been measured.

The transport properties of the spin injection devices depend on the device microstructure. The thickness of the insulating barrier and structure of the barrier determine how the current is injected. A uniform barrier thicker than ~100Å injects the current by capacitive coupling. In this case, there is no current flow from the spin-polarized source, so there is no net spin-polarization of the injected current. Thick barriers can directly transmit the current if there are defects in the barrier, such as pinholes, grain boundaries, or particulates. Thin barriers, less ~ 100Å can act as tunnel barriers. If the injected current tunnels into the superconductor, or is transmitted through defects in barrier, the spin polarization will likely be preserved. The maximum supercurrent suppression will occur when the spin polarization is 100%.

The effect of the injected current on the superconductor will depend on the uniformity of the superconductor and whether the injected current reaches a weak link. The critical current of the superconductor is determined by its weakest link. To measure a critical current suppression due to the spin injection, the spin injection suppression must be greater than that at the weakest link, or must occur at the controlling weak link. Metallic barriers between the ferromagnetic film and the superconductor would allow the injected current to "short-circuit" around weak links, or suppressed region of superconductivity, where the resistance was higher. Metallic barriers can only be used in spin injection devices if the superconductor is completely uniform, without weak links. This is not true for the high temperature superconductors, such as YBCO. To ensure that the injection current reaches the existing weak links in YBCO, an insulating barrier is needed. The insulating barrier spreads the injection current over the entire junction, so that no weak links are missed. If the insulating barrier has defects, such as pinholes, there must be a sufficient overlap between the barrier defects and the superconductor weak links, so that the injection reaches the weak links.

The devices used for this experiment had wide LSMO films (2mm) and narrow patterned YBCO (100μm) films. These junctions were designed to test the effects injection through barrier defects into a distribution of weak links. The resistivity of the barrier was much lower in the current-perpendicular to-plane direction than in the current-in-plane direction, 3 kΩ-cm vs. >1 MΩ-cm. This indicates that there were defect current paths through the barrier in the out-of-plane direction. Whether the defects were pinholes, particulates, or grain boundaries has not been determined. The gain of these devices at 78K was -0.46 for the 50Å STO device and -0.61 for the 100Å CO device.

In previous experiments [6,7], gains as high as -35 were reported. The devices used in those experiments had narrow patterned LSMO films and narrow patterned YBCO films, producing a junction area of 140μm x 140μm. The barriers were STO, 100-400Å thick. The difference in gain between the previous devices and those reported here, is in part the elimination of heating, but may also be due to differences in microstructure and the device geometry. Heating in the previous devices resulted from the high resistance of barrier, and higher resistivity of the LSMO (10^{-3} Ω-cm).

The optimum spin injection effect has not yet been observed. More research is needed to understand how the microstructure of the barrier and superconductor films affect the supercurrent suppression, so that the spin injection effect can be controlled. Time-response measurements are also need to help determine the cause of the suppression and for evaluating potential applications of the effect.

CONCLUSION

Supercurrent suppression due to the injection of spin-polarized current in YBCO/(STO,CO)/LSMO trilayer devices has been demonstrated. The microstructure of the films, particularly defects in the barrier, are believed to influence the transport properties of the devices. Compared to previously reported results for small area junctions $(140 \text{ } \mu\text{m})^2$, the results presented here for wide-area junctions, show smaller contributions from heating. However, the transport measurements of the wide-area junction barriers show evidence for non-uniform injection consistent with the presence of pinholes or grain boundary defects in the barrier film. Non-uniform injection was not apparent in measurements of the small area junctions. Continued research is necessary for determining the device geometry and film characteristics that maximize the spin injection effect.

ACKNOWLEDGMENTS

The authors thank Alberto Piqué for helpful discussion.

REFERENCES

1. J. Mannhart, Supercon. Sci. Tech. 9, p.49 (1996).

2. M. Johnson and R. H. Silsbee, Phys. Rev. B 77, p.5326 (1988).

3. M. Johnson, Appl. Phys. Lett. 65, p.1460 (1994).

4. D. B. Chrisey, M. S. Osofsky, J. S. Horwitz, R. J. Soulen Jr., B. Woodfield, J. Byers, G. M. Daly, P. C. Dorsey, J. M. Pond, T. W. Clinton and M. Johnson, IEEE Trans. Appl. Supercon. 7, p.2067 (1997).

5. V. A. Was'ko, V. A. Larkin, P. A. Kraus, K. R. Nikolaev, D. E. Grup, C. A. Nordman and A. M. Goldman, Phys. Rev. Lett. 78, p.1134 (1997).

6. R. M. Stroud, J. Kim, C. R. Eddy, D. B. Chrisey, J. S. Horwitz, D. Koller, M. S. Osofsky, R. J. Soulen Jr. and R. C. Y. Auyeung, J. Appl. Phys,, in press.

7. R. J. Soulen, M. S. Osofsky, D. B. Chrisey, J. S. Horwitz, R. Stroud, J. M. Byers, B. F. Woodfield, G. M. Daly, T. W. Clinton, M. Johnson, R. C. Y. Auyeung, Proc. of 1997 European Conf. on Supercond., in press.

8. Z. W. Dong, R. Ramesh, T. Venkatesan, M. Johnson,. Z. Y. Chen, S. P. Pai, V. Talyanski, R. P. Sharma, R. Shreekala, C. J. Lobb and R. L. Greene, Appl. Phys. Lett. 71, p.1718 (1997).

Part V

Physical Properties of
Metallic Magnetic Oxides

IN-PLANE GRAIN BOUNDARY EFFECTS ON THE TRANSPORT PROPERTIES OF La$_{0.7}$Sr$_{0.3}$MnO$_{3-\delta}$ THIN FILMS

J. Y. Gu[*§], S. B. Ogale[*], K. Ghosh[*], T. Venkatesan[*], R. Ramesh[*], V. Radmilovic[**], U. Dahmen[**], G. Thomas[**], and T. W. Noh[§]

[*]Center for Superconductivity Research, Department of Physics, University of Maryland, College Park, MD 20742, jygu@squid.umd.edu
[**]National Center for Electron Microscopy, Berkeley, CA 94720
[§]Department of Physics, Seoul National University, Seoul 151-742, Korea

ABSTRACT

C-axis oriented La$_{0.7}$Sr$_{0.3}$MnO$_{3-\delta}$ (LSMO) films were fabricated on the top of SrTiO$_3$/YBa$_2$Cu$_3$O$_7$ grown on MgO(001) substrates. From x-ray ϕ-scan and planar transmission electron microscopy measurements, the LSMO layer in the LSMO/SrTiO$_3$/YBa$_2$Cu$_3$O$_7$/MgO heterostructure is found to have coherent in-plane grain boundaries with a predominance of 45° rotations (between [100] and [110] grains) in addition to the cube-on-cube epitaxial relationship. Also, epitaxial LSMO/Bi$_4$Ti$_3$O$_{12}$/LaAlO$_3$ (001) and c-axis textured LSMO/Bi$_4$Ti$_3$O$_{12}$/SiO$_2$/Si(001) with random in-plane grain boundaries are introduced as the counterparts for comparison. The resistivity and magnetoresistance (MR) of LSMO layer were measured and compared in these three different heterostructures. The low field MR at low temperature shows a dramatic dependence on the nature of the grain boundary. An attempt is made to interpret these results on the basis of correlation between the magnetic properties and grain structures.

INTRODUCTION

Colossal magnetoresistance (CMR) in manganese oxide materials is being studied during the last few years not only for a better understanding of these metallic oxides but also to explore the possibility for various novel magnetic devices. However, the requirement of a large applied magnetic field and low temperature for getting a sizable magnetoresistance (MR) value in such systems has posed a hindrance to their applicability. In some recent studies[1-6] however, it has been demonstrated that the enhancement of MR at low fields could be achieved by controlling the crystallinity and defect constitution in CMR materials. The CMR effects in homogeneous crystalline manganites are most pronounced in the vicinity of the magnetic transition temperature[7]. However, local disturbance of the magnetic order at any temperature has a potential to alter the electrical conduction and induce a magnetoresistive effect. For example, defects such as grain boundaries have a strong perturbing effect on the local magnetization and can lead to an enhanced and controllable spin-dependent scattering[8]. The role of such internal interfaces and defect states is expected to be all the more important in CMR systems wherein the transport and magnetization are so intimately coupled[9]. The same interplay of magnetism and transport however complicates the analysis of experimental results especially when the interfaces and defects are included. This situation clearly calls for more controlled experiments on the subject wherein the departures from crystallinity are tailor-made and their consequences are examined.

In this work, we have made special grain boundaries in the CMR layer using the coincidence site lattice (CSL) type grain boundaries of YBa$_2$Cu$_3$O$_7$ (YBCO). Coherent in-plane grain boundaries were introduced in the La$_{0.7}$Sr$_{0.3}$MnO$_{3-\delta}$ (LSMO) layer by growing it on a SrTiO$_3$(STO)/YBa$_2$Cu$_3$O$_7$/MgO surface. The large in-plane lattice mismatch (8.8 %) between MgO and YBCO leads to the formation of CSL type in-plane grain boundaries[10]. In addition, to investigate the mechanism of the grain boundary effects on the transport properties, such as the resistivity and MR, we have also prepared other heterostructures with different in-plane grain

257

structures. One is LSMO/ $Bi_4Ti_3O_{12}(BTO)/SiO_2/Si(001)$ (LSMO(SiO_2/Si)) and the other is LSMO/BTO/LaAlO$_3$(001) (LSMO(LAO)) heterostructures. In our earlier paper,[4] we have reported that the low field MR sensitivity was significantly improved with in-plane grain boundaries of LSMO(SiO_2/Si) thin film. Here, LSMO(LAO) is introduced as an epitaxial counterpart for the comparison with thin films having in-plane grain boundaries. Thus, the main difference among these three different heterostructures is the degree of in-plane structural coherence.

EXPERIMENTS

X-ray θ-2θ and φ-scan measurements were used to determine the crystalline orientation and in-plane alignment. φ-scan measurement was performed on a Siemens four circle x-ray diffractometer. A standard four-probe method was used to measure the resistivity of the film. The dc magnetization was measured with a superconducting quantum interference device(SQUID) magnetometer. The low field magnetoresistance measurement was performed with a vibrating sample magnetometer. High-resolution electron microscopy (HREM) was carried out in the Berkeley atomic resolution microscope. HREM images were obtained under the conditions close to Scherzer defocus.

We deposited about 700 Å thick YBCO layer on MgO(001) single crystal substrate at 800 °C and 100 mTorr of oxygen by *in-situ* pulsed laser deposition. The pulse repetition rate was 5 Hz and the fluence was ~ 2.0 J/cm^2. Immediately following the YBCO deposition, the temperature was lowered to 650 °C and the oxygen pressure was maintained at the same value to deposit the STO layer on top of the YBCO film, where STO is used for insulation between the YBCO and LSMO layers. After the STO layer deposition, LSMO layer was grown at 700 °C and 400 mTorr oxygen atmosphere. The film was then slowly (5°C/min) cooled down to room temperature in 1 atmosphere of O$_2$. X-ray θ-2θ diffraction pattern of the LSMO/STO/YBCO/MgO(001) (LSMO(MgO)) heterostructure shows only the (00l) peaks of LSMO, STO, and YBCO indicating that the entire film on the MgO substrate has the c-axis normal to the substrate surface, *i.e.*, [001]$_{LSMO}$//[001]$_{STO}$//[001]$_{YBCO}$//[001]$_{MgO}$. The deposition process and x-ray θ-2θ diffraction patterns for two counterpart heterostructures of LSMO/BTO/LaAlO$_3$ and LSMO/BTO/SiO$_2$/Si were reported in earlier paper[4].

RESULTS

Figures 1 (a), (b), and (c) show the x-ray φ-scan spectrum of the LSMO{101} family of peaks in the three different heterostructures. LSMO(LAO) in Fig. 1 (a) shows only one set of peaks

(a)   (b)   (c)

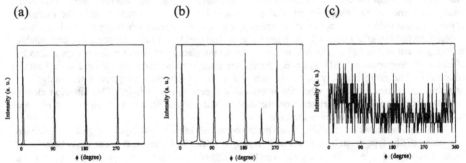

FIG. 1. Off axis x-ray φ-scans of the {101} family of peaks of LSMO in the (a) LSMO(LAO), (b) LSMO(MgO), and (c) LSMO(SiO$_2$/Si) heterostructures

spaced 90° apart indicating that the in-plane orientation relationship is $[100]_{LSMO}//[100]_{LaAlO_3}$. On the other hand, as shown in Fig. 1 (b), the existence of two principal orientation relationships was observed in the LSMO(MgO). In addition to the peaks of Fig. 1 (a), there are four peaks showing a rotation of 45° with respect to the grains with the primary orientation. These peaks correspond to the in-plane crystallographic relationship of $[110]_{LSMO}//[100]_{MgO}$. Ramesh *et al*,[10] reported that the thin film of YBCO grows with the *c*-axis normal to the single crystal MgO(001) substrate and the *a-b* axes lock into several preferred orientations. In our LSMO(MgO), the LSMO/STO layers are expected to follow the in-plane orientation of the underlying YBCO layer. This is also confirmed from the HREM images. Finally, in Fig. 1 (c), the LSMO(SiO₂/Si) is found to have no in-plane structural coherency.

Figure 2 shows an electron diffractogram and a planar section high resolution transmission electron micrograph (HRTEM) of the top LSMO layer in the LSMO(MgO). The electron diffractogram of Fig. 2 (a) shows a 8-fold symmetry diffraction pattern indicating that two neighboring grains in some region are rotated relative to each other by 45°. This 45° rotation is a well-known CSL rotation in cubic crystals. The transmission electron microscopy studies indicate that this 45° rotation is the primary type of CSL in these films, which is coincident with the φ-scan result, although other low-σ CSL's have also been observed. Of interest is the fact that these grain boundaries are structurally semi-coherent and the consequent absence of disorder and impurities that is commonly observed at grain boundaries. This coherent grain boundary feature can be also seen from the HRTEM image of Fig. 2 (b). Therefore, it can be said that this CSL type in-plane grain structure in the LSMO(MgO) corresponds to an intermediate state between in-plane epitaxial and in-plane random grain configurations by considering the degree of in-plane structural coherency.

In Fig. 3, the temperature dependence of the resistivity, ρ(T), of the LSMO layer in each heterostructure is compared. ρ(T) of the LSMO(MgO) is similar to the case of the epitaxial LSMO film even though the resistivity shows a more gentle slope and a large low temperature remnant resistivity due to the grain boundary effects. The coherent grain boundaries in LSMO(MgO) seem not to affect too much on the resistivity of LSMO layer compared to the epitaxial one. On the other hand, the LSMO(SiO₂/Si) shows several orders of magnitude larger resistivity and also a large difference between the resistivity peak temperature, T_p, and the Curie temperature, T_c. Based on the consideration of the difference in the magnetic configurations between the interface (grain boundary) and the bulk (core of each grain) regions, we can expect that the resistivity should decrease with

(a)   (b)

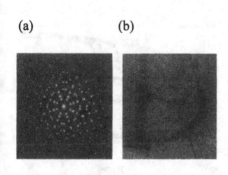

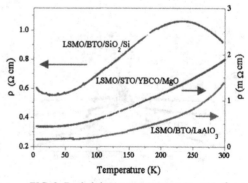

FIG. 2. Electron diffraction pattern and planar HRTEM image of LSMO layer grown on STO/YBCO/MgO.

FIG. 3. Resistivity vs. temperature curves of LSMO layer in the LSMO(LAO), LSMO (MgO), and LSMO(SiO₂/Si) heterostructures.

increasing temperature when the interface becomes paramagnetic, while the bulk phase within a grain is still in the ferromagnetic state. As a result, the resistivity peak appears at a lower temperature than the intrinsic Curie temperature of the bulk LSMO.

The magnetic field dependencies of the magnetoresistance, MR (%) = {[R(H)-R(0)]/R(0)} × 100, of the LSMO layer in the three heterostructures were measured at low fields. Figure 4 shows the MR vs. H curves measured at 77 K. The MR values at 77 K show a dramatic difference depending on the in-plane grain characteristics. Epitaxial LSMO(LAO) does not show any significant MR value. However, LSMO films with in-plane grain boundaries show the magnetic hysteresis behavior as well as enhanced MR values at the low field region. Especially, the resistance changes by ~ -14 % with respect to the maximum resistance value at ± 2000 Oe in the LSMO(SiO₂/Si). Of another interest is the fact that the samples with in-plane grain boundaries show the increase of the resistance from the value of zero magnetic field which is never observed in the epitaxial sample.

The magnetic hysteresis loops of the heterostructures were measured at 77 K to investigate the correlation between the MR behavior and the magnetization process at low field region. $4\pi M$ vs. H curves for each heterostructure are shown in Fig. 5. Hysteresis loops of LSMO(LAO) and LSMO(MgO) have almost the same coercive field and saturated $4\pi M$ value, however the curve of LSMO(SiO₂/Si) shows a much higher coercive field and a lower $4\pi M$ value.

To check the correlation between MR behavior and magnetization process MR vs. H and $4\pi M$ vs. H curves for the LSMO(MgO) sample are drawn at the same time in Fig. 6. It can be noticed that the point of maximum resistance in the MR curve is coincident with the point of zero-magnetization in the magnetic hysteresis loop. The zero-magnetization is the result of the cancellation of effective total magnetization from the complete random distribution of the magnetization direction of each magnetic domain. Therefore in this configuration, the resistance can be maximized due to a large scattering of the carriers through the randomly oriented magnetic domains. The resistance change at low field region is known to be correlated with the magnetic hysteresis loop that is determined by the magnetic domain rotation motion of this material[4]. From the base of this correlation observed in Fig. 6, we propose a model to explain the magnetotransport behavior in the film having in-plane grains. Without any external magnetic field, the magnetization direction of the each grain is randomly oriented even if it is ferromagnetic within the grain as shown in Fig. 7 (a). Also, the individual magnetic spins at the grain boundary region are randomly oriented because usually the grain boundary region is structurally defective. By applying a low magnetic field, the magnetization of the each grain starts to align toward the direction of the external magnetic field

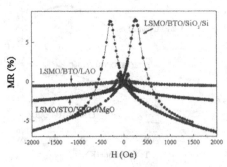

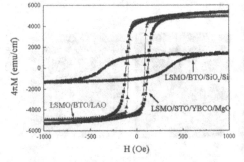

FIG. 4. Low field MR of LSMO layer in the LSMO (LAO), LSMO(MgO), and LSMO(SiO₂/Si) hetero-structures measured at 77K.

FIG. 5. Magnetic hysteresis loops in the LSMO (LAO), LSMO(MgO), and LSMO(SiO₂/Si) heterostructures measured at 77 K.

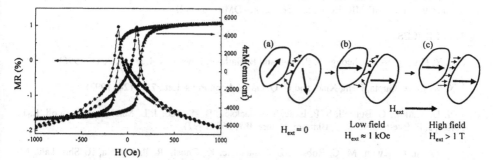

FIG. 6. Low field MR and hysteresis loop of LSMO in the LSMO(MgO) heterostructure measured at 77 K.

FIG. 7. A schematic model for the magneto-transport behavior of the film with grains.

as shown in Fig. 7 (b) and the resistance decreases due to the reduction of the scattering between the grains. However, a larger magnetic field is necessary to align the individual spins of the grain boundary region to the external magnetic field direction. The resistance change by the spin alignments at the grain boundary region will be related to the gentle slope of the MR curve at relatively high field region.

On the other hand, the difference of MR values between LSMO(SiO$_2$/Si) and LSMO(MgO) can be explained from the comparison of their grain characteristics. First, the magnetic correlation between the neighboring grains is much weaker in the incoherent grain configurations observed in LSMO(SiO$_2$/Si) as compared to the coherent grain structures in LSMO(MgO) since the incoherent grain boundaries are more defective. Therefore, the initial state of LSMO(SiO$_2$/Si) with no magnetic field has a large resistance due to no magnetic correlation between neighboring grains. However, in the coherent grain structures the neighboring grains have some correlation in their magnetic states since the magnetic and structure properties of this material are known to be highly correlated each other[9]. Indeed as shown in Fig. 3, the resistance of LSMO(SiO$_2$/Si) at zero magnetic field is several orders of magnitude larger than that of LSMO(MgO). Second, from the comparison of grain size LSMO(SiO$_2$/Si) gives more effective grain boundary regions rather than LSMO(MgO). At this point, the effect of defects on the magnetotransport in CMR materials have not been understood clearly. However, in our result, it can be pointed out that the magnetotransport properties of CMR manganite thin films are highly affected by the incoherence of the grain boundary region.

CONCLUSIONS

In conclusion, c-axis oriented La$_{0.7}$Sr$_{0.3}$MnO$_{3-\delta}$ films with coherent in-plane grain boundaries were fabricated on top of the SrTiO$_3$/YBa$_2$Cu$_3$O$_7$/MgO. Also, for comparison two different type of heterostructures, LSMO(LAO) and LSMO(SiO$_2$/Si), are prepared. Transport properties of LSMO layer were investigated and compared in the three different heterostructures. The low field MR at low temperature show a dramatic difference according to the grain boundary characteristics. The MR behavior of the sample with in-plane grain boundaries follows the magnetic hysteresis loop at low field region indicating that these two phenomena are closely correlated. Also, it is shown that the low field MR value can be improved using grain boundary effects.

ACKNOWLEDGMENTS

The present work was supported by the Office Naval Research under ONR-N000149510547 and partly by the NSF-MRSEC under contract no. DMR-9632521.

REFERENCES

1. H. Y. Hwang, S-W. Cheong, N. P. Ong, and B. Batlogg, Phys. Rev. Lett. **77**, 2041 (1996).

2. X. W. Li, A. Gupta, Gang Xiao, and G. Q. Gong, Appl. Phys. Lett. **71**, 1124 (1997).

3. N. D. Mathur, G. Burnell, S. P. Isaac, T. J. Jackson, B. -S. Teo, J. L. MacManus-Driscoll, L. F. Cohen, J. E. Evetts, and M. G. Blamire, Nature **387**, 266 (1997).

4. J. Y. Gu, C. Kwon, M. C. Robson, Z. Trajanovic, K. Ghosh, R. P. Sharma, R. Shreekala, M. Rajeswari, T. Venkatesan, R. Ramesh, and T. W. Noh, Appl. Phys. Lett. **70**, 1763 (1997).

5. A. Gupta, G. Q. Gong, Gang Xiao, P. R. Duncombe, P. Lecoeur, P. Trouilloud, Y. Y. Wang, V. P. Dravid, and J. Z. Sun, Phys. Rev. B **54**, 15629 (1996).

6. Ning Zhang, Weiping Ding, Wei Zhong, Dingyu Xing, and Youwei Du, Phys. Rev. B **56**, 8138 (1997).

7. C. Zener, Phys. Rev. **82**, 403 (1951).

8. K. Steenbeck, T. Eick, K. Kirsch, K. O'Donell, and E. Strinbeiβ, Appl. Phys. Lett. **71**, 968 (1997).

9. M. F. Hundley, M. Hawley, R. H. Heffner, Q. X. Jia, J. J. Neumeier, J. Tesmer, J. D. Thompson, and X. D. Wu, Appl. Phys. Lett. **67**, 860 (1995).

10. R. Ramesh, D. M. Hwang, T. S. Ravi, A. Inam, J. B. Barner, L. Nazar, S. W. Chan, C. Y. Chen, B. Dutta, and T. Venkatesan, Appl. Phys. Lett. **56**, 2243 (1990).

OBSERVATION OF GROWTH-RELATED MAGNETIC STRUCTURES IN $La_{0.67}Sr_{0.33}MnO_3$

M.E. HAWLEY, G.W. BROWN, AND C. KWON

Los Alamos National Laboratory, Los Alamos, NM 87545, USA, hawley@lanl.gov

ABSTRACT

Ambient observation of magnetic structures by magnetic force microscopy (MFM) in $La_{0.67}Sr_{0.33}MnO_3$ films has not yet been clearly correlated with stresses induced by the kinetic or thermodynamic growth processes or the compressive ($LaAlO_3$) or tensile ($SrTiO_3$) nature of the substrate lattice-mismatch. Although domain-like magnetic structures have been seen in some as-grown films on LAO and related to substrate-induced stress and film thickness, no magnetic structure has been seen for films on STO and other films grown under different kinetic conditions on LAO. In this study we have identified a set of pulsed-laser deposition conditions with the substrate temperature as a variable to determine the relationship between growth and stress-induced magnetic structures. Results from scanning tunneling, atomic force, and MFM microscopies, magnetization, and coercivity measurements will be presented.

INTRODUCTION

As industry evaluates new materials for utilization in next generation magnetoresistive devices, an issue that must be addressed is the presence of native and process-induced magnetic structure in the thin film candidates and the effects this has on their magnetoresistance (MR) properties. In particular, stress-induced magnetic structure will be very important since these materials will probably be used in thin film form, must be grown in ways compatible with other processing techniques (i.e. lower temperatures) and since they have relatively large magnetostriction constants ($\sim70^{-4}$). Stress in thin films can arise from substrate/film lattice mismatch, defects induced by substrate or growth details, growth mode as determined by deposition parameters, post growth cool down and annealing procedures, etc. A previous study [7] has examined the thickness dependence of surface magnetic structure on $La_{0.67}Sr_{0.33}MnO_3$ (LSMO) films grown on $LaAlO_3$ (LAO) and $SrTiO_3$ (STO) under a single set of deposition parameters and observed substrate dependent magnetic structure within a range of thicknesses. The lack of such a structure on thinner and thicker films with presumably the same c-axis lattice parameters as measured by x-ray suggests that the microscopic details of the growth may play more of a role than the lattice parameter. In addition, that study and another one examining LSMO on STO [2] obtained contradictory results for in-plane magnetic anisotropy which may lie in differences in the details of the growth. Our work was aimed at examining the effect of substrate-film lattice-mismatch and deposition conditions on the magnetic structure of LSMO films grown on LAO and STO using MFM, magnetization, and coercivity measurements. These observations are related to the physical structure observed by scanning probe microscopies.

EXPERIMENT

LSMO films were deposited on LAO and STO substrates at temperatures (T_s) of 550 °C, 650 °C, 700 °C, 750 °C, and 800 °C using a XeCl excimer laser ($\lambda = 308$ nm) incident on a stoichiometric target. Pulse repetition rate, width, and energy were 70 Hz, 20 ns, and 2 J/cm^2 respectively; depositions were carried out in 400 mTorr of oxygen. After deposition, samples were allowed to cool while the chamber with 7 atmosphere oxygen. No thermal treatment was performed on the samples after the deposition. The thickness of some of the films was measured

263

post-deposition. The crystallinity and out-of-plan lattice constant (c-axis lattice constant) were characterized by a Siemens four circle x-ray diffractometer. Hysteresis loops were obtained at room temperature (RT) by a vibrating sample magnetometer (VSM). The temperature dependent magnetization was measured by a Quantum Design SQUID magnetometer. For VSM and SQUID measurements, the field was applied along the in-plane direction of the sample.

The microstructure of the films was characterized by STM in air at ~7.4 V tip bias and 740 pA tunnel current. In-plane magnetization was measured with a squid magnetometer with 700 Oe applied field. Film coercivity was measured using a VSM. The c-axis lattice parameter of each film was measured by x-ray diffraction. Because of variations in deposition rate that depended strongly on the substrate position on the PLD sample holder chuck and changes in the laser output power, film thickness was determined after deposition by etching off part of the LSMO with a mixture of HNO_3 and HCl and measuring the step height with a Dektak profilometer.

MFM force gradient measurements using frequency shift detection were carried out with high coercivity cobalt-chrome coated Si tips. This is a lift mode used to minimize tip-sample interactions dominated by van der Waals forces in tapping mode. The MFM measurements provided us with qualitative measurements of the magnetic domain structure, relative strengths and orientations. Only relative qualitative information can be obtained since the tip's moment and geometry are not precisely known.

RESULTS

We observed significant variation in film thickness and properties with location on the sample holder and laser power. This was partly responsible for a variation in measured properties. A summary of x-ray, film thickness, Curie temperature (T_C), and coercive field (H_C) measurements for LSMO/LAO films are shown in Table I. As expected from the compressive substrate-film lateral lattice-mismatch, virtually all films grown on LAO had c-axis lattice parameters > the bulk value, 3.876 Å [3]; above a critical thickness, lattice-mismatch induced stress is expected to disappear, relaxing to the bulk value. However, elongation of the c-axis persisted up to ~770 nm thick films, probably due to stress induced by the details of the growth process. In contrast, STO films had c-axes < the bulk as expected from the tensile nature of the lattice-mismatch.

Table I. C-axis parameter, thickness, and Tc for films grown on LAO.

T_s	550 °C	650 °C	700 °C	750 °C	800 °C 3 min.	800 °C 5 min.	800 °C 5 min.
c-axis (Å)	3.9426	3.8750	3.8977	3.9736	3.8962	3.9122	3.9110
FWHM (degrees)	0.296	0.3362	0.2986	---	0.6393	0.5983	0.4332
% change	+7.72	-0.03	+0.39	+0.97	+0.52	+0.93	+0.90
Thickness (nm)	~ 60	~ 60	~ 60	~ 60	~130	~170	~170
$T_C \pm 5$ (°K)	325*	335**	345	355	355	355	---
H_C (Oe)	0	~2	~70	~30	---	~40	---

LSMO 3.876, *weak, **broad transition, --- not measured.

STM images of the microstructure of the LAO grown films as a function of temperature are shown in Figure 1. Films grown on STO (not shown) were qualitatively the same. The 550 °C films on both substrates consisted of large clusters (~40nm) with of small subgrains (< 5nm). The c-axis parameter suggests that the films are crystalline and somewhat epitaxial normal to the substrate surface. The 650 °C films were more island-like with evidence of layering within the

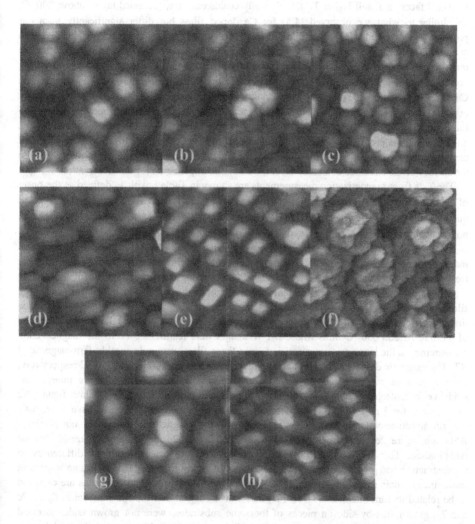

Figure 1. 200 nm x 200 nm STM images of films grown on LAO: (a) 550 °C (c = 3.9426 Å), (b) 650 °C (c = 3.8750 Å), (c) 700 °C (c= 3.8911 Å), (d) 750 °C (c = 3.9136 Å), (e) 800 °C - 3 min. (c = 3.8962 Å), (f) 800 °C -3 min. (c = 4.003 Å and 3.9156 Å), (g) 800 °C - 5 min. (c = 3.9122 Å), and h) 800 °C - 5 min. (c = 3.9110 Å).

islands. Although the island size didn't increase at 700 °C, the islands were more obviously layered and faceted. The islands in the 750 °C films were larger, even more faceted and layered in appearance. The T_s-dependent growth trends, i.e. the onset of epitaxy at about 650 °C followed by island faceting at still higher T_s's, and finally coalescence into extended layers above 800 °C, are similar to what we observed [4,5] for Ca doped films but differ significantly from our previous LSMO study [6] where films were grown in 200 mTorr oxygen at 5 Hz, the same conditions used for the Ca films. The evolution in structure was consistent with increased surface mobility with increasing T_s. The LSMO films, as-grown and annealed, from our previous study, however, had anomalously low ferromagnetic (FM) transition temperatures while our annealed Ca films, irrespective of T_s, had T_c's around 270 °K, comparable to the best results obtained by others. Although our Ca and Sr results appear inconsistent, they may be attributed to changes in the deposition systems used in these studies. As mentioned above, the microstructure of the STO-based films progressed virtually identically to the LAO films up to 750 °C. An unusual deviation occurred at 800 °C where square surface grains grew on grain boundaries rather than on the surface of the extended coalesced layers; we have no explanation for this observation.

The T_c's from magnetization measurements are shown in Table I, with T_c defined as the onset of the FM transition. All the films were ferromagnetic at RT, T_c increasing with improved crystal quality and epitaxy. The coercive field (H_c) measured at RT increased with increasing T_s; it was nonexistent for the 550 °C film increasing to ~ 40 Oe for the 800 °C film.

Since all the films in this study were ferromagnetic at RT, the MFM was used to reveal magnetic domain structure present. Due to the long range nature of magnetic forces, the magnetic tip, scanned ~40 nm above the surface, is capable of sensing the stray field out of the sample resulting from any magnetic structure present throughout the volume of the film down to the film-substrate interface. Figure 2 shows the magnetic structure observed with the MFM probe tip magnetized normal to the film surface for the same films shown in figure 1 with the exception of the 550 °C and 650 °C films whose MFM images were featureless. From the magnetization measurements, the latter was not surprising since these films were only weakly ferromagnetic at RT. The magnetic domain structures shown in figures 2a and 2b, 700 °C and 750 °C respectively, are very diffuse and broad, also typical of all films grown on STO, which we interpret as evidence for isotropic in-plane magnetization. Finally, three of the four 800 °C films, figures 2(c to f), grown for 3 min. (d and f) and 5 min. (c and e) as pairs, can be seen to have a magnetic domain structure with at least a large portion of the magnetization oriented out-of-plane. Although figure 2e can be described as maze-like, figures 2c and 2f are a maze-like/bubble combination. The subtle differences in magnetic structure are a result of differences in microstructure and growth or cool down processes, not exposure to magnetic fields, an important issue since domain structures are sensitive to magnetic history. The domain widths are expected to be related to film thickness [7]. If true, that would suggest that the films shown in figure 7e and 7g, grown side-by-side on pieces of the same substrates, were not grown under identical conditions, that another unknown growth variable was operative. The same can be said for the films in figure 7f and 7h, also grown side-by-side and leading to the absence of out-of-plane magnetic structure for one of the films, figure 2d, in this case probably leading to a lower T_c.

CONCLUSIONS

The evolution in microstructure was a predicable function of temperature, crystal quality and epitaxy improving with increasing temperature. The variations in c-axis parameter with substrate

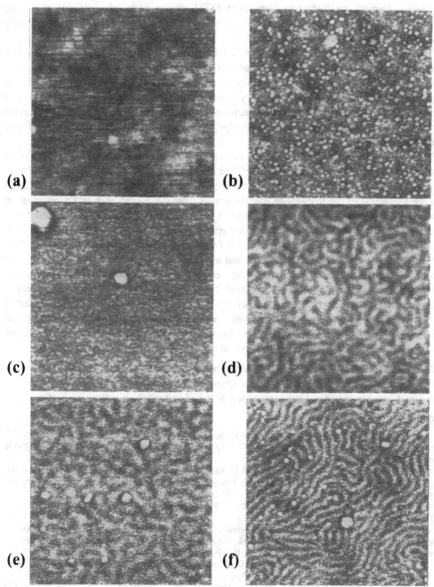

Figure 2. 4 μm x 4 μm MFM images of the magnetic structure of films grown on LAO: a) 700 °C (c = 3.8911 Å), b) 750 °C (c = 3.9136 Å), (c) 800 °C -3 min. (c = 4.003 A and 3.9156 Å), (d) 800 °C - 3 min. (c = 3.8962 Å), (e) 800 °C - 5 min. (c = 3.9122 Å), and f) 800 °C - 5 min. (c = 3.9110 Å)

lattice-mismatch, while consistent with the our notions about stress-induced elongation (LAO compressive in-plane mismatch) or contraction (STO tensile mismatch), did not show the predicted relaxation to the bulk value with thickness, suggesting that other not understood growth or post-deposition conditions, such as cool down rate, are preventing this relaxation.

The T_c's for these films also followed a predictable trend, T_c increasing with film quality. The observed magnetic domain structures in these soft magnetic films are the result of free-energy minimization resulting from balancing contributions to the magnetic energy primarily from a competition between magnetoelastic, crystalline anisotropy, magnetic exchange, and magnetostatic energies. Any one magnetic structure, although not globally unique, represents information about these energies, including shape effects (thin films), strain, and defects (e.g. domain wall pinning).

The elongated c-axes (LAO films) suggest the presence of magnetoelastic effects. However, in thin films demagnetizing fields due to shape effects are expected to win out, aligning spins in the film plane, figure 2c to f. The absence of a clear in-plane preferred orientation in the MFM data is probably a function of the inability to unambiguously distinguish exact spin or tip polarization directions. The weak diffuse structure is probably a combination of in-plane magnetization due to the shape anisotropy and lower Tc's, the latter resulting in incomplete magnetization and spin fluctuations effects. For the thicker 800 °C LAO films shape effects appear less important than stress effects manifested in c-axis elongation and a positive normal stress tensor and easy spin direction. Variations in the stripe domain structures are clues to variations in stress and defect structures in these films, hints at the nonuniformity of growth and post-deposition processes. Further work is needed to minimize these effects in order to sort out the effects of selected growth parameters on film properties so future work can concentrate on understanding and controlling the magnetic domain properties. Finally, as with microstructure, no significant difference in the Ts-dependent Hc's was found between the LSMO films on STO and LAO.

We would like to acknowledge Thomas Silva and Nicholas Rizzo, NIST, for the use of and help with of their VSM, and Yates Coulter and Quanxi Jia, Los Alamos National Lab, for the use of the Squid Magnetometer and support and use of the PLD deposition system, respectively.

References

7. C. Kwon, M. C. Robson, K.-C. Kim, J. Y. Gu, S. E. Lofland, S. M. Bhagat, Z. Trajanovic, M. Rajeswari, T. Venkatesan, A. R. Kratz, R. D. Gomez, and R. Ramesh, J. Magnetism and Magnetic Materials 772, 229 (7997).
2. Y. Suzuki, H.Y. Hwang, S-W. Cheong, and R.B. van Dover, ," in Epitaxial Oxide Thin Films III, D.G. Schlom, E.-C. Beom, M.E. Hawley, C.M. Foster, and J.S. Speck, eds., Mat. Res. Soc. Symp. Proceed. 474, 205 (Materials Research Society, Pittsburgh 7997).
3. M.C. Martin, G. Shirane, Y. Endoh, K. Hirota, Y. Moritomo, and Y. Tokura, Phys. Rev. B 5 53 (27), 74285 (7996).
4. M.E. Hawley, X.D. Wu, P.N. Arendt, C.D. Adams, M.F. Hundley, and R.H. Heffner, in Epitaxial Oxide Thin Films II, J.S. Speck, D.K. Fork, R.M. Wolf, and T. Shiosaki, eds., Mat. Res. Soc. Symp. Proceed. 407, 537(Materials Research Society, Pittsburgh 7996).
5. M.E. Hawley, C.D. Adams, P.N. Arendt, E.L. Brosha, F.H. Garzon, R.J. Houlton, M.F. Hundley, R.H. Heffner, Q.X. Jia, J. Neumeier, and X.D. Wu, J. Crystal Growth 774, pp.455-463 (7997).
6. G.W. Brown, Q.X Jia, E.J. Peterson, D.K. Hristova, M.F. Hundley, J. D. Thompson,. C.J Maggiore, J. Tesmer, and M.E. Hawley, in Epitaxial Oxide Thin Films III, D.G. Schlom, E.-C. Beom, M.E. Hawley, C.M. Foster, and J.S. Speck, eds., Mat. Res. Soc. Symp. Proceed. 474, (Materials Research Society, Pittsburgh 7997).
7. K. Shirai, K. Ishikure, N. Takeda, J. Appl. Phys. 82 (5), 2457 (7997).

The Effect of Elastic Strain on the Electrical and Magnetic Properties of Epitaxial Ferromagnetic SrRuO₃ Thin Films

Q. Gan *, R. A. Rao *, J. L. Garrett **, Mark Lee **, C. B. Eom *,
*Department. of Mechanical Engineering and Materials Science, Duke University, Durham, NC 27708, eom@acpub.duke.edu
**Department of Physics, University of Virginia, Charlottesville, VA 22903

ABSTRACT

We report the direct measurement of elastic strain effect on the electrical and magnetic properties of single domain epitaxial SrRuO₃ thin films, using a lift-off technique. The as-grown films on vicinal (001) SrTiO₃ substrates are subjected to elastic biaxial compressive strain within the plane and tensile strain normal to the plane. In contrast, the lift-off films prepared by chemical etching of SrTiO₃ substrates, are completely strain free with bulk like lattice. Our measurements indicate that the elastic strain can significantly affect the electrical and magnetic properties of epitaxial ferromagnetic SrRuO₃ thin films. For the strained films, the Curie temperature (T_c) was suppressed to 150K and the saturation magnetic moment (M_s) was decreased to 1.15μB/Ru atom as compared to a T_c of 160K and M_s of 1.45μB/Ru atom for the strain free films. These property changes are attributed to the structural distortion due to the elastic strain in the as-grown epitaxial thin films. Our results provide direct evidence of the crucial role of lattice strain in determining the properties of the perovskite epitaxial thin films.

INTRODUCTION

Perovskite oxide materials have a wide range of technologically important properties such as high temperature superconductivity, colossal magnetoresistance (CMR), and ferroelectricity. For device applications, these oxide materials have to be prepared in the form of epitaxial thin films or heterostructures. However, the lattice mismatch and the thermal expansion mismatch between the thin film layers and substrates can cause considerable strain in the thin films and heterostructures. As a result, the properties of epitaxial perovskite thin films can be quite different from those of the corresponding bulk materials or intrinsic properties.

The strain effect on the properties of epitaxial perovskite thin films has been actively studied recently. The most commonly used methods are to grow epitaxial thin films on substrates with different lattice mismatch and to vary the thickness of the films to induce the different strain states.[1-4] However, these methods can not rule out the property changes due to the change in crystalline quality of the films on different substrates, the different strain distribution in the films with different thickness, as well as the possible stoichiometry changes due to strain stabilization. To determine the intrinsic strain effect on the properties of epitaxial perovskite thin films, it is necessary to isolate the strain effect from the other effects mentioned above. The ideal method would be to compare the properties of the same thin film sample measured before and after strain relaxation. We have successfully prepared strain relaxed SrRuO₃ thin films from initially strained single domain thin films grown on vicinal (001) SrTiO₃ substrates, by using a lift off technique. The properties of the as-grown strained and lift-off strain-free films were compared to directly investigate the strain effect on the magnetic and electrical properties of epitaxial SrRuO₃ thin films.

SrRuO₃ is a conductive ferromagnetic (Curie temperature, T_c, = 160K) perovskite with a GdFeO₃-type structure and a bulk lattice parameter of 3.93Å. Furthermore, it is chemically very stable which allows us to selectively wet etch the SrTiO₃ substrates and prepare the lift-off thin films. The lift-off thin films should be strain-free as the lattice of the film is no longer constrained by that of the substrate.

EXPERIMENT

The SrRuO₃ thin films were deposited by 90° off-axis sputtering[5,6] using a 2" diameter stoichiometric composite target.[7] Vicinal (001) SrTiO₃ substrates with 2°, 4° and 8° miscut toward [010] axis were used. The deposition atmosphere consisted of 80 mTorr O₂ and 120 mTorr Ar, and the substrate block temperature was typically held at 600°C. After the deposition, the chamber was immediately vented in O_2 to a pressure about 300 Torr, and the samples were then cooled down. The thickness of the thin films was about 1000Å. X-ray analysis indicated that these epitaxial $SrRuO_3$ thin films are single domain with $SrRuO_3$ [110] normal to the substrate surface.[8]

Figure 1 shows a schematic of the procedure used to prepare lift-off films. After deposition, the as-grown $SrRuO_3$ thin films on 0.25"x0.25" $SrTiO_3$ substrates were diced into two rectangular shaped pieces of almost equal sizes. One piece was used as a reference as-grown sample. The other was used to prepare a lift-off thin film sample by selectively etching the $SrTiO_3$ substrate with a strong acid solution (50% $HF:70\%$ $HNO_3:H_2O\approx1:1:1$).[9] The lift-off thin film was then directly loaded on a (001) $LaAlO_3$ substrate. The top surface of the lift-off $SrRuO_3$ thin film was slightly roughened which is probably due to the strain relaxation. The complete three-dimensional strain state of the films were studied by using four circle x-ray diffraction. The magnetic properties were investigated with a SQUID magnetometer, and the resistivity as a function of temperature was measured in the Van der Pauw geometry.

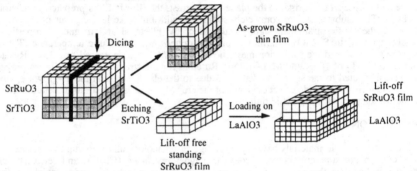

Fig. 1. The schematic diagram of the processing steps used to prepare lift off $SrRuO_3$ films and the strain state of the films during each step.

RESULTS AND DISCUSSION

<u>Strain States</u>

The strain states were characterized by measuring the lattice parameters of $SrRuO_3$ thin films. The normal $\theta-2\theta$ scans were used to determine the out-of-plane lattice parameter, $d_{(110)}$. The in-plane lattice parameters were measured using grazing incidence diffraction (GID) $\theta-2\theta$ scans and these measurements were further confirmed by the least squares method using several off-axis reflections. Fig. 2(a) shows a typical normal $\theta-2\theta$ scan and a GID scan of (002) or (220) plane for the as-grown thin film on 8° miscut (001) $SrTiO_3$. For comparison, the 2θ value corresponding to $d_{(220)}$ of the bulk material is shown by a vertical line. The in-plane lattice parameters of $SrRuO_3$ were found to be 3.90±0.02Å, very close to that of $SrTiO_3$ substrate (3.905Å). Therefore, epitaxial $SrRuO_3$ thin films grow coherently on $SrTiO_3$ substrates (see Fig. 1), which is consistent with cross-sectional TEM observations.[8,11] The smaller in-plane lattice parameters than that of bulk $SrRuO_3$ (3.93Å) indicate that the film is subjected to biaxial compressive strain in the plane ($\varepsilon_{xx}=\varepsilon_{yy}=-0.67\%$). From the 2θ value of the normal scan in Fig. 2(a), the out-of-plane lattice parameter was found to be 3.95Å, which is larger than that of bulk materials suggesting a uniaxial tensile strain along [110] direction ($\varepsilon_{zz}=0.50\%$) in the film (see Fig. 1). This tensile strain is due to the in-plane biaxial compressive stress.

Figure 2(b) shows both normal $\theta-2\theta$ scans and GID $\theta-2\theta$ scans for the lift-off thin film prepared from the same as-grown thin film shown in Fig. 2(a). Since the film was originally grown on 8° miscut (001) $SrTiO_3$ substrate, the [110] direction of the $SrRuO_3$ lift-off film was oriented 8° away from the [001] direction of the underlying exact (001) $LaAlO_3$ substrate. Therefore, after aligning $SrRuO_3$ [110] along x-ray scattering vector, **q**, only film peaks were observed in the normal θ–2θ scans. As can be seen, both the in-plane and out-of-plane lattice parameters of the lift-off film (~3.93Å) are the same as that of the bulk material. This implies that the as-grown SrRuO$_3$ thin film is subjected to an elastic strain which is fully relaxed in the lift-off thin film. Using these as-grown and lift-off samples, we have directly measured the intrinsic strain effect on magnetic and electrical properties of epitaxial SrRuO$_3$ thin films.

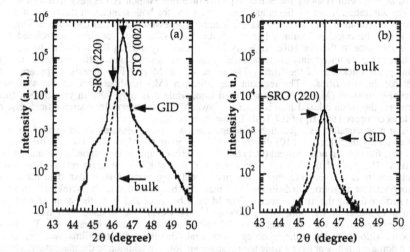

Fig. 2. Normal θ–2θ scans (solid line) and GID θ–2θ scans (broken line) for (a) as-grown SrRuO$_3$ thin film on 8° miscut (001) SrTiO$_3$ substrate and (b) lift-off strain relaxed SrRuO$_3$ thin film. The vertical line indicates the 2θ angle for d$_{(220)}$ of bulk SrRuO$_3$.

The surface morphology of the strained and strain relaxed thin films was studied by a scanning tunneling microscope (STM). Fig. 3(a) and (b) show the STM images of a SrRuO$_3$ thin film grown on a 2° miscut (001) SrTiO$_3$ substrate before and after lift-off. Both strained and strain relaxed films show periodic steps as a clear evidence of the step flow growth, and no significant microscopic change is observed. The root means square (RMS) surface roughness is about 20 Å over a 2μm x 2μm scan area for both films.

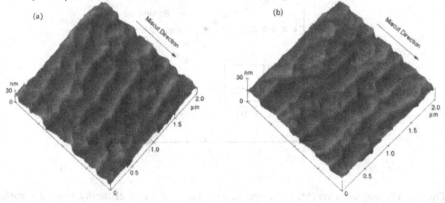

Fig. 3. The STM images over a 2μm x 2μm area for (a) as-grown, and (b) lift-off SrRuO$_3$ thin films grown on a 8° miscut (001) SrTiO$_3$ substrate.

Magnetic Properties

The magnetic field (H) dependence of the magnetization (M) for the as-grown and lift-off samples prepared from a film grown on a 8° miscut substrate was studied. The magnetization was measured at 10K with H along the SrRuO₃ [1̄10] direction, which is the easy axis in the film plane. The saturation magnetic fields are almost the same for both samples in the range of 2.5-3 Tesla but, the saturation magnetization is significantly suppressed (~20%) in the strained film. At H = 4 Tesla, the calculated saturated moments are $1.45\mu_B$/Ru atom for the strain relaxed film (which is the same as that for bulk single crystals[12]) and only $1.15\mu_B$/Ru atom for the strained film. There is a ±5% uncertainty in the saturated moment values due to the difficulties in accurately determining the thickness of the films. The shape of the M vs. H hysteresis loops are quite different for the two films. The remnant magnetization (M_r) is about 75% of the saturation magnetization moment (M_s) in the strain relaxed film, while it is only 50% in the strained film. Furthermore, the strain relaxed film is easier to "switch" due to its smaller coercive field, H_c, of 0.3 Tesla as compared to a H_c of 0.7 Tesla for the strained film.

The magnetization as a function of temperature was also measured in an applied field of 500 Gauss along the SrRuO₃ [1̄10] direction. Fig. 5 shows the M vs. T curves for the strained SrRuO₃ thin film grown on 8° miscut (001) SrTiO₃ substrates and the corresponding strain relaxed lift-off film. There is a sharp magnetization onset at about 160K for the strain relaxed film which does not saturate at low temperatures. This unsaturated magnetization is attributed to the strong magnetocrystalline anisotropy of SrRuO₃.[13] It may also be partly due to the wrinkles formed on the film surface during the lift-off process. The M vs. T behavior and T_c of the strain relaxed film are similar to those of the bulk single crystals.[12] However, the strained film has a broad ferromagnetic transition with an onset at ~150K and it saturates at low temperatures. The broad magnetic transition may due to the inhomogeneous strain distribution in the films, resulting in the magnetic domain alignment over a wide temperature window. In addition to the magnetocrystalline anisotropy, another extrinsic factor - the residual stress anisotropy - has to be considered in the strained thin films. This may explain why the magnetization saturates only in the strained film as shown in Fig. 4. The suppression of both T_c and magnetization moment in the strained thin films is believed to be due to the strain effect, as the spin-spin coupling is very sensitive to the interatomic distances. The change of the lattice parameters affects the overlapping of Ru:t2g and O:2p orbitals to which form a narrow π* band that is responsible for the ferromagnetism in SrRuO₃.[14]

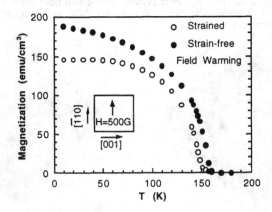

Fig. 4. The magnetization (M) vs. temperature (T) behavior for an elastically strained epitaxial SrRuO₃ thin film on a 8° miscut (001) SrTiO₃ substrate and the corresponding lift-off strain relaxed film. The inset shows the applied field direction with respect to the film in-plane directions.

The strain effect on the magnetic properties was also indirectly confirmed using SrRuO$_3$ thin films deposited on (001) LaAlO$_3$ substrates. Due to the large lattice mismatch between SrRuO$_3$ (3.93Å) and LaAlO$_3$ (3.793Å), the strain was relaxed during the film growth, which is confirmed by X-ray diffraction. These films have a multi-domain structure consisting of two [110] oriented domains, with [001] pointing in two orthogonal directions in the plane, and trace [001] oriented domain. Resistivity as a function of temperature and magnetization measurements showed a Curie temperature near 160K, and the magnetization did not saturate at low temperature, agreeing with that of the strain relaxed film. However, no magnetic anisotropy was observed in M vs. H loops because of the multi-domain structure in the films. The saturation moment for SrRuO$_3$ on LaAlO$_3$ was about 1.15±0.05 μB/Ru atom, lower than that of strain relaxed thin films. In this case, the decrease in saturation moment compared to strain relaxed samples is believed to be due to structural imperfections that result from growth on more highly lattice mismatched substrates. This clearly shows the limitation of using thin films on substrates with different lattice mismatch to study the strain effect.

Electrical Properties

We have also compared the electrical transport behavior of the strained and strain relaxed epitaxial SrRuO$_3$ thin films by measuring the film resistivity as a function of temperature. The ρ vs. T curves for strained epitaxial SrRuO$_3$ thin film on a 8° miscut (001) SrTiO$_3$ substrate and the corresponding strain relaxed lift-off film are shown in Fig. 5. The general features of resistivity behavior for both strained and strain relaxed thin films are the same, such as metallic behavior and a sharp kink corresponding to the Curie temperature T$_c$. The T$_c$ of the strained as-grown film is about 150K, while it is about 160K (same as bulk material) for the lift-off film due to complete strain relaxation. These values are consistent with the magnetization measurements. Above T$_c$, the resistivity for both strained and strain relaxed samples increased linearly with temperature with almost the same slope. Below T$_c$, the resistivity for the strain relaxed film decreased faster with decreasing temperature which is due to the stronger magnetization in that film. However, there is some uncertainty in the absolute resistivity values of the strain relaxed films as the lift-off films are very fragile and could have been damaged during loading on the LaAlO$_3$ substrates. In fact, micro cracks were observed on the strain relaxed film that was used in Fig. 5, which may be responsible for the slightly higher resistivity as compared to the strained film.

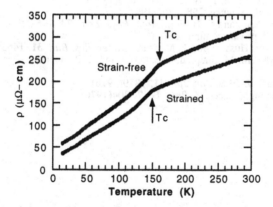

Fig. 5. The resistivity as a function of temperature for an elastically strained epitaxial SrRuO$_3$ thin film on a 8° miscut (001) SrTiO$_3$ substrate and the corresponding strain relaxed SrRuO$_3$ thin film. The arrows indicate the Curie temperature, T$_c$.

SUMMARY

We have directly measured the elastic strain effect on the magnetic and electrical properties of the single domain epitaxial $SrRuO_3$ thin films by using lift-off strain-relaxed thin films. The magnetic properties of strain relaxed $SrRuO_3$ thin films are essentially the same as that for bulk $SrRuO_3$ single crystal. The anisotropic elastic strain in epitaxial $SrRuO_3$ thin films can significantly change the magnetic properties by suppressing Curie temperature (by ~10K) and saturation magnetic moment (by ~20%) as compared to the strain relaxed films. Our results provide conclusive evidence of the crucial role of the strain effect in determining the properties of the perovskite epitaxial thin films used in electronic and magnetic heterostructures and devices. The chemical lift-off technique can also be used to prepare strain relaxed single domain epitaxial thin films for the study of intrinsic properties of other complex oxide materials which are difficult to be synthesized in bulk single crystal form.

ACKNOWLEDGMENT

We would like to thank L. Klein, T. H. Geballe, and A. F. Marshall for helpful discussions. This work is supported by the ONR Grant No. N00014-95-1-0513, the NSF Grant No. DMR 9421947.

REFERENCES

1. S. Jin, T. H. Tiefel, M. McCormack, R. A. Fastnacht, R. Ramesh, and L. H. Chen, *Science* **264**, 413(1994); S. Jin, T. H. Tiefel, M. McCormack, H. M. O'Bryan, L. H. Chen, R. Ramesh, D. Schurig, *Appl. Phys. Lett.* **67**, 557(1995).
2   Y. Suzuki, H. Y. Hwang, S-W. Cheong, and R. B. van Dover, *Appl. Phys. Lett.* **71**, 140(1997).
3   H. Sato and M. Naito, *Physica C* **274**, no. 3-4, 221(1997).
4   C. H. Kwon *et al.*, *J. Magn. Magnetic Materials*, **172**, 229(1997).
5   C. B. Eom, J. Z. Sun, K. Yamamoto, A. F. Marshall, K. E. Luther, S. S. Laderman, T. H. Geballe, *Appl. Phys. Lett.* **55**, 595(1989).
6   C. B. Eom, J. Z. Sun, S. K. Streiffer, A. F. Marshall, K. Yamamoto, B. M. Lairson, S. M. Anlage, J. C. Bravman, T. H. Geballe, S. S. Laderman, and R. C. Taber, *Physica C* **171**, 351(1990),
7   R. A. Rao *et al.*, *Appl. Phys. Lett.* **70**, 3035(1997).
8   C. B. Eom *et al.*, *Science,* **258**, 1766(1992).
9   Q. Gan, R. A. Rao, and C. B. Eom, *Appl. Phys. Lett.* **70**, 1962(1997).
10 A. F. Marshall, private communication.
11 A. Gupta, B. W. Hussey, and T. M. Shaw, *Mater. Res. Bull.* **31**, 1463(1996)
12 G. Cao, S. McCall, M. Shepard, J. E. Crow, and R. P. Guertin, *Phys. Rev. B* **56**, 321(1997)
13 A. Kanbanyasi, *J. Phys. Soc. Jpn.* **41**, 1879(1976).
14 J. B. Goodenough, *Czech. J. Phys.* **B 17**, 304(1976)

EFFECTS OF LOCALIZED HOLES ON CHARGE TRANSPORT, LOCAL STRUCTURE AND SPIN DYNAMICS IN THE METALLIC STATE OF CMR La$_{1-x}$Ca$_x$MnO$_3$

R. H. HEFFNER,[1] M.F. HUNDLEY,[1] C. H. BOOTH[1,2]
[1]Los Alamos National Laboratory, Los Alamos, NM 87545
[2]Physics Department, University of California, Irvine, CA 92697

ABSTRACT

We review resistivity, x-ray-absorption fine-structure (XAFS) and muon spin relaxation (μSR) data which provide clear evidence for localized holes causing polaron distortion and unusual spin dynamics *below* T_C in "colossal magnetoresistive" (CMR) La$_{1-x}$Ca$_x$MnO$_3$. Resistivity measurements for $x=0.33$ under an applied field H have shown that $\ln[\rho(H,T)] \propto -M$, where M is the magnetization. The XAFS data show a similar functional dependence for the polaron distortions on M. The data from these two measurements are interpreted in terms of some fraction of the available holes x remaining localized and some increasing fraction becoming delocalized with increasing M. Finally, this polaron-induced spatial inhomogeneity yields anomalously slow, spatially inhomogeneous spin dynamics below T_C, as shown in the μSR data. These experiments individually probe the charge, lattice and spin degrees of freedom in this CMR system and suggest that the polarons retain some identity even at temperatures significantly below T_C.

INTRODUCTION

In 1954 Volger [1] measured a large magnetoresistance effect near the ferromagnetic transition in La$_{0.8}$Sr$_{0.2}$MnO$_3$. At that time the double-exchange theory was being developed by Zener [2], Anderson and Hasegawa [3], and deGennes [4], so the magnetoresistance was loosely attributed to this mechanism. At the time, deGennes [4] pointed out that the theory of double exchange (DE) really should include coupling to the lattice degrees of freedom. About 40 years later, the consequences of such a coupling are now being appreciated. Recently, the magnetoresistance in several related materials was studied [5] in more detail and found to be quite large, prompting the term "colossal" magnetoresistance (CMR). Millis *et al.* [6] recognized CMR was too large to be explained by the DE model without invoking lattice distortions (polarons) in the paramagnetic state. Many subsequent experiments confirmed the existence of both polaron transport [7,8] and local lattice distortions [9,10] that are partially removed in the ferromagnetic state, as well as the expected isotope effect [11]. Moreover, there is mounting evidence that the local lattice distortions play an even bigger role in the transport, possibly dominating the electronic properties over a wide range of temperatures both above and below any magnetic transitions. We review some of these measurements, including those that demonstrate the strong dependence of magnetic field on the resistivity and the local polaron distortions. These data have a remarkably similar functional dependence on the magnetization, as we will show. Moreover, μSR experiments [12] indicate that there are anomalously slow and spatially inhomogeneous spin dynamics in these systems, which can be associated with considerable inhomogeneity in the local magnetic environment. Taken together, these data suggest that there exist regions of the sample in the ferromagnetic state that are essentially still conducting via polaron transport, and that the electronic properties are governed primarily by how much the magnetization can change the relative size of these regions. Regions of various

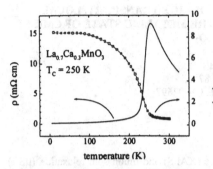

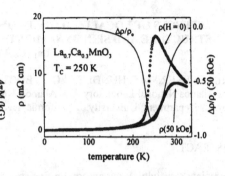

Figure 1: $M(T)$ and $\rho(T)$ for a CMR film with $T_C = 250$ K.

Figure 2: $\rho(T)$ in zero and 50 kOe applied fields (left axis) and transverse MR(T) in 50 kOe (right axis).

sizes can in principle give rise to different spin fluctuation rates. These different regions can also support a wide range of polaron binding energies, consistent with the resistivity and local structure measurements.

DEPENDENCE OF MAGNETIZATION ON RESISTIVITY

The temperature-dependent magnetization and resistivity of an epitaxial thin film of $La_{0.7}Ca_{0.3}MnO_3$ grown via pulsed laser deposition are depicted in Fig. 1. The T-dependent resistivities in zero field and in 50 kOe as well as the resulting magnetoresistance are shown in Fig. 2. The Curie temperature for this specimen was determined to be 250 K from magnetization Arrott plots. For temperatures above T_C, ρ exhibits Arrehenius behavior with an activation energy of 0.1 eV. Hall measurements at 300 K indicate that the carrier concentration is consistent with 0.3 carriers per formula unit and that the drift mobility is only 0.02 cm^2/V-sec, corresponding to a mean-free path that is a fraction of a lattice constant. This clearly indicates that conduction proceeds via nearest-neighbor adiabatic small polaron hopping above T_C. Thermopower and Hall effect measurements confirm that this is indeed the case [13,14]. By applying a 50 kOe magnetic field the peak in ρ at T_C is drastically suppressed and a substantial magnetoresistance is achieved: $\Delta\rho/\rho_0(50 \text{ kOe}) = -85\%$ at T_C.

The data in Figs. 1 and 2 show that both $d\rho/dH$ and $d\rho/dT$ are largest near T_C. This region is exactly where dM/dT is a maximum and where an applied magnetic field has the greatest effect on the microscopic magnetism. Thus there is clearly a close connection between charge transport and magnetic order in the manganite CMR compounds. To more fully explore the relationship between ρ and M, careful measurements of $\rho(H,T)$ and $M(H,T)$ were made on the same thin-film sample used to produce the data shown in Figs. 1 and 2. The data were measured at nine temperatures from 272 K to 10 K in fields sufficient to saturate the domain structure so that the measured M would correspond to a microscopic magnetization. The results are presented in Fig. 3 where $\rho(H,T)$ is plotted versus $M(H,T)$ rather than as a function of H or T. The data indicate a correlation that encompasses two orders-of-magnitude variation in ρ and can be parameterized as

$$\rho(H,T) = \rho_m \exp\{-M(H,T)/M_\rho\}, \qquad (1)$$

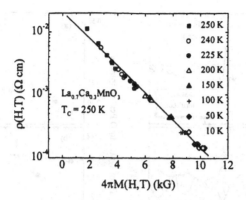

Figure 3: $\rho(H,T)$ plotted as a function of $M(H,T)$.
At each temperature points are shown for applied
fields of 10, 20, 30, 40, and 50 kOe. The solid
line is a least-squares fit to the data.

with ρ_m = 21 mΩ-cm and $4\pi M_\rho$ = 2.0 kG. This resistivity-magnetization correlation was first
reported by Hundley *et al.* [7] and has since been confirmed by many others in both thin
film [13,15] and bulk samples [12] of CMR compounds with ordering temperatures less than
room temperature. Because this correlation appears to hold throughout the temperature range
below T_C it is evident that the CMR compounds are not conventional ferromagnetic metals even
at low temperatures. Clearly, electronic transport is influenced by magnetic order in a highly
unusual way.

By making a few simple assumptions, the phenomenology expressed by Eq. 1 can
provide insight into the transport process below T_C. One approach is to assume that the adiabatic
small polaron hopping description that is valid above T_C will still apply over a limited
temperature range below the ordering temperature. In this scenario, the developing
magnetization acts to reduce the polaron binding energy W_P that characterizes the degree to
which charge carriers are localized. By following this approach [16] it is straightforward to show
that Eq. 1 leads to the simple result that W_P is reduced linearly by the magnetization that
develops from either reducing the temperature below T_C or by applying a magnetic field.
However, this analysis cannot directly validate the physical picture implied by the adiabatic
small polaron theory, namely that W_P is a measure of the spatial extent of the polaron distortion.
Consequently, more local experiments, such as a local structure or a local magnetic probe, are
necessary to provide a microscopic picture of the transport mechanism.

LOCAL STRUCTURE MEASUREMENTS OF MnO$_6$ OCTAHEDRA, AND RELATION TO MAGNETIZATION

We now report on local-structure measurements of the MnO$_6$ distortions in CMR
La$_{1-x}$Ca$_x$MnO$_3$ and how they relate to the zero-field magnetization. (These measurements are also
discussed in Refs. [17,18].) The Mn-O bond length distribution width is measured using the
x-ray-absorption fine-structure (XAFS) technique. The measured changes in this width $\Delta\sigma$ due
to removal of the polaron distortions below T_C are found to relate to the zero-field magnetization
simply as $\ln(\Delta\sigma) \propto M$. Using a simple model whereby each doped hole is either localized

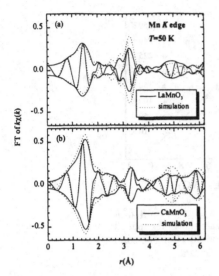

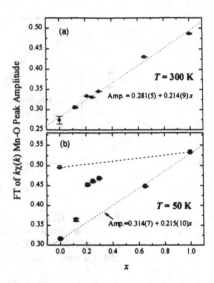

Figure 4: Mn K edge Fourier transforms (FT) of $k\chi(k)$ vs. r for (a) LaMnO₃ and (b) CaMnO₃ at 50 K. The oscillating curve is the real part and the envelope is the amplitude of the complex transform ($[Re^2+Im^2]^{1/2}$). Dotted lines are from simulations using structures from diffraction, as described in the text.

Figure 5: (a) Amplitude of the Mn-O peak in the FT of $k\chi(k)$ at 300 K as a function of x for La₁₋ₓCaₓMnO₃. (b) As in (a), except at 50 K. The dotted lines in (a) and (b) are linear fits to the data. The (upper) dashed line is approximately the amplitude one expects if the individual Mn ions were actually mixed valent.

(causing a polaron distortion) or delocalized (causing no polaron distortion), we find that the delocalized hole concentration n_{dh} also changes as $\ln(n_{dh}) \propto M$. This result is in agreement with the dependence of the resistivity on magnetization $\ln(\rho) \propto -M$ shown in Fig. 3. These local structure measurements and the transport measurements therefore provide empirical relations between the spin, charge and lattice degrees of freedom in the Ca-doped CMR manganite perovskites.

The XAFS experiments yield the local structure around a central species of atom, chosen by tuning an incident x-ray beam to a particular absorption edge. For these experiments we measure the absorption region ~900 eV above the Mn K edge. The absorbing photon forces the ejection of a photoelectron with wave vector k given by $E-E_0=\hbar^2k^2/2m_e$, where E_0 is the edge energy and m_e is the electron mass. As the photoelectron is back-scattered from near-neighbor atoms, it interferes with itself at the absorbing atom, causing the absorption coefficient μ to oscillate as $\sin(2kr)$, where r is the distance to the neighboring atom. By isolating the oscillatory part of the absorption (defined as $\chi(k)$), a Fourier transform (FT) of $k\chi(k)$ yields peaks in the amplitude that correspond to the distance (r) between and the number (N) and distribution width (σ) of neighboring atoms. Since the photoelectron back-scattering amplitude $F(k)$ is also a function of k, the FT of $k\chi(k)$ is only *related* to the radial distribution function (RDF). However, $F(k)$ can be calculated to high precision by multiple scattering codes such as FEFF [19], and therefore fits to the FT of $k\chi(k)$ can yield N, r, and σ for the near-neighbor ($< \sim 4.5$ Å) atoms.

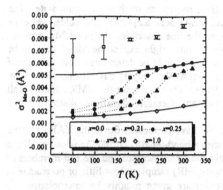

Figure 6: The square of the Mn-O bond length distribution width σ as a function of temperature. The solid line is a fit of the data for $CaMnO_3$ to the correlated-Debye model with Debye temperature $\Theta_D=940\pm30$ K. This same temperature dependence is also shown with a static of 0.07 Å (upper solid line).

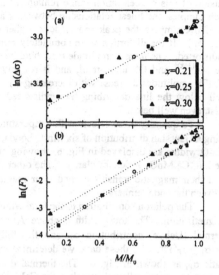

Figure 7: (a) $\ln(\Delta\sigma)$ as a function of the magnetization. $\Delta\sigma$ is the amount of Mn-O bond length distribution width that is removed in the ferromagnetic state for the CMR samples. The magnetization is normalized to the magnetization at T=5 K. (b) $\ln(F)$ as a function of the normalized magnetization, as derived for the changes in the local structure. F is the fraction number of delocalolized holes, $F=n_{dh}/x$ (see text).

Fig. 4 shows the FT of $k\chi(k)$ for the end-member compounds $LaMnO_3$ and $CaMnO_3$, together with simulations calculated with FEFF [19], using crystal structures measured by Mitchell et al. [20] and Poeppelmeier et al. [21], respectively. Considering the large distortions present in both of these compounds (the La/Ca site sits off-center within the perovskite cube, creating a wide range of Mn-La/Ca bond lengths near 3.3 Å), the agreement between the diffraction and XAFS data is remarkably good. One should note that the Mn-O peak at ~1.5 Å (corresponding to Mn-O bond lengths near 1.95 Å) is much lower in amplitude in $LaMnO_3$ than in $CaMnO_3$. This decreased amplitude is a consequence of the Jahn-Teller (JT) distortion that occurs around Mn^{3+} ions. In contrast to the undistorted MnO_6 octahedra in $CaMnO_3$ which gives six Mn-O pairs at ~1.90 Å, the JT distortion causes three distinct bond lengths in $LaMnO_3$, with two oxygen neighbors to the manganese at 1.91 Å, two at 1.97 Å and two at 2.15 Å.

Now we must consider the effect of Ca substitution for La in $LaMnO_3$ on the local Mn-O environment. If each added calcium converts a Mn^{3+} (distorted) into a Mn^{4+} (undistorted) ion, one expects that the amplitude of the Mn-O peak in the FT of $k\chi(k)$ will be linear in the calcium concentration x. This assertion can be verified by simulating such a local environment with FEFF. This linear relation indeed is shown to exist experimentally, as seen in Fig. 5a. Therefore, above T_C, the Mn-O amplitude is an accurate predictor of the *localized* hole concentration.

As we drop below T_C, samples with $0.2<x<0.5$ undergo an insulator-to-metal (CMR) transition, together with a (partial) removal of the JT distortion of the MnO_6 octahedra. Samples outside of this concentration range remain insulating, regardless of their magnetic state. For such samples, the linear relationship between x and Mn-O peak amplitude still holds, while for the CMR samples, the peak amplitude is higher than the line describing the non-CMR samples (Fig. 5b). If the distortion were completely removed, one might expect the Mn-O peak to be undistorted. The *lowest* amplitude that could be expected would be determined from the sum of the $Mn^{(3+x)+}$ and O^{2-} ionic radii, and is roughly shown as the (upper) dashed line in Fig. 5b. (XAFS amplitudes decrease with increasing r.) The real amplitudes for the CMR materials fall well below the line describing the undistorted case, and therefore significant distortions still exist, even well below T_C.

To quantify these effects with temperature, we have performed fits to the XAFS spectra, using a Gaussian distribution of six Mn-O bonds, with a bond-length distribution width σ. These Mn-O widths are displayed in Fig. 6, showing only the CMR samples and the end-members of the $La_{1-x}Ca_xMnO_3$ series for clarity. The other (non-CMR) samples show little or no change in σ at their magnetic transitions, and their magnitudes are given roughly by interpolating $1/\sigma$ between the end members.

The polaron contribution to the total width can be isolated and compared to the zero-field magnetization. The total width σ above T_C can be written as the sum (in quadrature) of a thermal (Debye) contribution σ_D and the fully-developed (static) polaron contribution σ_P: $\sigma=(\sigma_D^2+\sigma_P^2)^{1/2}$. For these data, we determine σ_D by fitting the data for $CaMnO_3$ and adding a static σ_P, as shown in Fig. 6. The thermal dependence of σ above T_C in the CMR materials appears to agree well with the data for $CaMnO_3$. Below T_C, the change in the width due the finite magnetization can be written as $\Delta\sigma=(\sigma^2-\sigma_D^2+\sigma_P^2)^{1/4}$. The zero-field magnetization can be estimated by the SQUID-measured magnetization M normalized to the low temperature magnetization M_0. (Magnetization measurements are reported in Ref. [17,18].)

Fig. 7a demonstrates that $\ln(\Delta\sigma) \propto M/M_0$, with all the data from this small sample set falling on the same line. This relationship strongly supports the existence of the JT distortions below T_C, and shows how they develop with decreasing magnetization, and thus increasing temperature. These data are in agreement with PDF data on the related $La_{1-x}Sr_xMnO_3$ system which demonstrate the existence of JT distortions below T_C for $x < \sim 0.35$ [22]. It is important that such a relationship is calculable within a double-exchange theory that includes polarons [23], and thus may provide an important test of such theories [24,25].

A "two-fluid" model of the Mn-O distortions and the conductivity

A simple "two fluid" model can describe these data as well. Consider that in the metallic state each of the doped holes x can be either in a localized (non-conducting) or a delocalized (conducting) state. Such a model ignores the effect of changes in the carrier mobility on the structure, such as may occur if the states become only partially extended. With this assumption, we can model the effect of the distortions that we measure by assuming that each localized charge will create a Mn^{3+} site and a JT distortion around that site. This model exactly describes the linear relationship in the paramagnetic state shown in Fig. 5a. When the material becomes ferromagnetic, some fraction of the holes become delocalized and no longer contributes to the lattice distortion. In the limit of no distortion, the width of the peak should be the same as $CaMnO_3$. With such a simple distribution of distortions, we can immediately write down the delocalized hole concentration n_{dh}. In terms of distribution amplitudes (which are slightly different that the amplitude of the FT of $k\chi(k)$ in Fig.5 [18]), n_{dh} is given by the difference between the measured amplitude A and the amplitude one expects if the distortion remains fully developed A_D, normalized by the difference between the amplitude expected for no distortion

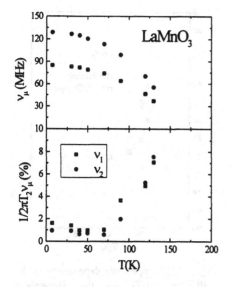

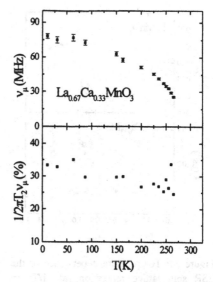

Figure 8: Temperature dependence of the µSR frequency (upper panel), proportional to the sublattice magnetization, in LaMnO₃. Temperature dependence of the inhomogeneous linewidth 1/T₂ divided by the precession frequency ωμ.

Figure 9: Temperature dependence of the µSR frequency (upper panel), proportional to the sublattice magnetization, in La₀.₆₇Ca₀.₃₃MnO₃. Temperature dependence of the inhomogeneous linewidth 1/T₂ divided by the precession frequency ωμ.

A_{ND} and A_D. These amplitudes are given by the number of Mn-O neighbors (six) times $(2\pi\sigma^2)^{-\frac{1}{2}}$. We may therefore write

$$n_{dh} = \frac{A(x,T,M) - A_D(x,T)}{A_{ND}(x,T) - A_D(x,T)}$$
$$= \frac{1/\sigma - 1/\sigma_D}{1/\sigma_{ND} - 1/\sigma_D}$$

(2)

A plot of the fractional number of delocalized holes $F = n_{dh}/x$ determined in this fashion is shown in Fig. 7b. Again, $\ln(F) \propto M$. Using fits to these data, one also finds that even at $T=0$, not all the holes in these samples are completely delocalized: $n_{dh}(T=0)/x$ goes from $0.77 \rightarrow 0.86$ for $x=0.21 \rightarrow 0.30$. We do not have a good explanation of this change in $n_{dh}(0)$, but it may be related to the number of cation vacancies in these samples.

ANOMALOUSLY SLOW, INHOMOGENEOUS SPIN DYNAMICS BELOW T_C FROM µSR

The XAFS data described above present strong evidence for significant average lattice distortions at temperatures far below T_C, which in turn is associated with charge localization.

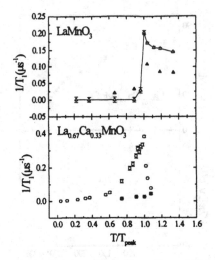

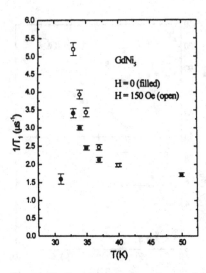

Figure 10: Temperature dependence of the µSR spin lattice relaxation rate $1/T_1$ in LaMnO$_3$ (upper panel) and La$_{0.67}$Ca$_{0.33}$MnO$_3$ lower panel. The open symbols are for zero applied field and the closed symbols are for 3 kOe applied field.

Figure 11: Temperature dependence of the µSR rate $1/T_1$ in the Heisenberg ferromagnet GdNi$_5$ ($T_C \cong 32$ K).

The clear connection between the residual distortion, the overall sample magnetization and the resistivity relates the microscopic lattice degrees of freedom to the *macroscopic* charge and spin degrees of freedom. This correlation with the spin degrees of freedom can be explored *microscopically* using the muon spin relaxation (µSR) technique. Here 100% spin polarized muons are implanted interstitially and the relaxation of the muons' polarization in time is detected by monitoring the anisotropy in the direction of the emitted decay positrons. This leads to a relaxation function that is given [12] by the formula

$$G_z(t) = A_1 \exp(-(t/T_1)^K) + A_2 \exp(-t/T_2) \cos(2\pi\nu_\mu t + \phi). \qquad (3)$$

Here T_1^{-1} and T_2^{-1} are the respective homogeneous and inhomogeneous linewidths, $\omega_\mu = 2\pi\nu_\mu = \gamma_\mu |B|$ is the muon Larmor precession frequency in the internal field B, and A_1 and A_2 are the relative amplitudes of the fluctuating and precessing components of the local field, respectively. The above assumes a single muon magnetic environment, so that $A_1 + A_2 = 1$.

Figs. 8 and 9 show the temperature dependence of the muon frequency ν_μ and the fractional linewidth $1/(2\pi T_2 \nu_\mu)$ in both LaMnO$_3$ and La$_{0.67}$Ca$_{0.33}$MnO$_3$. The ν_μ values are proportional to the local sublattice magnetization. There are two frequencies found in LaMnO$_3$, corresponding to two magnetically inequivalent muon sites in the antiferromagnetic structure. Only a single muon line is observed in the doped material, indicating a single interstitial site. In an earlier publication [12] we showed that the temperature dependence of ν_μ in La$_{0.67}$Ca$_{0.33}$MnO$_3$ is well described by the functional form expected for a second order phase transition $\nu_\mu = \nu_0(1-T/T_C)^{1/3}$. Both of the lines for LaMnO$_3$ are fit by this same expression, but with an

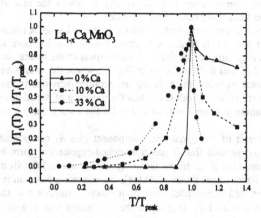

Figure 12: Temperature dependence of the spin lattice relaxation rate $1/T_1$ in $La_{1-x}Ca_xMnO_3$. The x-axis has been normalized to the magnetic ordering temperature and the y-axis to the maximum relaxation rate in each sample.

exponent of about 0.29. The lower panel in Fig. 8 shows that the inhomogeneous linewidth in the undoped sample is quite narrow, except near T_N. By contrast, in $La_{0.67}Ca_{0.33}MnO_3$ (lower panel, Fig. 9) the linewidth tracks the sublattice magnetization and is at least an order of magnitude larger than in the undoped material. In this case both the magnitude and temperature dependence (proportional to the magnetization) of the linewidth suggest that its origin comes from a distribution of grain shapes in the polycrystalline sample, giving rise to a distribution of demagnetization factors. We return to this point below.

Fig. 10 shows the temperature dependence of the spin lattice relaxation rates in $LaMnO_3$ and $La_{0.67}Ca_{0.33}MnO_3$. The open and closed symbols are for zero and 3 kOe applied field, respectively. One expects almost no relaxation below the ordering temperature in magnets with appreciable spin stiffness constants [26]. This is because $1/T_1$ is usually dominated by two-magnon relaxation from spin waves, which is proportional to $T(\ln T)/D^3$, where D is the spin wave stiffness [26]. The stiffness constant in the manganites is indeed large enough that no relaxation from the spin waves is expected ($D \cong 155$ meV Å^2 in $La_{0.67}Ca_{0.33}MnO_3$). This is what is observed in $LaMnO_3$; i.e., no appreciable μSR rate is observed below T_N. However, in the Ca-doped material the relaxation below T_C is appreciable and is attributed to the effects of spin-lattice polarons, as discussed below. An important point to note is that $1/T_1 \propto \Delta^2 \tau$, where Δ is the amplitude of the fluctuating hyperfine field and τ is the correlation time of the fluctuations. Thus anomalously slow fluctuation rates or long correlation times τ are associated with the relaxation below T_C in $La_{0.67}Ca_{0.33}MnO_3$.

These anomalously slow fluctuations produce a μSR rate that is very sensitive to an applied field, as seen in Fig. 10. Whereas a field of only 3 kOe completely destroys the peak in the relaxation rate for $La_{0.67}Ca_{0.33}MnO_3$, one still sees the relaxation peak in the antiferromagnet $LaMnO_3$. Although one would expect an applied field to couple more strongly to a ferromagnet than an antiferromagnet, fields of comparable sizes in other ferromagnetic materials do not have such a drastic effect. This is seen in the case of $GdNi_5$, a typical Heisenberg ferromagnet, shown in Fig. 11. The application of a small field actually enhances the relaxation rates near T_C presumably because the field slows down the fluctuations.

We have also performed measurements on the material $La_{0.90}Ca_{0.10}MnO_3$, which lies in a region of the phase diagram where canted antiferromagnetic order has been suggested [27] (i.e., ferromagnetism is just beginning to develop). A comparison of the relaxation rates for the three systems with $x = 0.0$, 0.10 and 0.33 is shown in Fig. 12, where the axes have been normalized. One notes that the relaxation below the ordering temperature increases in magnitude relative to its peak rate as the system is doped with Ca. Thus the presence of doped holes (and presumably their associated spin-lattice polarons) is responsible for the anomalously slow relaxation below T_C.

A detailed analysis of the dynamical component (the A_1 component in Eq. 3) of the relaxation function $G_z(t)$ reveals that it has a stretched-exponential form below and slightly above T_C, with an exponent $K \leq \frac{1}{2}$. In the ferromagnetic compound $GdNi_5$ discussed above, as well as in the random ferromagnet $PdMn$ (2%) [28], the relaxation function is exponential above and below T_C. A stretched exponential relaxation function implies that there exists a broad distribution of relaxation rates $1/T_1$, leading to the conclusion that Δ and/or τ are distributed quantities.

To understand how a distribution in Δ affects the measured relaxation function we convoluted both a Gaussian and a Lorentzian broadening function $P(\Delta)$ with the exponential relaxation function $\exp(-t/T_1)$, i.e.,

$$\exp(-(t/T_1)^K) = \int d\Delta \, P(\Delta) \exp(-t/T_1). \tag{4}$$

Using the value obtained from the inhomogeneous linewidth shown in Fig. 9 (lower panel) as an upper limit for Δ, we find that $K \geq 0.6$. Obviously, the smaller the value of Δ the larger the value of K, i.e., $K = 1$ for $\Delta = 0$. Recall that the physical meaning of Δ is the amplitude of the fluctuating hyperfine field produced by the Mn spins at the muon site. Thus the actual value of Δ for our experiment in $La_{0.67}Ca_{0.33}MnO_3$ is likely to be much smaller than the value obtained from the inhomogeneous linewidth $1/T_2$, which is presumably dominated by a distribution of demagnetizing fields from the various grain shapes in the sample. Consequently, the measured exponent K cannot be explained by a spread in the intrinsic Δ values alone, strongly suggesting that the local-field correlation time τ is also a distributed quantity. Because μSR is a local probe one may therefore infer that the fluctuations are spatially inhomogeneous.

In summary, the principle findings from μSR are that the spin fluctuations below T_C in $La_{0.67}Ca_{0.33}MnO_3$ are anomalously slow, spatially inhomogeneous and very sensitive to small-applied fields. These results can be related to the combined XAFS, resistivity and magnetization experiments by assuming that the small polarons observed previously in transport measurements above T_C retain some identity in the ferromagnetic state; i.e., the system is not homogeneous below T_C. The presence of structural inhomogeneity below T_C is quite clear from the XAFS measurements, and as discussed above, this can be related to the existence of both localized and delocalized charges. The μSR measurements likewise indicate inhomogeneity, but in the fluctuations of the Mn spins.

To reconcile these results we hypothesize that the regions of localized and delocalized charge (or alternatively, regions of large and small local lattice distortion) produce spin clusters of varying sizes and that the fluctuation rates within a cluster depend on the cluster size. This gives rise to a distribution of relatively slow fluctuation rates (we find $1/\tau \leq 10^{-11}$ s^{-1} near and below T_C). One may estimate an upper limit for the cluster size by noting that when the spin cluster or polaron reaches the size of the ferromagnetic coherence length it can support spin waves, which do not relax the muon spin, as discussed above. Estimates of this correlation length are $\geq$ 15- 20 Å [29]. A reasonable lower limit for the cluster size is 1-2 lattice spacings, slightly larger that the expected size of the small polarons in the paramagnetic state. The unusual field dependence observed for the μSR relaxation rate could thus be explained by postulating that a small field will enlarge the cluster size a few lattice spacings (by enhancing the charge transport which is mediated through the double exchange) until spin waves are again supported locally. Note that a much larger field is required to significantly reduce the resistivity, where charge must be transported across the entire sample.

In conclusion, we have presented three different types of measurements that separately probe the charge, lattice and spin degrees of freedom in the Ca-doped CMR manganites. As expected, the transport, local structure, and μSR measurements show clear evidence for the presence of small polarons both in the electronic properties and from the Jahn-Teller distortions in the paramagnetic state. What is surprising is that these measurements provide compelling evidence that even well below T_C, there are still signatures of local lattice distortions and other polaron-like effects, albeit diminished from those observed above T_C. Taken together, these experiments are consistent with a model wherein regions of the sample display small spin-lattice polaron dynamics and that these regions continuously change into regions with free-carrier-like dynamics below the magnetic ordering temperature. The presence of polarons and the dramatic effect they have on the transport, structure and magnetic properties indicates that these compounds are *not* simple ferromagnetic metals well below T_C and any description of the underlying physical mechanism responsible for their unique properties must account for this fact. These measurements therefore provide an important glimpse into the nature of the ground-state in CMR compounds.

ACKNOWLEDGEMENTS

The authors thank A. Millis for useful discussions. Work at Los Alamos National Laboratory (LANL) was performed under the auspices of the U.S. Department of Energy (DOE), and supported in part by funds provided by the University of California for the conduct of discretionary research by LANL. The XAFS experiments were performed at the Stanford Synchrotron Radiation Laboratory, which is operated by the DOE, Division of Chemical Sciences, and by the NIH, Biomedical Resource Technology Program, Division of Research Resources.

References
1. J. Volger, Physica **20**, 49 (1954).
2. C. Zener, Phys. Rev. **82**, 403 (1951).
3. P. W. Anderson and H. Hasegawa, Phys. Rev. **100**, 675 (1955).
4. P. G. deGennes, Phys. Rev. **118**, 141 (1960).
5. R. M. Kusters, J. Singleton, D. A. Keen, R. McGreevy, and W. Hayes, Physica B **155**, 362 (1989); K. Chahara, T. Ohuo, M. Kasai, and Y. Kozono, Appl. Phys. Lett. **63**, 1990 (1993); R. von Helmholt, J. Wecker, B. Holzapfel, L. Schultz, and K. Samwer, Phys. Rev. Lett. **71**, 2331 (1993); S. Jin, M. McCormack, T. H. Tiefel, R. M. Fleming, J. Phillips, and T. Ramesh, Science **264**, 413 (1994).
6. A. J. Millis, P. B. Littlewood, and B. I. Shraiman, Phys. Rev. Lett. **74**, 5144 (1995).
7. M. F. Hundley, M. Hawley, R. H. Heffner, Q. X. Jia, J. J. Neumeier, J. Tesmer, X. D. Wu, and J. D. Thompson, Appl. Phys. Lett. **67**, 860 (1995).

8. M. Jaime, M. B. Salamon, M. Rubinstein, R. E. Treece, J. S. Horwitz, and D. B. Chrisey, Appl. Phys. Lett., **68**, 1576 (1996).

9. S. J. L. Billinge, R. G. DiFrancesco, G. H. Kwei, J. J. Neumeier, and J. D. Thompson, Phys. Rev. Lett. **77**, 715 (1996).

10. C. H. Booth, F. Bridges, G. J. Snyder, and T. H. Geballe, Phys. Rev. B **54**, R15606 (1996).

11. G. Zhao, K. Conder, H. Keller, and K. A. Müller, Nature **381**, 676 (1996).

12. R.H. Heffner, L.P. Le, M.F. Hundley, J.J. Neumeier, G.M. Luke, K. Kojima, B. Nachumi, Y.J. Uemura, D.E. MacLaughlin, and S-W. Cheong, Phys. Rev. Lett. **77**, 1869 (1996).

13. M. Jaime, M.B. Salamon, M. Rubinstein, R.E. Treece, J.S. Horwitz, and D.B. Chrisey, Phys. Rev. B **54**, 11914 (1996).

14. M. Jaime, H.T. Hardner, M.B. Salamon, M. Rubinstein, P. Dorsey, and D. Emin, Phys. Rev. Lett. **78**, 951 (1997); M.F. Hundley and J.J. Neumeier, Phys. Rev. B **55**, 11511 (1997).

15. J.Z. Sun, L. Krusin-Elbaum, S.S.P. Parkin, and Gang Xiao, Appl. Phys. Lett. **67**, 2726 (1995); B.X. Chen, C. Uher, D.T. Morelli, J.V. Mantese, A.M. Mance, and A.L. Micheli, Phys. Rev. B **53**, 5094 (1996); B. Martinez, J. Fontcuberta, A. Seffar, J.L. Garcia-muñoz, S. Piñol, and X. Obradors, Phys. Rev. B *54*, 10001 (1996).

16. M.F. Hundley, J.J. Neumeier, R.H. Heffner, Q.X. Jia, X.D. Wu, and J.D. Thompson in Epitaxial Oxide Thin Films III, edited by D.G. Schlom, C-B. Eom, M.E. Hawley, C.M. Foster, and J.S. Speck (Mater. Res. Soc. Proc. 474, Pittsburgh, PA 1997), p. 167.

17. C. H. Booth, F. Bridges, G. H. Kwei, J. M. Lawrence, A. L. Cornelius, and J. J. Neumeier, Phys. Rev. Lett., in press.

18. C. H. Booth, F. Bridges, G. H. Kwei, J. M. Lawrence, A. L. Cornelius, and J. J. Neumeier, submitted to Phys. Rev. B.

19. S. I. Zabinsky, A. Ankudinov, J. J. Rehr, and R. C. Albers, Phys. Rev. B **52**, 2995 (1995).

20. J. F. Mitchell et al., Phys. Rev. B **54**, 6172 (1996).

21. K. R. Poeppelmeier, M. E. Leonowicz, J. C. Scanlon, and J. M. Longo, J. Solid State Chem. **45**, 71 (1982).

22. D. Louca, W. Dmowski, T. Egami, E. L. Brosha, H. Röder, and A. R. Bishop, Phys. Rev. B **56**, R8475 (1997).

23. A. J. Millis and H. Röder, private communication.

24. A. J. Millis, R. Mueller, and B. I. Shraiman, Phys. Rev. B **54**, 5405 (1996).

25. H. Röder, J. Zang, and A. R. Bishop, Phys. Rev. Lett. **76**, 1356 (1996).

26. A. Yaouanc and P. Dalmas de Réotier, J. Phys.: Cond. Matter **3**, 6195 (1991).

27. E. O. Wollen and W. C. Koehler, Phys. Rev. **100**, 548 (1995).

28. S. A. Dodds, G. A. Gist, D. E. MacLaughlin, R. H. Heffner, M. Leon, M. E. Schillaci, G. J. Nieuwenhuys and J. A. Mydosh, Phys. Rev. B **28**, 6209 (1983).

29. J. W. Lynn, R. W. Erwin, J. A. Borchers, Q. Huang, A. Santoro, J.-L. Peng, and Z. Y. Li, Phys. Rev. Lett. **76**, 4046 (1996).

THE EFFECT OF RADIATION INDUCED DISORDER ON
La$_{0.7}$Ca$_{0.3}$MnO$_{3-\delta}$ THIN FILMS

R.M. STROUD, V.M. BROWNING, J.M. BYERS, D.B. CHRISEY, W.W. FULLER-MORA, K.S. GRABOWSKI, J.S. HORWITZ, J. KIM, D.L. KNIES and M.S.OSOFSKY,
Naval Research Lab., 4555 Overlook Ave. SW, Washington, DC 20375

ABSTRACT

The effects of disorder on the transport and magnetic properties of pulsed laser deposited La$_{0.7}$Ca$_{0.3}$MnO$_3$ thin films were studied. Ion irradiation with 10 MeV I and 6 MeV Si ions was used to produce controlled levels of defects ranging from 0.006 to 0.024 displacements per atom. The peak resistance temperature of the I-irradiated films decreased from 264K for the undamaged film to 0K for the 0.016 dpa film. The magnetic ordering temperature decreased from 270K (un-damaged) to 130K (0.011 dpa), and remained nearly constant at 130K for higher damage levels. This demonstrates a decoupling of the magnetic and metal-insulator transitions. A characteristic relationship between the peak resistance temperature and activation energy of resistance with the form $T_p = 285(1-\Delta_\rho/115)^{0.13}$, was observed for all films. This characteristic relationship indicates that the range of resistivity and magnetoresistivity values for observed La$_{0.7}$Ca$_{0.3}$MnO$_3$ can be explained in terms of disorder-limited polaron hopping.

INTRODUCTION

The metal-insulator transition temperatures and magnetoresistance values of doped manganites vary widely between samples. For La$_{0.7}$Ca$_{0.3}$MnO$_3$ (LCMO) in particular, the peak resistance temperatures can range from 280K to 0K, and magnetoresistance ((R(T,H)-R(T,O))/R(T,H)) from 10% to 10^6%.[1, 2, 3] The double exchange model [4] explains the occurrence of the transition from the high temperature paramagnetic phase to the low temperature ferromagnetic phase. However, it does not explain either the colossal magnetoresistance observed in some manganite thin films[5], or the large variation of properties between samples. Explaining this variation of transport and magnetic properties is one of the important outstanding problems in manganite research.

Many studies have been conducted in which an electronic, structural, or magnetic perturbation was applied to a manganite, and the effects on the metal-insulator transition were measured. The transition to the metallic phase has been induced by applying a magnetic field, or irradiating the sample with x-rays.[6] The transition temperature has been altered by cation substitution [7], using electric fields [8], changing oxygen stoichiometry [9] and applying high pressure [10]. It is clear from the results of these studies that the metal-insulator transition is subject to many influences. It is not clear from these results how the range of properties from ordinary magnetoresistance to colossal magnetoresistance arises from the interaction of these influences. In this paper, we demonstrate how a range of metal-insulator transition temperatures, and thus magnetoresistance values, can result from extrinsic disorder.

EXPERIMENT

Films, 260 nm thick, with nominal composition La$_{0.7}$Ca$_{0.3}$MnO$_3$ were deposited on (100) LaAlO$_3$ substrates by pulsed laser deposition. The deposition conditions were: 750°C substrate temperature, 2.0 J/cm^2 laser fluence at 248 nm, 200 mT oxygen atmosphere. The films were an-nealed ex-situ for 24 hours at 1000°C to ensure full oxygenation and minimize structural defects.

Quantifiable, uniformly distributed defects were produced in the films using ion irradiation. Two types of ions were used: 6 MeV Si and 10 MeV I. The energy of the ions was sufficiently high that all ions passed through the films, and were deposited at a depth of 2 μm into the substrate. The damage produced by the ions was simulated using SRIM[11]. A fluence of 1.1×10^{14} 6 MeV Si ions/cm^2 or 5.0×10^{12} I ions/cm^2 was predicted to result in 0.006 displacements per atom (dpa). Four I and four Si fluences were used to produce defect concentrations in the films ranging from 0.006 dpa to 0.024 dpa. The 10 MeV I irradiations were all performed on film pieces cut from one film. The silicon irradiations were performed on a series of different films.

The films were characterized as a function of defect concentration using x-ray diffraction (XRD), transmission electron microscopy (TEM), resistivity, magnetization, and thermopower measurements. The xrd measurements were made with a Rigaku diffractometer with a rotating anode Cu source and a graphite monochromater. Two transmission electron microscopes were used: a 300 KeV Philips CM-30 with a windowless EDS detector for analytical measurements, and 300 KeV Hitachi S-9000 for high-resolution measurements. Four-point probe resistivity measurements were made over the temperature range of 4K to 300K in a closed cycle refrigerator, with measuring currents ranging from 10 nA to 100 μA depending on the sample resistance. The Curie temperatures were determined using a SQUID magnetometer by measuring the sample magnetization as a function of temperature in a 1000 Oe applied magnetic field. For thermopower measurements, a temperature gradient was applied across the films and the resulting Seebeck voltage was measured.

RESULTS

Structural characterization of the films before irradiation showed them to be well-ordered. The XRD data showed single-phase growth, single direction growth, with the LCMO (100) axis parallel to the (100) $LaAlO_3$ out-of-plane axis. The lattice parameter was 3.86Å. The TEM studies revealed strain at the film-substrate interface and columnar growth defects. High resolution TEM studies showed well-ordered regions of film extending approximately 500Å * 500Å, and good epitaxy between the films and substrate.[12]

The effect of the radiation damage on the structure of the films was small. Although the (200) XRD rocking curve peak widths of the films increased from 0.3° to 0.45°with damage, no change in the order was observed by high resolution TEM.

For the I irradiation series, the peak resistance temperature (T_p) decreased from 264K for the undamaged film to 75K for the 0.011 dpa film (Fig. 1). The room temperature resistivity increased with damage. At the highest damage levels, 0.016 dpa and 0.021 dpa, activated resistivity was observed until the resistance of the film exceeded the measurement limit of $\sim$ 5 MΩ. At 4K, the resistance of the 0.016 dpa and 0.021 dpa films remained above 5 MΩ. The peak resistance temperature of these films is taken to be 0K. For the Si irradiation series, quantitatively similar changes of the temperature dependent resistivity were observed[12].

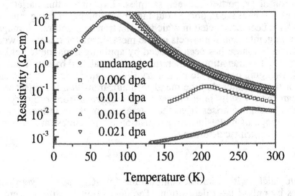

Figure 1: Resistivity versus temperature for 10 MeV I-irradiated $La_{0.7}Ca_{0.3}MnO_3$.

The effect of disorder due to the two different ions, 10 MeV I and 6 MeV Si, can be quantitatively compared by plotting the peak resistance temperature and magnetic ordering temperature (T_c) as a function of damage (Fig. 2). The decrease in peak resistance temperature for Si-induced damage and I-induced damage is similar. However, the threshold for elimination of the metal-insulator transition, $T_p = 0$, is lower for 10 MeV I (0.016 dpa) than for 6 Mev Si (0.024 dpa). Unlike T_p, T_c does not decrease to zero with damage. After the initial decrease in T_c with damage from 270K to $\sim$ 130K, T_c remains fixed. This demonstrates a disorder-induced decoupling of the magnetic and metal-insulator transitions. For comparison the peak resistance temperature of oxygen deficient films as a function of radiation- induced damage is also shown. The oxygen deficiency results in depressed peak resistance temperature before irradiation, and a decreased damage threshold for elimination of the M-I transition (0.011 dpa).

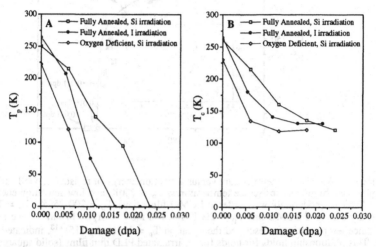

Figure 2: A.) Peak resistance temperature versus radiation damage. B.) magnetic ordering temperature versus radiation damage. The peak resistance and magnetic ordering temperatures diverge for large damage levels > 0.01 dpa.

The effect of disorder on the M-I transition can be analyzed by plotting T_p as a function of the activation energy of the insulating state (Fig. 3). The activation energy, Δ_ρ, is calculated by fitting the resistivity data above T_p to the function $\rho = \rho_0 exp(\Delta_\rho/k_B T)$. This analysis has been performed for resistivity data obtained from annealed, radiation damaged, damaged and re-annealed, and oxygen deficient samples, grown under varying deposition conditions and on different substrates. Together, these data define a characteristic relationship between T_p and Δ_ρ, $T_p = 285(1 - \Delta_\rho/115)^{0.13}$, as indicated by the solid line. The iodine irradiation series is a representative subset of the data, and quantitatively demonstrates how disorder modulates Δ_ρ and determines T_p. The undamaged film has a high T_p and low Δ_ρ, $T_p = 264K$, $\Delta_\rho = 60$ meV. A small amount of disorder, 0.006 dpa, results in a small decrease in T_p to 207K, and a large increase in Δ_ρ to 107 meV. For 0.011 dpa, T_p decreases to 75K and Δ_ρ increases to 110 meV. At 0.016 dpa and 0.021 dpa, $T_p = 0K$ and $\Delta_\rho = 115$ meV and 118 meV respectively. This characteristic relationship between activation energy and T_p holds for the resistivity activation energy only. There is no simple relationship between T_p and the thermopower activation energy, Δ_s (Fig. 3 inset).

DISCUSSION

The purpose of using radiation damage in this study was to simulate the effects of the disorder on the properties of LCMO under controlled conditions. Many LCMO samples contain disorder as a result of less than ideal growth conditions. Oxygen deficiency and strain at the substrate-film interface are particularly common in thin films. These extrinsic defects can have dramatic effects on the M-I transition and the magnetoresistance, but they do not change the fundamental physics governing the properties of the material. Similarly, radiation damage can produce disorder that affects the transport properties of LCMO without fundamentally altering the physics of the damaged samples. The proposition that radiation damage in these samples is comparable to growth defects is supported by the data. The T_p-Δ_ρ data (Fig. 3) shows complete overlap of the radiation damaged sample data with that for samples grown under varying deposition conditions. The T_p

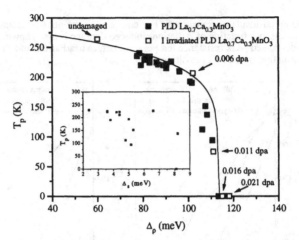

Figure 3: Peak resistance temperature versus activation energy of resistance. Well-ordered LCMO samples have high peak resistance temperatures, $T_p > 250K$, and low activation energy, $\Delta_\rho < 70$ meV. Typical, moderately well-ordered LCMO films have $T_p = 200-250K$, $\Delta_\rho = 100-80$ meV. Critically damaged LCMO has $T_p < 200K$ and 100 meV $< \Delta_\rho < 115$ meV. The range of peak resistance temperatures is described the equation $T_p = 285(1-\Delta_\rho/115)^{0.13}$, indicated by the solid line. This relationship holds for holds for un-irradiated PLD thin films (solid squares), as well as irradiated PLD thin films (open squares). Inset: Peak resistance temperature versus thermopower activation energy. No consistent relationship between T_p and Δ_s is observed.

and T_c data from radiation damaged oxygen deficient samples (Fig. 2) is qualitatively the same as that for fully-oxygenated radiation damaged samples. There is small offset between the oxygen deficient and oxygenated T_p and T_c values that indicates oxygen deficiency can be mapped to displacement damage.

The primary effect of the radiation-induced disorder on the LCMO transport properties is to lower the mobility of carriers. In the insulating state of LCMO, the carriers are polarons. The disorder limits the mobility of the polarons by increasing the average barrier to polaron hopping. The experimental manifestation of the increased barrier to hopping is the increased activation energy of resistance, Δ_ρ. Both carrier concentration and carreir mobility affect Δ_ρ, however the thermopower measurements provide an activation energy, Δ_s, that is sensitive to carrier concentration only. The thermopower activation energy Δ_s is more than an order of magnitude smaller than Δ_ρ (Fig. 3), and shows no obvious trend with damage. Thus, the variation in Δ_ρ results primarily from changes in polaron mobility.

One way in which the radiation damage decreases the polaron mobility is by decreasing the amount of magnetic order. The decreased magnetic order is evidenced by the lower magnetic ordering transition temperature (Fig. 2b), a decrease in saturation magnetization and increased coercivity[13]. Instead of the purely ferromagnetic ordering of the undamaged samples, the radiation damaged samples show either ferrimagnetic, or superparamagnetic ordering. The decrease in magnetic ordering leads to the decoupling of the magnetic and M-I transitions shown in Fig. 2. A magnetic transition persists after $T_p \rightarrow 0$, but the magnetic ordering is relatively weak. The increase in polaron mobility due to the weak magnetic ordering is not enough to produce a metal-insulator transition. A similar decoupling of the magnetic and M-I transitions has been observed previously for oxygen-deficient $La_{0.7}Sr_{0.3}MnO_{3-\delta}$.[14]

The range of metal-insulator transition temperatures and resistivity values for LCMO can be understood in terms of disorder. The transition onset temperature, T_p, is determined by Δ_ρ, which is in turn determined by the amount disorder. The characteristic $T_p-\Delta_\rho$ relationship for LCMO in zero-field is shown in Fig. 3. For well-ordered samples, $T_p > 250K$, and $\Delta_\rho < 80$ meV. The majority of thin film samples produced for this study, and reported in the literature, have $T_p \sim 200\text{-}250K$, $\Delta_\rho \sim 80\text{-}100$ meV. These samples are moderately well-ordered films, and contain more disorder than the bulk materials that show $T_p = 280K$. The precise level of disorder is most important for the critical range of activation energies, 110 meV $< \Delta_\rho < 115$ meV, over which T_p decreases from 200K to 0K. Samples disordered to the point that $\Delta_\rho > 115$ meV generally show no transition, only activated resistance.

The range of magnetoresistance properties of LCMO can now be explained in terms of the characteristic $T_p\text{-}\Delta_\rho$ relationship and disorder. For well-ordered LCMO, the activation barrier for polaron hopping, Δ_ρ, is low. An applied magnetic field can only produce a small increase in the order and polaron mobility, resulting in only a small change in the resistivity. The magnetoresistance of these samples is small. Colossal magnetoresistance occurs when the samples are disordered, producing large Δ_ρ, and low T_p. For these samples, a large magnetic field can greatly enhance the order and polaron mobility, but only at low temperatures. The difference between activated resistivity and metallic resistivity is greatest at lower temperature, so the effect of the magnetic ordering on the resistivity is the greatest.

CONCLUSION

There is a characteristic relationship between the metal-insulator transition temperature and the resistivity activation energy that is a function of disorder. Disorder increases the barrier to polaron hopping, which increases the resistivity activation energy, and decreases the peak resistance temperature. The relationship $T_p = 285(1\text{-}\Delta_\rho/115)^{0.13}$ holds for disorder produced by radiation damage, oxygen deficiency and growth defects in thin films. This relationship provides a mechanism for understanding the range of resistivities, metal-insulator transition temperatures and magnetoresistances observed for LCMO. Magnetoresistance is determined by the amount of disorder, the transition temperature and the resistivity activation energy. There is a critical range of activation energies (110 meV - 115 meV) for which the magnetoresistance is greatest. For sufficiently high levels of disorder at which the activation energy exceeds 115 meV, the magnetic ordering and metal-insulator transitions are decoupled.

We speculate that the form of the characteristic relationship observed for LCMO will hold for other colossal magnetoresistive manganites. The value of the critical range of activation energies for maximum magnetoresistance will vary with the composition of the manganite, but the effects of disorder on polaron mobility should be the same as for LCMO.

ACKNOWLEDGMENTS

Financial support from ONR for the completion of this work is gratefully acknowledged. Additional support was received in the form of a NRC/NRL Cooperative Research Associateship for RMS.

REFERENCES

1 J. B. Goodenough, Phys. Rev. **100**, p. 564 (1955).

2 E. O. Wollan and W. C. Koehler, Phys. Rev., **100** (2) p. 545 (1955).

3 S. Jin, T. H. Tiefel, M. McCormack, R. A. Fastnacht, R. Ramesh, and L. H. Chen, Science **264**, 413 (1994).

4 C. Zener, Phys. Rev. **82**, 403 (1951).

5 A. J. Millis, P. B. Littlewood and B. I. Shraiman, Phys. Rev. Lett. **74**, 5144 (1995).

6 V. Kiryukhin, D. Casa, J. P. Hill, B. Keimer, A. Vigilante, Y. Tomioka and Y. Tokura, Nature **386**, 813 (1997).

7 J. Fontcuberta, B. Martinez, A. Seffar, S. Pinol, J. L. Garcia-Monuz and X. Obradors, Phy. Rev. Lett. **76**, 1122 (1996).

8 S. Jin, JOM **49**, p. 61 (1997).

9 J. S. Horwitz, P. C. Dorsey, N. C. Koon, M. Rubenstein, J. M. Byers, M. S. Osofsky, V. G. Harris, K. S. Grabowski, D. L. Knies, R. E. Treece and D.B. Chrisey, SPIE **2703**, 526 (1996).

10 H. Y. Hwang, T. T. M. Palstra, S.-W. Cheong and B. Batlogg, Phys. Rev. B **52**, 15046 (1995).

11 J. F. Ziegler, J. P. Biersack and U. Littmark, The Stopping Range of Ions in Solids, (Pergamon Press, New York, 1985).

12 R. M. Stroud, V. M. Browning, J. M. Byers, D. B. Chrisey, W. W. Fuller-Mora, K. S. Grabowski, D. L. Knies and M. S. Osofsky, in Epitaxial Oxide Thin Films III , edited by C. Foster, J. S. Speck, D. Schlom, C.-B . Eom and M. E. Hawley (Mater. Res. Soc. Proc. **474** Pittsburgh, PA 1997), in press.

13 V. M. Browning, R. M. Stroud, W. W. Fuller-Mora, J. M. Byers, M. S. Osofsky, D. L. Kneis, K. S. Grabowski, D. Koller, J. Kim, and D.B. Chrisey and J. S. Horwitz, submitted to J. Appl. Phys..

14 R. Mahendiran, R. Mahesh, A. K. Raychaudhuri and C. N. R. Rao, Solid. State. Comm. **99**, 149 (1996).

VOLUME-BASED CONSIDERATIONS FOR THE
METAL-INSULATOR TRANSITION OF CMR OXIDES

J. J. Neumeier[1,2], A. L. Cornelius[2], M. F. Hundley[2], K. Andres[3], and K. J. McClellan[2]
[1]Department of Physics, Florida Atlantic University, Boca Raton, FL 33431, USA
[2]Material Science and Technology Division, Los Alamos National Laboratory, Los Alamos, NM 87545, USA
[3]Walther-Meissner-Institut fuer Tieftemperaturforschung, Walther-Meissner-Str. 8
D85748 Garching, Germany

ABSTRACT

The sensitivity of ρ to changes in volume which occur through: (1) applied pressure, (2) variations in temperature, and (3) phase transitions, is evaluated for some selected CMR oxides. It is argued that the changes in volume associated with (2) and (3), which are referred to as *self pressures*, are equivalent in magnitude to changes in volume resulting from pressures in the range of 0.18 to 0.45 GPa. Through consideration of thermal expansion and electrical resistivity data, it is shown that these *self pressures* are responsible for large features in the electrical resistivity and are an important component for the occurrence of the metallicity below T_c. It is suggested that this is a manifestation of a strong volume dependence of the electron phonon coupling in the CMR oxides.

INTRODUCTION

Ferromagnetic transition metal oxides of the chemical composition $A_{1-x}A_xMnO_3$ (where A = La. Pr, Gd, Pb, Nd and A = Ca, Sr, Ba, Cd) have been investigated since the early 1950's [1-2]. At the ferromagnetic transition temperature T_c the electrical resistivity displays a metal-insulator transition. This is quite different from common metallic ferromagnets for which the electrical resistivity exhibits a modest change in slope at T_c due to a decrease of spin-disorder scattering as ferromagnetism sets in. Zener [3] proposed an explanation for the ferromagnetic interaction in the manganites called double exchange (DE). In the double exchange picture, electrical conduction can occur only if neighboring Mn ions are ferromagnetically aligned. The conduction lowers the ground-state energy of the system and mediates the ferromagnetic exchange.

The interplay between magnetic, structural, and electrical transport properties in this system is exceptionally strong, and leads to the largest negative magnetoresistivity (MR) observed in any material near room temperature; the effect is often referred to as colossal magnetoresistance (CMR). If one defines MR = $|\rho(H) - \rho(0)|/\rho(0)$, it can have a magnitude larger than 90% [4-8]. This exceedingly large MR opens the door for a number of potential applications of $A_{1-x}A_xMnO_3$ in sensor technology. In addition to the technological promise of the CMR oxides, they provide an example of how subtle aspects of the crystal structure play an important role in determining the electrical and magnetic properties of complex oxides.

A number of diverse experiments indicate that electrical conduction in CMR materials takes place via magnetic polarons. A magnetic polaron consists of a conduction electron and a group of magnetic spins which propagates through the lattice by aligning neighboring spins along its path. Diffuse magnetic scattering of neutrons above T_c displays an activation energy consistent with that observed in electrical resistivity measurements [9], in support of electrical conduction via magnetic polarons. Measurements of thermoelectric power [10,11] and electrical resistivity [8] illustrate that the energy gap determined by electrical resistivity measurements is about 10 times larger than that determined by thermoelectric power, this is considered the hallmark of conduction via polarons. Local structure analysis of powder neutron diffraction data [12] and x-ray absorption measurements [13,14] reveal local structural distortions which are consistent with lattice polaron formation. Thus, a reasonable scenario is that the polarons manifest themselves as both a magnetic and lattice distortion. Theoretical investigation has noted that the DE model alone cannot explain the large MR [15] and the significance of magnetic polarons as well as the structural distortion associated with the Jahn-Teller effect [15-17]. Thus, both experimental and theoretical

Mat. Res. Soc. Symp. Proc. Vol. 494 © 1998 Materials Research Society

investigations indicate that lattice distortions, presumably associated with magnetic polarons, play an important role in determining the properties of CMR materials.

The formation of polarons in the manganese oxides is a result of large electron-phonon coupling [18]. The polarons begin to delocalize as T is decreased below T_c, thus permitting formation of the metallic state. The significance of electron-phonon coupling for the ferromagnetic exchange has been demonstrated by the observation of a large oxygen isotope effect on T_c [19]. In addition, a series of high pressure experiments on specimens, where the structure has been altered by chemical substitution, indicates that electron-phonon coupling plays an important role in the influence of pressure on T_c and the electrical resistivity [20]. In this scenario, the delocalization of the polarons below T_c is a result of a decrease in electron-phonon coupling as T is reduced below T_c.

The focal point of this work is the thermodynamic properties of the CMR materials. These experiments deal with measurements of chemical compositions at the boundary between ferromagnetic metal and ferromagnetic insulator [21]. The sensitivity of CMR materials to pressure (i.e. a controlled reduction of the volume) is examined to evaluate the influence of temperature-dependent volume changes on the electrical resistivity. These results are discussed within the framework of electron-phonon coupling.

EXPERIMENTAL

A 10 g polycrystalline specimen of $La_{0.79}Ca_{0.21}MnO_{3-y}$ was prepared from 99.99% purity La_2O_3 (dried at 600 °C), $CaCO_3$, and MnO_2 which were mixed and reacted at 1100 °C in air for 20 h. The specimen was always removed from the furnace at T > 1000 °C to reduce the number of La vacancies σ that form by the uptake of oxygen [22]. After the initial reaction, the powder was reground, reacted at 1200 °C for 20 h, reground and reacted at 1350 °C for 20 h. The specimen was then reground, pelletized, and reacted at 1350 °C for 20 h; this step was repeated once. After regrinding, it was pressed, reacted at 1375 °C for 40 h, removed from the furnace at 1200 °C, and labeled specimen A (or simply A). A portion of this specimen (specimen B or B) was heated in air to 1375 °C and slow-cooled at 60 °C/h to 26 °C. Iodometric titration under the assumption of +3, +2, +2, and -2 valences for La, Ca, Mn, and O, respectively, in acidic solution yielded Mn valences of 3.239 ± 0.004 ($\sigma = 9.70 \times 10^{-3}$ vacancies/formula-unit) for A and 3.266 ± 0.004 ($\sigma = 18.7 \times 10^{-3}$ vacancies/formula-unit) for B. Thus, A has about 10% fewer doped holes than B. A monoclinic cell was used for lattice parameter calculations based on powder x-ray diffraction data with a Si standard yielding a=3.890(2), b=3.889(2), c=3.880(2) and β=89.94(5) for A and a=3.888(2), b=3.885(2), c=3.877(2) β = 89.96(5) for B whose unit-cell volume is 0.25 % smaller than A. Four-probe electrical resistivity and ac magnetic susceptibility measurements at high pressure are described elsewhere [23]. Magnetization studies were conducted with a commercially available SQUID magnetometer. Thermal expansion measurements utilized a capacitive dilatometer. A single crystal specimen of $La_{0.83}Sr_{0.17}MnO_3$ was grown using the optical float zone method. Polycrystalline feed and seed rods were melted in air at ambient pressure. Rotation rates of 50 rpm were used for the feed and seed and growth rates were 3-5 mm/h.

THERMODYNAMICS OF A SECOND-ORDER PHASE TRANSITION

When a system undergoes a second-order phase transition, its volume V changes in a continuous fashion. In contrast, when a system experiences a first-order phase transition, V is discontinuous at the phase transition. For a second-order phase transition, the derivative of V, the volume thermal expansion $\beta(T)$, is discontinuous at T_c. The continuity of the Gibb's free energy and its first derivative leads to the Ehrenfest relation given by

$$\Delta\beta = \frac{\Delta C_p}{V_{mol}T_c}\left[\frac{dT_c}{dP}\right],$$ (1)

where $\Delta\beta$ is the discontinuity in $\beta(T)$, ΔC_p is the discontinuity in the specific heat at constant pressure, V_{mol} is the molar volume, and dT_c/dP is the derivative of the phase transition temperature as a function of pressure near $P = 0$ [24]. Thus, Equation (1) provides a clear relation among a number of experimentally determinable quantities. The Ehrenfest relation will allow analysis of the ferromagnetic to paramagnetic phase transition in two of the CMR specimens investigated below.

MAGNETIC AND ELECTRICAL PROPERTIES AT AMBIENT PRESSURE

In Fig. 1 the magnetization M at H = 4000 Oe in units of μ_B/Mn-ion is plotted versus temperature for the $La_{0.79}Ca_{0.21}MnO_{3-y}$ specimens. M versus H at 5 K appears in the inset. Both

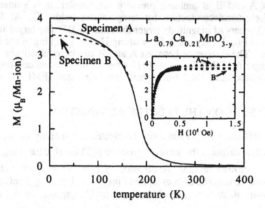

Fig. 1. Magnetization versus temperature for A and B at H = 4000 Oe. In the inset, the magnetic moment at 5 K is plotted versus H.

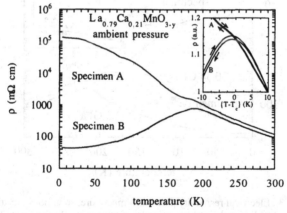

Fig. 2. Electrical resistivity versus temperature for Specimens A and B at ambient pressure. In the inset, the electrical resistivity (in arbitrary units) is plotted versus $T-T_c$ for warming and cooling.

the hole deficient specimen A and specimen B are clearly ferromagnetic with magnetic saturation moments M(5 K, 35 kOe) of $3.95\mu_B$/Mn-ion and $3.71\mu_B$/Mn-ion for A and B, respectively. Arrott plots (i.e. M^2 vs H/M) of magnetization data taken above and below the ferromagnetic transition temperatures and electrical resistivity data yielded T_c values of 188.1 K and 192.6 K for A and B, respectively. These observations illustrate that the magnetic properties of A and B are similar.

In Fig. 2 the electrical resistivity ρ versus temperature for A (the hole deficient specimen) and B is shown at ambient pressure. At T_c, A (the reduced specimen) does not exhibit a metal-insulator transition. Specimen B, on the other hand, exhibits a metal-insulator transition typical for CMR materials. Warming and cooling through the transition at 0.17 K/min revealed minimal hysteresis for A and B of 0.4 K and 1.0 K, respectively, as illustrated in Fig. 1. Although the magnetic properties of A and B at ambient pressure are similar, it is evident that they differ markedly in terms of electrical conduction for temperatures below T_c. At 4 K, the electrical resistivity of A is over 3 orders of magnitude larger than that of B! The larger magnetic saturation moment of A, revealed in Fig. 1, may result from strong Hund's coupling of the localized electrical carriers to the Mn-ions. The difference between A and B can be qualitatively appreciated if one considers the phase diagram of Schiffer et al. [21] wherein the calcium composition 0.21 resides near the phase boundary separating the ferromagnetic insulating state (FMI) from the ferromagnetic metallic state (FMM).

INFLUENCE OF PRESSURE ON THE ELECTRICAL RESISTIVITY

In Fig. 3, the influence of pressure on the electrical resistivity of $La_{0.79}Ca_{0.21}MnO_{3-y}$ is presented. The dashed line displays the ambient pressure $\rho(T)$ for B; data at higher pressures for a similar specimen appear elsewhere [23]. Application of pressure to A dramatically decreases ρ revealing curves comparable to those of B at ambient pressure. $Pr_{0.7}Ca_{0.3}MnO_3$ displays a similar pressure-induced metal-insulator transition [25]. Releasing the pressure on A from P = 1.5 GPa

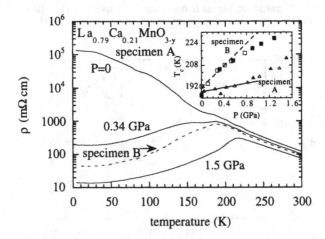

Fig. 3. Electrical resistivity ρ versus temperature, solid lines are for A, dashed line is the ambient pressure curve for B. In the inset, T_c versus pressure for A (triangles) and B (squares), filled symbols are from ρ data, open symbols are from ac magnetic susceptibility. Straight lines have slopes of 7.79 K/GPa and 40.3 K/GPa for A and B, respectively.

restored the insulating behavior. Thus, pressure facilitates a reversible crossover from the FMI to the FMM regions of the Schiffer et al. phase diagram [21]. Furthermore, the pressure sensitivity of ρ is largest in the ferromagnetic state for both specimens. The pressure dependence of T_c is presented in the inset of Fig. 3, the solid symbols were determined from the feature in ρ at T_c while the open symbols were determined by ac magnetic susceptibility χ_{ac}; the feature in $\rho(T)$ occurs slightly below T_c at ambient pressure. The pressure derivatives $dT_c/dP(P=0)$ from χ_{ac} for these two specimens differ dramatically. Specimen A exhibits $dT_c/dP(P=0) = 7.79$ K/GPa while B displays $dT_c/dP(P=0) = 40.3$ K/GPa; dT_c/dP of B agrees with that reported earlier [23]. The slope dT_c/dP for A increases at pressures greater than 0.6 GPa and attains values comparable to B.

VOLUME THERMAL EXPANSION AT AMBIENT PRESSURE

In Fig. 4 ambient pressure measurements of the volume expansion $\delta V_T \equiv \{V(T)-V(0)\}/V(0)$ are displayed. As is readily apparent, the volume changes in a continuous fashion at T_c for both specimens. This indicates that the phase transition is second-order in nature, according to our discussion above. However, the electrical resistivity data in Fig. 2 illustrate that a modest hysteresis occurs near T_c. It is suspected that the phase transition may possess a weak first-order component, perhaps associated with a minor change in crystal structure. At any rate, the hysteresis observed here is minimal compared to that reported for a specimen of $Pr_{0.7}Ca_{0.3}MnO_3$ [25]. The volume thermal expansion coefficient $\beta(T) = d(\delta V_T)/dT$ is plotted in the inset. The solid lines are for specimen A while the dashed lines are for specimen B. The jump in β associated with the ferromagnetic transitions are $\Delta\beta = 7.6 \times 10^{-6}$ K^{-1} and 3.0×10^{-5} K^{-1} for specimens A and B, respectively; note the surprisingly large difference in the $\Delta\beta$ values. Since T_c is comparable for A and B, and if ΔC_P is assumed to be similar [26], the Ehrenfest relation predicts a smaller value of dT_c/dP for A. This is indeed observed in the inset of Fig. 3. ΔC_P values calculated with Eq. (1) (5.1 J/mole-K and 4.0 J/mole-K for A and B, respectively) are comparable to ΔC_P reported [27]

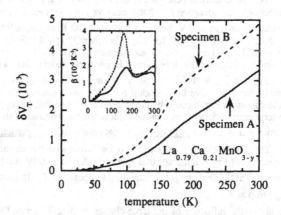

Fig. 4. $\delta V_T \equiv \{V(T) - V(0)\}/V(0)$ versus temperature for A (solid line) and B (dashed line). In the inset, the thermal volume expansion $\beta \equiv d(\delta V_T)/dT$ is plotted versus temperature.

for a $La_{0.9}Ca_{0.1}MnO_3$ specimen ($\Delta C_P \approx 4.1$ J/mole-K). This simple analysis indicates that our observation, namely the large difference in dT_c/dP (and $\Delta\beta$) between A and B, appears to be consistent with the thermodynamics of a second-order phase transition.

VOLUME AND ITS ROLE IN THE METAL-INSULATOR TRANSITION

Specimens A and B are quite similar in terms of composition, that is A has about 0.03 less doped holes than B. Surprisingly, their electrical properties at ambient pressure are dramatically different. The temperature dependence of δV_T as well as the volume dependence of T_c (i.e. dT_c/dP) are also dissimilar for these two specimens. If we consider Specimen A, a small reduction of volume (through the application of hydrostatic pressure) results in electrical properties similar to B and an increase in dT_c/dP.

To develop a better qualitative understanding of the data, the bulk modulus $B_M = -V(dP/dV)$ is needed to estimate the volume change $\delta V_P \equiv \{V(0)-V(P)\}/V(0)$ resulting from a given pressure. A room-temperature value of $B_M = 182\pm18$ GPa was estimated [28] from crystal chemistry considerations. Resonant ultrasound measurements (RUS) [29] of a $La_{0.83}Sr_{0.17}MnO_3$ single crystal ($T_c = 265$ K) reveal $B_M(295$ K, H=0$) = 177 \pm 11$ GPa (in good agreement with our estimate), $B_M(270$ K, H=0$) = 138 \pm 8$ GPa, and $B_M(200$ K, H=0$) = 126 \pm 7$ GPa. The $La_{0.83}Sr_{0.17}MnO_3$ specimen is different than $La_{0.79}Ca_{0.21}MnO_{3-y}$ in that it undergoes a rhobohedral-orthorhombic structural transition at 285 K. Presumably the large decrease in B_M is associated with this transition. For this reason, we utilize the estimated $B_M = 182$ GPa for $La_{0.79}Ca_{0.21}MnO_{3-y}$ in the following discussion.

The importance of the subtle difference in crystallographic volume between A and B at room temperature can be appreciated through use of B_M. A given relative volume change $\Delta V/V$ (i.e. δV_P or dV_T) is multiplied by B_M to give the pressure which would be required to realize that volume change. At ambient pressure and 295 K the unit-cell volume of A is 0.25% larger than B, a pressure of 0.45 GPa would be required to reduce the volume of A to that of B at P=0; in fact, at this pressure A exhibits a M-I transition. This observation suggests that the contrast in temperature dependence of ρ for A and B can be qualitatively attributed to the differing unit-cell volumes.

In considering the temperature dependence of the δV_T curves, it is worthwhile to bring the data into perspective through consideration of what is observed in a less exotic material. Consider copper which exhibits an expansion of δV_T between 4 K and 300 K of about 9.6 x 10^{-3}, nearly a factor of two larger than what we observe for $La_{0.79}Ca_{0.21}MnO_{3-y}$ [30]. Furthermore, the bulk modulus of copper is 137 GPa [31] compared to our estimate of 182 GPa for the manganite. Thus the manganite is "harder" than copper (i.e. it requires a larger pressure to realize a particular change in volume), and the changes in δV_T with T are smaller. If the total change in δV_T between 4 K and 300 K were multiplied by the respective bulk moduli for these two materials (as a normalization), the decrease in volume shown in Fig. 4 could be considered similar to that of copper.

Although the temperature-dependent volume effects are comparable to those of a common metal, a major difference exists in that ρ(T) of the manganese oxides is particularly sensitive to pressure [20,23,25]. Thus, one might expect $\rho(T)$ to be influenced by temperature-dependent volume changes. Inspection of Fig. 4 indicates that specimen A, the poorly conducting specimen, exhibits a weak change in β at the ferromagnetic transition temperature T_c. In contrast, specimen B exhibits a large change in β at T_c.

In order to evaluate the influence of the large change in β at T_c on $\rho(T)$ for specimen B which exhibits a metal-insulator transition, let us consider some $\rho(P)$ isotherms. In Fig. 5 ρ is plotted as a function of P at three temperatures chosen above, near and below T_c. The upper axis of Fig. 5 indicates estimated values of $\delta V_P = \{V(0)-V(P)\}/V(0)$ which were obtained using the bulk modulus of 182 GPa. The electrical resistivity is particularly sensitive to changes in volume near and below T_c; at temperatures about 60 K above T_c the sensitivity is much reduced. Note that values of δV <u>as small as a few parts in one thousand</u> are sufficient to reduce ρ by 50% or more when T is near or below T_c. From these data and the data of Fig. 3 it is clear that a reduction of volume in the presence of ferromagnetism (or strong ferromagnetic fluctuations) can have a dramatic influence on ρ.

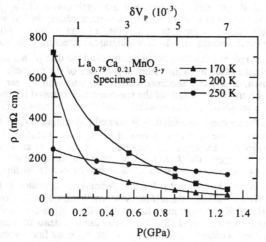

Fig. 5. Electrical resistivity as a function of pressure (lower axis) for specimen B at 170 K (triangles), 200 K, squares, and 250 K. T_c for this specimen is 192.6 K. The upper axis indicates the change in volume estimated with the bulk modulus of 182 GPa. Solid lines are guides to the eye.

With the volume sensitivity of ρ in mind, let us assess the temperature dependence of δV_T illustrated in Fig. 4 for its possible influence on ρ. In the temperature region immediately below T_c where the majority of the decrease in ρ occurs, the data in Fig. 2 (i.e. $T_c < T < T_c$-100 K) provides a value for $\delta V_T(T_c) - \delta V_T(T_c$-100 K) $\approx 2.5 \times 10^{-3}$. Multiplication of this value by B_M yields a pressure of 0.45 GPa. We call this the *self pressure* to indicate that it is the "pressure" generated by the volume change, in this case, associated with the paramagnetic to ferromagnetic transition. Note that the data in Fig. 5 indicate that a pressure of 0.45 GPa, if it were exerted on the specimen at 200 K (i.e. above T_c), is large enough to depress ρ significantly.

Similar considerations for specimen A provide $\delta V_T(T_c) - \delta V_T(T_c$-100 K) $\approx 1.3 \times 10^{-3}$. Multiplication of this value by B_M yields a *self pressure* of 0.24 GPa. Consideration of the $\rho(T)$ curves in Fig. 3 illustrate that a pressure of this magnitude is not sufficient to induce a metal-insulator transition. A report of δV_T data similar to those of specimen A were presented for a ferromagnetic $La_{0.88}Ca_{0.12}MnO_3$ specimen which also did not exhibit a metal-insulator transition [32].

The above arguments highlight volume as a parameter important in understanding the occurrence of a metal-insulator transition in CMR materials. It is by virtue of the extreme sensitivity of the electrical resistivity to changes in volume, that consideration of *self pressure* becomes relevant. In common metals, the sensitivity of the electrical resistivity is small in this pressure regime, thereby making any consideration of *self pressure* irrelevant. On the other hand, this is not the case for <u>all</u> metals. INVAR alloys are an example where the absence of a volume contraction below T_c results in a positive contribution to ρ [33]. High temperature superconductors are another example where ρ is sensitive to pressure. In this case, the temperature-dependent volume results in a sizable contribution to ρ which adds curvature to the $\rho(T)$ curve [34].

In Fig. 6 below, we show an additional example of the influence of *self pressure* on the electrical resistance of CMR materials. This specimen is a single crystal of $La_{0.83}Sr_{0.17}MnO_3$. This particular composition is known to exhibit an orthorhombic to rhombohedral structural transition near 300 K [34] whereby the low-temperature phase has a larger volume. All measurements in Fig. 6 were done along the rhombohedral 110 crystallographic direction. The solid curve displays the change in specimen length as a function of temperature and the data points are $\rho(T)$ normalized to the value at 330 K. In the inset, the temperature dependence of the magnetization in a magnetic field of 4000 Oe is shown; clearly this specimen is ferromagnetic with T_c near 275 K. Above T_c there is a discontinuous jump in the length of the specimen as illustrated by the solid line in the figure. This is a first-order phase transition which results in an expansion of the length by an amount $\Delta l/l = 1 \times 10^{-3}$ as the specimen is cooled. The normalized volume δV_T of this specimen (not shown) expands by 1×10^{-3} as the specimen is cooled through the transition; this is similar, but about a factor of 2 smaller than the expansion reported by Asamitsu et al. [35]. The structural transition is nearly 25 K above the ferromagnetic transition temperature. At a coincident temperature, a discontinuous increase in the electrical resistivity occurs as illustrated by the data points in the figure. It is well-known that ρ for this particular composition is highly sensitive to applied pressure [36]. We can estimate the self pressure which the specimen experiences by virtue of volume expansion at 300 K. A bulk modulus of 177 GPa [29] indicates that this expansion corresponds to a *negative* pressure of 0.18 GPa. There are two items of interest with regard to these data. As the specimen is cooled from room temperature, we see first a volume expansion that is not related to the ferromagnetic transition. This results in an increase in ρ. About 25 K below this structural transition, ferromagnetism sets in with a rapid reduction in volume due to the second-order paramagnetic to ferromagnetic phase transition. This results in a positive *self pressure* and a reduction in ρ similar to what is observed in specimen B above. Thus, through this example we see that both negative and positive *self pressures* can result in an increase and decrease in ρ, respectively. The electrical resistivity of this specimen is isotropic, so we believe

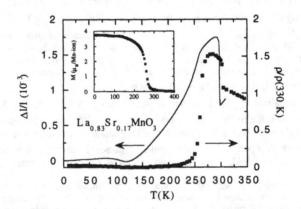

Fig. 6. Specimen length (solid curve) and normalized electrical resistivity (squares) plotted versus temperature. The magnetization at 4000 Oe as a function of temperature is plotted in the inset.

that the change in volume, and not a uniaxial length change, is responsible for the feature in ρ(T) at 300 K. Further measurements on this specimen will be discussed in detail in a forthcoming publication [37].

IMPORTANCE OF ELECTRON-PHONON COUPLING

To develop an understanding for the surprising contrast between specimens A and B, one must consider the influence of pressure and composition on the electron-phonon (e-p) coupling. It is clear from the oxygen isotope experiments that an e-p interaction exists in these materials [19] and that it is important for the occurrence of ferromagnetism. Furthermore, Laukhin et al. [20] have recently argued that the strong pressure effect on ρ and T_c, particularly in the region of doping where specimens A and B exist, is largely a result of a pressure-induced *reduction* of the e-p interaction. Other experiments have also suggested the important role of e-p coupling [38]. A theoretical treatment in 1987 considered the importance of e-p coupling and DE for the metal-insulator transition in a doped ferromagnetic semiconductor [39]. Millis, Shraiman and Mueller [18] presented a theory for the CMR manganites which considered the e-p coupling parameter λ in conjunction with the DE hopping matrix element t_{DE}. The strong e-p interaction decreases as T is reduced below T_c [18,39] and it is also influenced by the level of doping [18]. Consideration of the results for specimen A, it is clear that the occurrence of ferromagnetism <u>alone</u> (and the associated increase of t_{DE}) is not sufficient to result in a metal-insulator transition. When comparison is made between A and B, it becomes apparent that the temperature-dependence of the volume plays an important role in the occurrence of a metal-insulator transition. It is also valuable to note that when pressure is applied to a specimen which exhibits a metal-insulator transition, it decreases the large peak in β(T) [40]. This illustrates that as the volume and e-p coupling are decreased through hydrostatic pressure, <u>both</u> the peak in ρ(T) and β(T) are decreased. Thus, the large peak in β(T) observed in specimen B would probably vanish if one could reach pressures high enough to create metallicity at all temperatures. This further illustrates the intimate relation between the parameter volume and the electrical resistivity.

As to the physical basis for the surprising difference in δV_T and dT_c/dP between specimens A and B, we suspect that the phase diagram in Ref. 18 provides a clue. These two specimens straddle the FMI/FMM phase boundary in the phase diagram. The ambient pressure volume of A and B differs by about 0.25% and a reduction of the volume in A (through hydrostatic pressure) results in a M-I transition at T_c. The room temperature volume of A is presumably such that its e-p coupling is simply too large to be overcome by the reduction in volume which occurs as T is reduced to 4 K. Thus, the slightly larger volume of A seems to have upset the delicate balance between λ and t_{DE} required to result in a metal-insulator transition.

CONCLUSIONS

The sensitivity of ρ to hydrostatic pressure (a reduction of volume) is a crucial component in understanding the origin of the metal-insulator transition in CMR materials. This was illustrated with three ferromagnetic specimens which have various ρ(T) and δV_T behaviors. The changes in volume which occur at a structural or magnetic transition can produce *self pressures* in the range of 0.18 to 0.45 GPa. These *self pressures* are responsible for large features in the electrical resistivity and are an important component for the occurrence of the metallicity below T_c. Presumably, this arises because of a strong volume dependence of electron phonon coupling in these particular manganese oxide perovskites. The influence of *self pressure* on the magnetoresistance will be discussed elsewhere. Near and below T_c approximately 50% of the magnetoresistance can be attributed to the magnetovolume [41].

ACKNOWLEDGMENTS

We acknowledge T. Darling, R. H. Heffner, F. Garzon, A. Migliori, E. J. Peterson, J. S. Schilling, J. D. Thompson, and S. Von Molnar for interesting discussions. Work at Los Alamos was performed under the auspices of the USDOE. One of us (ALC) was supported (in part) by funds provided by the University of California for discretionary research at Los Alamos National Laboratory. JJN and KA are thankful to NATO for providing a Collaborative Research Grant (CRG 970260). Work at Florida Atlantic University was supported in part by a FAU-sponsored Research Initiation Award. JJN is thankful to the Walther Meissner Institut and the Bavarian Academy of Science for providing a stipend during his visit and to M. Kund for his help with the measurements.

REFERENCES

1. G. H. Jonker and J. H. Van Santen, Physica **16**, 337 (1950).
2. J. H. Van Santen and G. H. Jonker, Physica **16**, 599 (1950).
3. C. Zener, Phys. Rev. **81**, 440 (1951); **82**, 403 (1951).
4. J. Volger, Physica **20**, 49 (1954).
5. R. M. Kusters, J. Singelton, D. A. Keen, R. McGreevy, and W. Hayes, Physica B **155**, 362 (1989).
6. R. von Helmholt, J. Wecker, B. Holzapfel, L. Schultz, and K. Samwer, Phys. Rev. Lett. **71**, 2331 (1993).
7. S. Jin, T. H. Tiefel, M. McCormack, R. A. Fastnacht, R. Ramesh, and L. H. Chen, Science **264**, 413 (1994).
8. M. F. Hundley, M. Hawley, R. H. Heffner, Q. X. Jia, J. J. Neumeier, J. Tesmer, J. D. Thompson, and X. D. Wu, Appl. Phys. Lett. **67**, 860 (1995).
9. K. N. Clausen, W. Hayes, D. A. Keen, R. M. Kusters, R. L. McGreevy, and J. Singleton, J. Phys. Condens. Matter **1**, 2721 (1989).
10. M. F. Hundley and J. J. Neumeier, Phys. Rev. B 55 (1997) 11511.
11. M. Jaime, M. B. Salamon, M. Rubinstein, R. E. Treece, J. S. Horwitz, and D. B. Chrisey, Phys. Rev. B **54**, 11914 (1996); M. Jaime, M. B. Salamon, K. Pettit, M. Rubinstein, R. E. Treece, J. S. Horwitz, and D. B. Chrisey, Appl. Phys. Lett. **68**, 1576 (1996); M. Jaime, H. T. Hardner, M. B. Salamon, M. Rubinstein, P. Dorsey, and D. Emin, Phys. Rev. Lett. **78**, 951 (1997).
12. S. J. L. Billinge, R. G. DiFrancesco, G. H. Kwei, J. J. Neumeier, and J. D. Thompson, Phys. Rev. Lett. **77**, 719 (1996).
13. T. A. Tyson, J. Mustre de Leon, S. D. Conradson, A. R. Bishop, J. J. Neumeier, H. Roder, and Jun Zang, Phys. Rev. B **53**, 13985 (1996).
14. C. H. Booth, F. Bridges, G. H. Kwei, J. M. Lawrence, A. L. Cornelius, and J. J. Neumeier, to appear in Phys. Rev. Lett. **80** (1998).
15. A. J. Millis, P. B. Littlewood, and B. I. Shraiman, Phys. Rev. Lett. **74**, 5144 (1995).
16. A. J. Millis, Phys. Rev. B **53**, 8434 (1996).
17. H. Röder, J. Zang, and A. R. Bishop, Phys. Rev. Lett. **76**, 1356 (1996).
18. A. J. Millis, B. I. Shraiman, and R. Mueller, Phys. Rev. Lett. **77**, 175 (1996).
19. G.-M. Zhao, K. Conder, H. Keller, and K. A. Müller, Nature **381**, 676 (1996).
20. V. Laukhin, J. Fontcuberta, J. L. García-Muñoz, and X. Obradors, Phys. Rev. B **56**, R10009 (1997).
21. P. Schiffer, A. P. Ramirez, W. Bao, and S.-W. Cheong, Phys. Rev. Lett. **75**, 3336 (1995).
22. In this report, we neglect the formation of cation vacancies incoporate differing Mn valance states as an oxygen deficiency. For details on the defect chemistry of $LaMnO_3$ see: J. A. M. Van Roosmalen and E. H. P. Cordfunke, J. Solid State Chem. **110**, 109 (1994).
23. J. J. Neumeier, M. F. Hundley, J. D. Thompson, and R. H. Heffner, Phys. Rev. B **52**, R7006 (1995).
24. P. Ehrenfest, Communications Leiden, Vol. **XX**, Suppl. 75b, 1993.
25. H. Y. Hwang, T. T. M. Palstra, S.-W. Cheong, and B. Batlogg, Phys. Rev. B **52**, 15046 (1995).

26. $\Delta C_P \propto T_c dM(T)^2/dT$ and is expected to be comparable for A and B since T_c and $M(T)$ are similar.
27. A. P. Ramirez, P. Schiffer, S.-W. Cheong, C. H. Chen, W. Bao, T. T. M. Palstra, P. L. Gammel, D. J. Bishop, and B. Zegarski, Phys. Rev. Lett. **76**, 3188 (1996).
28. A. L. Cornelius, S. Klotz, and J. S. Schilling, Physica C **197**, 209 (1992).
29. T. W. Darling, A. Migliori, E. G. Moshopoulou, Stuart A. Trugman, J. J. Neumeier, J. L. Sarrao, A. R. Bishop, and J. D. Thompson, to appear in Phys. Rev. B **57** (1998).
30. F. R. Kroeger and C. A. Swenson, J. Appl. Phys. **48**, 853 (1977).
31. K. Geschneidner, Jr., Solid State Physics **16**, 275 (1964).
32. G. H. Kwei, D. N. Argyriou, S. J. L. Billinge, A. C. Lawson, J. J. Neumeier, A. P. Ramirez, M. A. Subramanian, and J. D. Thompson, proceedings of the Spring MRS meeting (1997).
33. T. Soumura, J. Phys. Soc. Jpn **42**, 826 (1977); E. F. Wasserman in Ferromagnetic Materials, Vol. 5 edited by K.H.J. Buschow and E. P. Wohlfarth (Elsevier, Amsterdam, 1990) p. 238.
34. K. Sundqvist, Solid State Commun. **66**, 623 (1988); B. Sundqvist and B. M. Andersson, ibid. **76**, 1019 (1990).
35. A. Asamitsu, Y. Moritomo, Y. Tomioka, T. Arima, and Y. Tokura, Nature 373 (1995) 407.
36. Y. Moritomo, A. Asamitsu, and Y. Tokura, Phys. Rev. B **51**, 16491 (1995).
37. J. J. Neumeier and K. Andres, in preparation.
38. J.-S. Zhou, W. Archibald, and J. B. Goodenough, Nature **381**, 770 (1996).
39. D. Emin, M. S. Hillary, and N.-L. H. Liu, Phys. Rev. B **35**, 641 (1987).
40. Z. Arnold, K. Kamenev, M. R. Ibarra, P. A. Algarabel, C. Marguina, J. Blasco, J. García, Appl. Phys. Lett. **67**, 2875 (1995).
41. J. J. Neumeier, A. L. Cornelius, and M. F. Hundley, in preparation.

RAMAN INVESTIGATION OF THE LAYERED MANGANESE PEROVSKITE La$_{1.2}$Sr$_{1.8}$Mn$_2$O$_7$

D.B. ROMERO[*,***], V.B. PODOBEDOV[*,1], A. WEBER[*], J.P. RICE[*], J.F. MITCHELL[**], R.P. SHARMA[***], and H.D. DREW[***]
[*] Optical Technology Division, NIST, Gaithersburg, MD 20899, USA
[**] Materials Science Division, Argonne National Laboratory, Argonne, IL 60439, USA
[***] Department of Physics, University of Maryland, College Park, MD 20742, USA
[1] Institute of Spectroscopy, Russian Academy of Science, Troitsk, Moscow 142092, Russsia

ABSTRACT

We report the results of a detailed polarization and temperature dependence study of the Raman scattering from La$_{1.2}$Sr$_{1.8}$Mn$_2$O$_7$. The Raman spectra reveal three general spectral features. First, there are sharp peaks due to long-wavelength optic phonons. Phonons, attributed to the distortion of the MnO$_6$ octahedra, reveal an anomalous behavior which correlates with the transition from a paramagnetic-insulating (PI) to a ferromagnetic-metallic (FM) phase at a critical temperature T$_c$. Second, there is an electronic continuum that is suppressed for $\omega <$ 500 cm^{-1} at T $<<$ T$_c$. Third, broad peaks between 400 cm^{-1} to 800 cm^{-1} seen in the PI state, surprisingly, disappear in the FM state. The implications of these results are discussed.

INTRODUCTION

The interplay of the lattice and electronic degrees of freedom is necessary to explain the phenomenon of colossal magnetoresistance in the mixed-valence manganese perovskites. While the double-exchange mechanism first proposed by Zener [1] correctly pointed out the connection between electrical conduction and ferromagnetism in these compounds, it is not sufficient to account for the phenomenon [2]. By incorporating the localization of charge-carriers into small polarons in the paramagnetic state, a quantitative explanation of certain aspects of the phenomenon was achieved [3]. On the experimental side, there is growing body of evidence for small polaron formation in the cubic manganites [4].

La$_{1.2}$Sr$_{1.8}$Mn$_2$O$_7$ is a quasi two-dimensional manganese-oxide perovskite. As shown in Fig. 1, it has a layered-type structure consisting of two MnO$_2$ sheets separated by a rock-salt (La,Sr)$_2$O$_2$ layer. Its electronic properties are anisotropic as well with resistivity ratio of $\rho_c/\rho_{ab} \approx 10^2$ and magnetization ratio of M$_{ab}$/M$_c$ $\approx$10 [5]. In this work, Raman spectroscopy is used as a probe of the lattice and electronic excitations in La$_{1.2}$Sr$_{1.8}$Mn$_2$O$_7$. We observed anomalous behaviors in some phonons that could provide possible signatures of small polaron formation in the layered manganites. We find evidence for electronic Raman scattering with spectral features that are

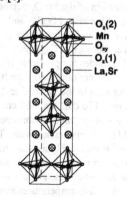

Fig. 1. Crystal structure of La$_{1.2}$Sr$_{1.8}$Mn$_2$O$_7$.

Mat. Res. Soc. Symp. Proc. Vol. 494 © 1998 Materials Research Society

suggestive of the presence of short-range magnetic order in the PI state and the opening of a spin-gap in the FM state.

EXPERIMENT

The $La_{1.2}Sr_{1.8}Mn_2O_7$ single-crystals investigated in this work undergo a paramagnetic to ferromagnetic phase-transition concomitantly with an insulator to metal transition at a critical temperature $T_c \approx 120$ K [6]. The measured resistivity has a peak at $T_p \approx 135$ K, above which it is activated with $E_{ac} \approx 67$ meV. Polarized Raman scattering experiments were performed on these single crystals over the temperature range of 5 K to 320 K using a commercial triple-stage spectrometer equipped with a liquid-nitrogen cooled charge-coupled device detector. The pre-monochramator stage of the spectrometer served as a filter that prevents parasitically scattered light from reaching the detector. In our measurements, the scattered light was polarized along the vertical direction while the polarization of the incident light was rotated between the vertical and horizontal directions using a half-wave plate. In this way, we obtain an accurate comparison of the Raman intensities in the parallel and perpendicular scattering configurations. All of the spectra reported in this work were taken in a back-scattering geometry using the 514.5 nm line of an Ar^+ laser as the exciting radiation with nominally 50 mW power incident on the sample. Similar room temperature Raman spectra were obtained on three different samples. The detailed results reported here is from the biggest sample with approximate dimensions of 6 mm x 4 mm x 0.7 mm with the shortest length along the crystal c-axis. In the following discussions, x and y refer to light polarization parallel to the in-plane crystallographic axes while x' and y' are rotated by 45^0 and z is along the c-axis.

RESULTS

Phonons

The sharp features in the room temperature spectra of Fig. 2 are due to first-order Raman-scattering by optic phonons. The crystal-structure of $La_{1.2}Sr_{1.8}Mn_2O_7$ belongs to the space-group I4/mmm (D_{4h}^{17}) [7]. Neglecting the distortion of the MnO_6 octahedra, only the motion of the apical $O_z(1)$- and (La,Sr)- atoms can give rise to Raman-active phonons. These atoms are located at C_{4v}-sites for which ($A_{1g}+E_g$) modes are predicted [8]. The E_g-modes were not observed. The two A_{1g} phonons seen in the xx, x'x' and zz spectra of Fig. 2 near 170 cm^{-1} and 450 cm^{-1} are assigned to the c-axis symmetric vibration of the (La,Sr)- and $O_z(1)$- atoms, respectively. The peak near 170 cm^{-1} is noticeably asymmetric as it consists of two peaks corresponding to the vibrations of the La- and Sr-atoms. These two peaks are resolved at low temperatures as seen in Fig. 3.

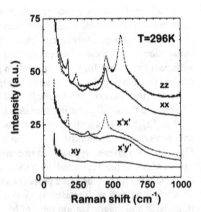

Fig. 2. Polarization dependence of the room temperature spectra. (i,s) indicates the polarization of the incident (i) and and scattered (s) light. The unprimed coordinates are along the crystal-axes. The primed coordinates are at 45^0 from the in-plane crystal-axes.

306

The strongest A_{1g} phonon in the zz-spectrum around 570 cm^{-1} is apparently a symmetry forbidden mode.

Neutron diffraction studies [6] of La$_{1.2}$Sr$_{1.8}$Mn$_2$O$_7$ have shown evidence of the distortion of the MnO$_6$ octahedra. Such distortion lowers the site-symmetry of the Mn- and in-plane O$_{xy}$-atoms. X-ray refinements [7] of the structure at room temperature of related layered manganites yielded C$_{4v}$ and C$_{2v}$^{v} site-symmetries for the Mn- and O$_{xy}$-atoms, respectively. Consequently, (A_{1g}+E$_g$) modes are allowed for the Mn-atoms while (A_{1g}+B$_{1g}$+2E$_g$) modes are allowed for the O$_{xy}$-atoms [8]. From these results, the A_{1g} phonon in the zz-spectrum at 243 cm^{-1} is assigned to the in-phase c-axis vibrations of the Mn-atoms and the weak B$_{1g}$ phonon in the x'y'-spectrum at 324 cm^{-1} is identified as the out-of-phase c-axis motion of the O$_{xy}$-atoms.

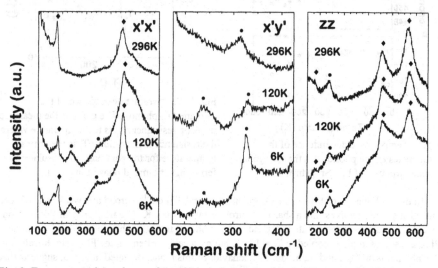

Fig. 3. Temperature dependence of the phonons in La$_{1.2}$Sr$_{1.8}$Mn$_2$O$_7$. Those marked with ♦ are phonons associated with undistorted tetragonal crystal-structure while those marked with • refer to phonons of the distorted MnO$_6$ octahedra.

The spectra at three representative temperatures are shown in Fig. 3. No remarkable changes are observed in the phonons (marked as ♦ in Fig. 3) associated with the tetragonal crystal-structure of La$_{1.2}$Sr$_{1.8}$Mn$_2$O$_7$. Figure 4 shows that the frequencies of these phonons are nearly temperature independent. However, the Mn- and O$_{xy}$- phonons (marked as • in Fig. 3) of the distorted MnO$_6$ octahedra reveal anomalous behaviors as the sample undergo the PI to FM phase-transition.

Above T$_c$ = 120 K, the Mn-phonon is seen only in the zz-spectrum. Its symmetry is pure A_{1g}. This result implies that, at temperatures far above T$_c$, the distortion of the MnO$_6$ octahedra must involve a displacement of the Mn-atoms along the c-axis, in order to maintain the same site-symmetry as the (La,Sr)- and apical O$_z$(1)- atoms. However, for temperatures close and below T$_c$, the Mn-phonon acquires a mixed (A_{1g}+B$_{1g}$) character as peaks at ca. 237 cm^{-1} in x'x' and at ca. 240 cm^{-1} in x'y' appear simultaneously. In this latter case, the Mn-atoms are displaced perpendicular to the c-axis as well.

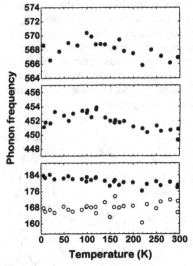

Fig. 4. Temperature dependence of the frequencies of the phonons of the tetragonal crystal-structure of $La_{1.2}Sr_{1.8}Mn_2O_7$.

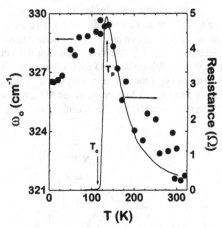

Fig. 5. Temperature dependence of the in-plane O_{xy}-phonon. The curve is the measured in-plane resistance. T_p is the temperature where the resistance is maximum. T_c is the critical temperature for the paramagnetic-insulator to ferromagnetic-metal phase-transition.

In the x'y'-spectra of Fig. 3, the O_{xy}-phonon around 324 cm^{-1} is broad and weak at T = 296K but subsequently evolves into a sharp and strong peak at T = 6 K. The frequency of this phonon reveals an anomalous temperature dependence. As shown in Fig. 5, this phonon hardens in the PI state as the sample is cooled to $T_p \approx 135$ K but noticeably softens in the FM state. Recall that this phonon is an O_{xy} bending mode. The behavior of this mode, depicted in Fig. 5, suggests that the in-plane O_{xy}-Mn-O_{xy} bond-angle decreases from 180^0 as T_p is approached but subsequently relaxes back towards 180^0 as the temperature is lowered from T_p.

Electronic Raman Scattering

Two additional features in the spectra of Fig. 1 are attributed to electronic Raman scattering in $La_{1.2}Sr_{1.8}Mn_2O_7$. One is a continuum that displays a quasielastic peak at low frequencies becoming flat and structureless at higher frequencies. The quasielastic peak is present in the xx, x'x', x'y', and zz spectra. It is weakly discernible in the xy spectrum. Note the coincidence of the electronic continuum in the x'x' and x'y' spectra. Also, this background in the xx ($A_{1g}+B_{1g}$) spectrum is nearly the sum of those in the x'x' (A_{1g}) and x'y' (B_{1g}) spectra. These results indicate that the electronic continuum is an intrinsic Raman effect. The other feature is a weak broad peak between 400 cm^{-1} to 800 cm^{-1}. It is observed in the scattering geometries where the quasielastic peak is present. This broad peak is clearly discerned in the x'y' spectrum. However, it is obscured by the presence of a strong phonon peak at 450 cm^{-1} in the xx and x'x' spectra as well as phonon peaks at 450 cm^{-1} and 570 cm^{-1} in the zz spectrum.

The two spectral features described above reveal dramatic changes as the temperature is decreased through the PI to FM phase-transition, as illustrated in Fig. 6 for the x'y' scattering geometry. The quasielastic peak has noticeably narrowed and reduced in intensity close to $T_c = 120$ K. The feature between 400 cm^{-1} to 800 cm^{-1} is now prominently strong and resolved into two peaks near 515 cm^{-1} and 630 cm^{-1}. For $T \ll T_c$, both the quasieleastic peak and the two broad peaks have disappeared. Furthermore, the electronic scattering continuum is suppressed for $\omega < 500$ cm^{-1}.

The temperature dependence of the electronic continuum depends on the Bose-Einstein occupation factor and on the Raman response function. Upon normalization of the spectra in Fig. 6, we find that the quasielastic peak is largely due to the Bose-Einstein contribution. However, the Raman response function manifests an intrinsic temperature dependence. This is clearly seen in the spectra at $T = 50$ K and 5 K in which the effect of the Bose-Einstein factor is negligible in the frequency range shown in Fig. 6.

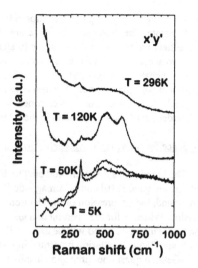

Fig. 6. Temperature dependence of the electronic Raman scattering in $La_{1.2}Sr_{1.8}Mn_2O_7$. The tick marks on the vertical indicates the zero-intensity position for the $T = 120$ K (lower) and $T = 296$ K (upper) spectra.

It is possible that the broad peaks between 400 cm^{-1} to 800 cm^{-1} are due to second-order phonon scattering. However, we believe this assignment is unlikely for two reasons. First, higher-order scattering is generally much weaker than one-phonon scattering. This is not the case for the above features particularly at temperatures close to T_c. Second, in the cubic manganites, where second-order scattering is reportedly seen [9], the pertinent spectral features are only weakly temperature dependent. In contrast, the above-described broad features are strongly and anomalously temperature dependent. For these reasons, we argue that these peaks are electronic in origin.

CONCLUSION

Evidence for Small Polaron Formation ?

The anomalous changes in the Mn- and O_{xy}- phonons point to a structural phase-transition involving the double-perovskite layers within the two MnO_2 sheets. The mechanism which drives such transition must be related to the PI to FM phase-transition in $La_{1.2}Sr_{1.8}Mn_2O_7$. One possible mechanism is the formation of small polarons in the PI state. Far above T_p, the perovskite layers are nearly cubic accounting for the weak and broad phonon peaks. On lowering the temperature, the layers responds to the localization of the e_g-electrons with the collapse of the in-plane O_{xy}-ions towards the Mn^{3+}-ions, leading to the buckling of the O_{xy}-Mn-O_{xy} bonds. Such distortion is optimum at T_p in which the electrons are strongly localized around the Mn^{3+}-sites. Below T_c, the distortion relaxes due to the delocalization of the e_g-electrons in the FM state. The accompanying structural phase-transition reflects the difference in symmetry of the electronic wavefunctions in the PI and FM states.

Two aspects of our results are suggestive of the above picture. First, we note that the unusual behaviors are exhibited only by the Mn- and O_{xy}- atoms of the MnO_2 sheets. This is consistent with the reported [5] quasi two-dimensional electronic properties of $La_{1.2}Sr_{1.8}Mn_2O_7$ in which the electron density is confined within the MnO_2 layers. Second, as depicted in Fig. 5, the hardening of the O_{xy}- phonon in the PI state follows the activated behavior of the resistance.

Evidence for Short-range Magnetic Order and Spin-gap ?

The broad peaks between 400 cm^{-1} to 800 cm^{-1} may be related to the observed short-range antiferromagnetic [10] and ferromagnetic [5,6] order in the PI state in $La_{1.2}Sr_{1.8}Mn_2O_7$. On the other hand, the suppression of the electronic scattering below 500 cm^{-1} at low temperatures is possible evidence for the opening of a gap in the spin-excitation spectrum. These preliminary results are interesting and will be explored in the future. As a further note, the features of the electronic Raman scattering presented in this work have striking similarity to those found [11] in $CuGeO_3$, a quasi one-dimensional spin-1/2 antiferromagnet with strong electron-phonon interactions. In this latter compound, a quasielastic peak dominates the electronic Raman spectrum at high temperatures. A regime of short-range antiferromagnetic order develops at low temperatures. Below the spin-Peierls temperature, the antiferromagnetic order is suppressed and a spin-Peierls gap opens in the spinon spectrum. It is intriguing to speculate whether these features are general properties of low-dimensional spin systems with strong spin-lattice interactions.

ACKNOWLEDGEMENTS

We thank Dr. R. Datla for his interest in and support of this work. JFM acknowledges the support of the U.S. Department of Energy, Basic Energy Sciences-Materials Sciences, under contract no. W-31-109-ENG-38.

REFERENCES

1. Zener, Phys. Rev. **82**, p. 403, (1951).
2. A.J. Millis, P.B. Littlewood, and B.I. Shraiman, Phys. Rev. Lett. **74**, p. 5144, (1995).
3. A.J. Millis, B.I. Shraiman, and R. Mueller, Phys. Rev. Lett. **77**, p. 175, (1996).
4. See the references cited in J.B. Goodenough, J. Appl. Phys. **81**, p. 5330, (1997).
5. Y. Moritomo, A. Asamitsu, H. Kuwahara, and Y. Tokura, Nature **380**, p. 141, (1996).
6. J.F. Mitchell, D.N. Argyriou, J.D. Jorgensen, D.G. Hinks, C.D. Potter, and S.D. Bader, Phys.Rev. B **55**, p. 63, (1997).
7. P. Laffez, G. Van Tendeloo, R. Seshadri, M. Hervieu, C. Martin, A. Maignan, and B. Raveau, J. Appl. Phys. **80**, p. 5850, (1996).
8. D.L. Rousseau, R.P. Bauman, and S.P.S. Porto, J. Raman Spectroscopy **10**, p. 853, (1981).
9. V.B. Podobedov, A. Weber, D.B. Romero, J.P. Rice, and H.D. Drew, Solid State Commun., in press.
10. T.G. Perring, G. Aeppli, Y. Moritomo, and Y. Tokura, Phys. Rev. Lett. **78**, p. 3197, (1997).
11. P.H.M van Loosdrecht, J.P. Boucher, and G. Martinez, Phys. Rev. Lett. **76**, p. 311, (1996).

HIGH FREQUENCY MAGNETO-ELECTRODYNAMICS OF La$_{1-x}$Sr$_x$MnO$_3$ SINGLE CRYSTALS

H. Srikanth, B. Revcolevschi, S. Sridhar [*], L. Pinsard, A. Revcolevschi [**]
*Department of Physics, Northeastern University, Boston, MA 02115
**Laboratoire de Chimie des Solides, Université Paris-Sud, 91405 Orsay Cedex, France

ABSTRACT

The radio frequency (RF) response of La$_{1-x}$Sr$_x$MnO$_3$ single crystals reveal a variety of features associated with the structural, electronic and magnetic properties of the system. The resonance technique operating at ~ 4 MHz employed in this study is sensitive to small changes in both the magnetic susceptibility and resistivity of the samples. Very sharp changes in frequency are observed at the ferromagnetic (FM) and structural phase transitions in both the metallic (x = 0.175) and insulating (0.125) crystals studied.

In addition to the known transitions identified as FM and orthorhombic distortions, our experiments show rich structures which are not observed in conventional DC magnetization and transport experiments. Our results demonstrate that RF experiments are ideally suited to investigate the complex phase diagram in the manganites.

The colossal frequency change that we observe at the FM transition in the La$_{1-x}$Sr$_x$MnO$_3$ crystals is indicative of the enormous potential for using these materials in high frequency switching applications.

INTRODUCTION

The discovery of colossal magnetoresistance (CMR) [1] and other interesting effects have made the manganese-based perovskite oxides a unique class of materials of current interest [2-3]. Subtle interplay between electronic, magnetic and structural properties in these materials give rise to a complicated phase diagram and lead to novel phenomena associated with different phases.

In the La$_{1-x}$Sr$_x$MnO$_3$ system, the parent compound, LaMnO$_3$, is an antiferromagnetic Mott insulator and contains Mn^{3+} ions with t$_{2g}$^{3}e$_g$^{1} (spin quantum number S = 2) configuration. Strong intra-atomic ferromagnetic coupling acts between the localized t$_{2g}$ spins and the itinerant e$_g$ electrons which gives rise to a parallel alignment of their spins. Chemical substitution of divalent Sr in place of La3+ introduces mobile holes into the e$_g$ orbitals which mediate the interatomic ferromagnetic interaction between the Mn atoms. For Sr content (x > 0.17), the system becomes metallic with a ferromagnetic ground state caused by the so-called "double exchange" mechanism between the Mn^{3+} and neighboring Mn^{4+} ions [4].

The important role of the Jahn-Teller (JT) effect in the transport and magnetic properties of doped manganese oxides has been pointed out [5] and indeed recent experiments have provided evidence for strong coupling of the charge carriers to the JT distortions and polaron formation [6-7]. Very recent experiments have also revealed that the charge carriers responsible for the conduction mechanism in the La$_{1-x}$Sr$_x$MnO$_3$ system have significant oxygen 2p character [8].

A consequence of the magnetic and structural transitions and the interplay between these effects leads to a very rich phase diagram for these materials. Elucidation and characterization of the phase diagram has indeed been an area of extensive experimental research in La$_{1-x}$Sr$_x$MnO$_3$ based on measurements like transport [9-10], magnetization [11-12] and neutron diffraction [13-

311

14]. While a broad picture has emerged, many issues governing the nature of the phase transitions and finer details about the various phases themselves are yet to be resolved.

In this paper, we report on some of the electrodynamic properties determined from the RF response at 4 MHz, of lightly doped $La_{1-x}Sr_xMnO_3$ single crystals. Our experimental technique has a unique advantage of having the ability to look at the combined response due to charge transport and magnetic properties. At the outset, our data show very sharp frequency changes at characteristic temperatures which match well with the magnetic and structural phase transitions identified by other methods. Closer examination reveals finer details associated with some of the transitions and also the observation of a novel structure not seen using conventional DC experiments. Signatures associated with the thermodynamic nature of the transitions are also seen in the hysteretic response.

In addition to the sensitivity of our experiments in exploring the transitions themselves, the enormous frequency changes seen in the vicinity of the transitions suggest that these materials hold a lot of promise as potential candidates for high frequency switching devices.

Overall, our observations indicate that the RF electrodynamics is a very useful probe of the ground state properties of manganese oxides. To our knowledge, this is the first investigation of its kind in the $La_{1-x}Sr_xMnO_3$ system.

EXPERIMENT

Single crystals of $La_{1-x}Sr_xMnO_3$ were grown by a floating zone method associated with an image furnace [15]. For this study, two crystals were chosen: one which is metallic (x = 0.175) and the other insulating (x = 0.125).

The sample is mounted inside a coil which forms the inductive part of an LC tank circuit driven by a stable tunnel diode oscillator (TDO) self-resonant at ~ 4 MHz. The resonance frequency f is measured with a commercial frequency counter (Phillips) and changes in f can be measured accurately as a function of temperature T.

Changes in the effective reactance X of the sample, caused by varying temperature leads to changes in the coil inductance which in turn results in shifts in the resonant frequency f given by the relation:

$$\delta X(T) \equiv -g \, [f(T) - f_0(T)] \tag{1}$$

where g is a geometrical factor set by the sample and coil dimensions and f_0 is the resonance frequency of the empty coil. (Note that the reactance can also be expressed in terms of a magnetic screening length $\delta\lambda = \delta X/\mu_0\omega$).

The very high stability of the circuit, typically 1 Hz in 4 MHz, leads to a very high resolution in detection of changes in screening length of the order of a few Å. The technique has been extensively validated through precise measurements of vortex parameters such as H_{c1} and pinning force constants in cuprate and borocarbide superconductors [16-17].

In the case of a magnetic material, the observed effective reactance reflects changes in both the conductivity and susceptibility of the sample and can be written as :

$$\delta X(T) \propto (\mu_{eff}/\sigma)^{1/2} \tag{2}$$

At any particular transition temperature where both the conductivity and susceptibility of a material change sharply, the shift in resonant frequency and consequently the change in magnetic screening length is determined by the dominant parameter.

RESULTS

The temperature dependence of the RF data for the metallic sample (Sr = 0.175) is shown in Fig. 1.

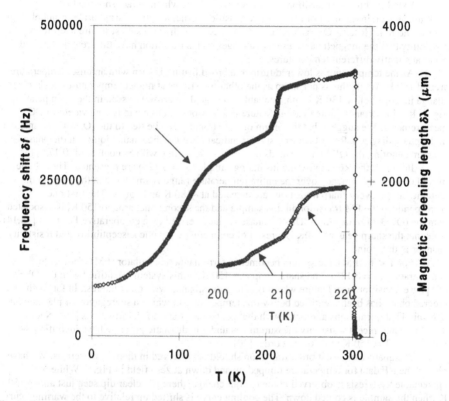

Fig. 1 The resonance frequency shift (f_{310K} - f) (where f_{310K} = 2.7 MHz) plotted as a function of the temperature (T). Change in RF screening length extracted from this data is also shown on the right axis. Note the very sharp jump at ~ 300K (FM transition) and the onset at ~215K associated with the (R3c → O) transition. The arrow indicates the onset of a broad feature around 130K which is not seen in conventional DC measurements. Inset shows an exploded view of the (R3c → O) transition which reveals the fine structure indicated by the arrows.

The data reveal several interesting features at characteristic temperatures. An extremely sharp transition is seen at 300 K which matches well with the ferromagnetic transition identified by other experiments. The change in frequency is about 400 kHz out of a resonance frequency (f) of about 2.7 MHz. This colossal frequency shift at the FM transition also takes place over a very narrow temperature range as seen from Fig. 1. The implication of our observation in terms of potentially using this effect for high frequency switching applications is quite obvious. The fact

313

that this transition takes place at room temperature for this Sr concentration makes it appealing for practical applications.

It should be noted that as T is decreased below the FM Curie temperature (T_c), the magnetic susceptibility increases and the resistivity decreases as one crosses over to the FM state. The data in Fig. 1 in this region indicates that the RF susceptibility term completely dominates over the resistivity.

A slightly broader transition occurs below ~ 215 K which in fact shows a two-slope structure as seen in the inset of Fig. 1. This transition coincides with a structural rhombhohedral to orthorhombic (R3c → O) transformation. We associate the two-step structure itself to the possibility that the magnetic and resistive changes at this transition have different widths and occur at slightly different temperatures.

As the temperature is lowered further, a broad feature is seen with an onset temperature around 130 K. While this is not sharp as the other transitions at higher temperatures, a clear change in slope below 130 K is unmistakable (indicated by arrow) as seen in the main panel of Fig. 1. It is important to note that this feature is a new observation and is not evident in DC measurements. We suggest that the origin of this structure can be tied to the (O → O'') transition in the phase diagram. Recent neutron studies indicate that the the small "dome" in the phase diagram denoting the O'' phase extends to the metallic regime with Sr content of 0.175 [14].

In Fig. 2, the RF data for the insulating sample (Sr = 0.125) are presented. Though the overall change in frequency over the entire temperature range is smaller than that seen in the metallic sample, very sharp transitions are detected at ~180 K and 150 K. The first one corresponds to the FM transition in this sample and the second one, around 150 K, is associated with the (O' → O'') transition. This O' phase is characterized by a co-operative JT effect [18,14]. Note that the sharp peak at 180K is a result of competing magnetic susceptibility and resistivity changes at this point.

For T < T_c (FM), the system becomes a ferromagnetic insulator (FMI) for this Sr concentration (x = 0.125). The sharp jump at ~ 150 K in this system (identified with the O' → O'' phase transition) is a feature which is being investigated with great interest. In fact a charge - ordered phase has been identified below this temperature in which a segregation of planes with dynamic JT deformations alternates with that containing static JT deformations [19]. Neutron diffraction and recent resistivity measurements under hydrostatic pressure have identified the transition at T = 150 K to be of first order [20].

A signature of a first order transition should be observed in thermal hysteresis. We have plotted the RF data for temperature ramped up and down at zero field in Fig. 2. While no appreciable hysteresis is observed in the overall change, there is a clear dip seen just above 150 K when the sample is cooled down. The cooling curve is shifted up relative to the warming curve for clarity and the arrow in Fig. 2 indicates this dip. We conclude that this unusual hysteretic behavior is characteristic of a first order transition.

It is to be noted that the DC resistivity data also show a resistivity minimum for 150K < T < 180 K with a width of about 20 to 30 K [18]. However, the width of the dip seen in the RF data is much smaller and it is not clear if a corresponding analogy can be drawn in this case.

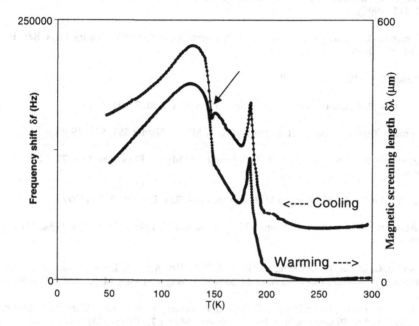

Fig. 2 The RF data for the $La_{1-x}Sr_xMnO_3$ (x = 0.125) crystal. Both the zero field warming and cooling curves are shown. The cooling curve is shifted up by 50000 Hz for clarity. The 180 K transition corresponds to FM ordering and and sharp increase at 150 K is associated with (O' → O'') transition. Note the sharp dip (marked by arrow) seen in the cooling curve.

CONCLUSION

In conclusion, we have demonstrated the effectiveness of RF electrodynamics as a probe of the electronic and magnetic properties of doped manganese oxides. Finer details are observed in the transitions which are not visible in DC measurements. A colossal frequency shift seen in metallic samples holds promise for switching applications.

ACKNOWLEDGEMENTS

Work at NU supported by NSF-DMR-9623720. The authors would like to thank Durga Choudhury and Z. Zhai for assistance and discussions.

REFERENCES

1. R. von Helmolt, J. Wecker, B. Holzapfel, L. Schultz and K. Samwer, Phys. Rev. Lett. **71**, 2331 (1993).

2. S. Jin, T. H. Tiefel, M. McCormack, R. A. Fastnacht, R. Ramesh and L. H. Chen, Science **264**, 413 (1994).

3. A. Urushibara, Y. Moritomo, T. Arima, A. Asamitsu, G. Kido and Y. Tokura, Phys. Rev. B **51**, 14103 (1995).

4. C. Zener, Phys. Rev. **82**, 403 (1951).

5. A. J. Millis, P. B. Littlewood, B. I. Shraiman, Phys. Rev. Lett. **74**, 5144 (1995).

6. Guo-meng Zhao, K. Conder, H. Keller and K. A. Muller, Nature **381**, 676 (1996).

7. A. Shengelaya, Guo-meng Zhao, H. Keller and K. A. Muller, Phys. Rev. Lett. **77**, 5296 (1996).

8. H. L. Ju, H.-C. Sohn and Kannan M. Krishnan, Phys. Rev. Lett. **79**, 3230 (1997).

9. A. Asamitsu, Y. Moritomo, R. Kumai, Y. Tomioka and Y. Tokura, Phys. Rev. B **54**, 1716 (1996).

10. A. Anane, C. Dupas, K. Le Dang, J.-P Renard, P. Veillet, A. M. de Leon Guevara, F. Millot, L. Pinsard, A. Revcolevschi and A. G. M. Jansen, J. Mag. and Mag. Mater. **165**, 377 (1997).

11. A. Anane, C. Dupas, K. Le Dang, J. P. Renard, P. Veillet, A. M. de Leon Guevara, F. Millot, L. Pinsard and A. Revcolevschi, J. Phys.: Condens. Matter **7**, 7015 (1995).

12. H. Kawano, R. Kajimoto, M. Kubota and H. Yoshizawa, Phys. Rev. B **53**, 2202 (1996).

13. H. Kawano, R. Kajimoto, M. Kubota and H. Yoshizawa, Phys. Rev. B **53**, R14709 (1996).

14. L. Pinsard, J. Rodriguez-Carvajal and A. Revcolevschi, J. Alloys and Compounds (in press) (1997).

15. A. Revcolevschi and G. Dhalenne, Adv. Mater. **5**, 657 (1993).

16. D. H. Wu and S. Sridhar, Phys. Rev. Lett. **65**, 2074 (1990).

17. S. Oxx, D. P. Choudhury, Balam A. Willemsen, H. Srikanth, S. Sridhar, B. K. Cho and P. C. Canfield, Physica C **264**, 103 (1996).

18. L. Pinsard , J. Rodriguez-Carvajal, A. H. Moudden, A. Anane, A. Revcolevschi and C. Dupas, Physica B **234-236**, 856 (1997).

19. Y. Yamada, O. Hino, S. Nolido, R. Kanao, T. Inami and S. Katano, Phys. Rev. Lett. **77**, 904 (1996).

20. J. -S. Zhou, J. B. Goodenough, A. Asamitsu and Y. Tokura, Phys. Rev. Lett. **79**, 3234 (1997).

Magnetic and Electronic Transport Properties of Single Crystal $La_{0.64}Pb_{0.36}MnO_3$

Jihui Yang*, Siqing Hu and Ctirad Uher
Department of Physics, University of Michigan, Ann Arbor, MI 48109

P. D. Han and D. A. Payne
Department of Material Science and Engineering Science and Technology Center for Superconductivity,
University of Illinois at Urbana-Champaign, Urbana, IL 61801

ABSTRACT

We studied the magnetic and electronic transport properties of a single crystal sample of $La_{0.64}Pb_{0.36}MnO_3$ in the temperature range 5 K to 350 K and magnetic field up to 5.5 T. A magnetic transition is found at 210 K. The single crystal sample is ferromagnetic below the transition temperature (T_c) and becomes paramagnetic at temperatures $T > T_c$. Magnetization measurements along three different orthorhombic crystal axes show no significant difference. The magnetoresistance approaches a maximum value of about -60% at T_c in 5 T magnetic field strength and has qualitatively different field dependence below and above T_c. The scaling behavior between resistivity and magnetic moment is examined for temperatures both below and above the transition. A low temperature (T<15 K) $d\rho / dT < 0$ effect is attributed to possible quantum tunneling of carriers between neighbouring distortions.

INTRODUCTION

The properties of perovskite-based compounds, the series of the type $R_{1-x}B_xMnO_3$, have been investigated both experimentally and theoretically since the 1950s[1,2]. Here R represents a rare-earth ion and B represents a divalent ion. The parent compound $RMnO_3$ is usually an antiferromagnetic insulator in which R and Mn are both trivalent. Upon doping with the divalent ion B, a corresponding number of Mn^{4+} ions are created on the Mn^{3+} sites in order to maintain stoichiometry. For the optimal doping concentration x ranging from 0.2 to 0.5, the compound becomes a ferromagnetic metal. The theory of double exchange has been developed to explain the phenomenon[2-4]. Recently discovered colossal magnetoresistance (CMR) effect in $La_{1-x}B_xMnO_3$ has lead to a renewed wide interest in these compounds because of the challenging physics as well as its great application potential. Lately, calculations indicated that Jahn-Teller distortion may be required to explain the magnetoresistance within the double exchange model[5]. Chen et al.[6], measured the magnetoresistance and the magneto-thermal transport of a polycrystalline sample, and concluded that the temperature and field dependence of the thermal conductivity are attributed to the scattering of the phonons by spin fluctuations. Experimental works using single crystal samples are however limited. Measurements on single crystals are highly desirable because various subtle features of transport may be easier to resolve and such measurements would also allow to explore the role of anisotropy. Here we report our magnetic and electronic transport studies on a single crystal sample of $La_{0.64}Pb_{0.36}MnO_3$.

EXPERIMENTAL TECHNIQUES

The single crystal sample used in this experiment was prepared by a flux method similar to the one described in ref. 7. The weight ratio of the solute and solvent was 15% with

PbO/PbF$_2$ = 1. High purity chemicals, La$_2$O$_3$ (5N), MnO$_2$ (5N) and CaO (5N), were loaded into a Pt crucible. Crystal growth was performed in a vertical furnace at slow cooling rate ~ 2-4 °C from the soaking temperature 1325 °C (soaking 2 hours). Cubic-like crystals were separated mechanically. The dimensions of the sample are 0.35 x 2.0 x 1.6 mm^3. The sample was cut so that the edges are along the [001], [100] and [010] crystal directions respectively. The lattice parameters along the [100], [010], [001] crystal directions are 5.4741Å, 5.5102 Å, and 7.7633Å respectively. Lattice parameters were obtained by XRD on a Rigaku D-Max diffractometer using Cu Kα radiation. The magnetic properties were measured by a Quantum Design SQUID Magnetometer from 5 K to 350 K. The maximum magnetic field strength is 5.5 T. Galvanomagnetic measurements were conducted using conventional four-probe method with a low frequency Linear Research AC resistance bridge.

RESULTS AND DISCUSSION

Our magnetization data show no significant difference of magnetization along the three different orthorhombic crystal directions. The coercive fields in the hysteresis curve measurements are all less than the resolution of our magnetometer. Figure 1 shows the magnetization curves of the sample from 5 K to 350K with the magnetic field along [100] direction. The saturation moment is 3.1 μ$_B$/Mn. This saturation moment is close to the expected value from high spin manganese in octahedral coordination. For spin only (orbital quenched) moment,

$$M_s = g [0.64 \, S_{3+} \, \mu_B + 0.36 \, S_{4+} \, \mu_B] \tag{1}$$

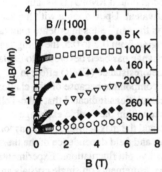
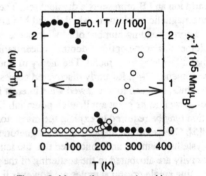

Figure 1. Magnetization curves for the single crystal sample from 5 K to 350 K. B is along [100].

Figure 2. Magnetic moment and inverse magnetic susceptibility of the single crystal sample measured in magnetic field B=0.1T applied along [100].

where M$_s$ is the saturation moment, g = 2 is the electron g-factor, S$_{3+}$ = 2 and S$_{4+}$ = 3/2 are the spins of Mn^{3+} and Mn^{4+} respectively, and μ_B is the Bohr magneton. According to equation (1), M_s = 3.67 μ_B. Because oxygen concentration has a noticeable effect on T_c and M_s[8], some deviation from the calculated value is expected. In figure 2, magnetic moment and inverse magnetic susceptibility are plotted from 5 K to 350 K with B = 0.1 T along [100]. At low temperature, the magnetic moment is large and it decreases rapidly above 100 K, almost reaching zero at 210 K. At the same temperature (210K), χ^{-1} starts to increase linearly with temperature. Figures 1 and 2 indicate a magnetic transition from a ferromagnetic state to paramagnetic state at 210 K. A plot of M^2 vs B/M further confirms the transition temperature $T_c = 210K$.

Resistivity of the sample is plotted in Figure 3 from 5 K to 350 K for values of magnetic field B = 0, 1, 3, and 5 T. Current is along [100] and B is applied along [001]. At temperatures lower than 15 K, a $d\rho / dT < 0$ is observed. This unusual effect is possibly due to the formation of a *Holstein* Polaron (HP) when the carrier together with its associated crystalline distortion is comparable in size to the cell parameter. Electronic conduction can then occur via quantum tunneling between neighbouring distortions and $\rho(T)$ is expected to increase with decreasing T[9, 10]. At temperatures above 15 K and up to T_c, the observed slope $d\rho / dT$ is positive. For temperature below 100K, the large magnetic moment indicates strong ferromagnetic coupling between magnetic ions that prevents the formation of a magnetic polaron or spin disorder. The resistivity is relatively low and magnetic field has little effect on the resistivity. For T >100 K, the magnetic moment decreases rapidly and resistivity increases correspondingly. At T=210 K, the resistivity reaches a maximum. This temperature coincides with the transition temperature derived from the magnetic measurements. The high resistivity state is attributed to the formation of magnetic polarons, spin disorder scattering and possibly a modified Jahn-Teller effect[11,5].

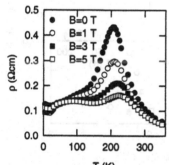

Figure 3. The resistivity of the single crystal sample from 5 K to 350 K with B=0, 1, 3 and 5 T. Magnetic field B and the sample current I are along [001] and [100] respectively.

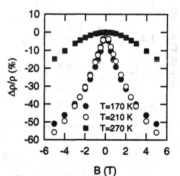

Figure 4. Magnetoresistance of the single crystal sample at 170, 210 and 270 K. B and I are along [001] and [100] respectively.

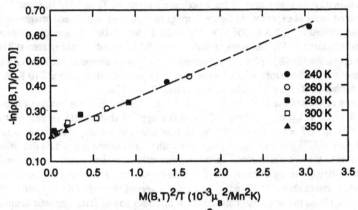

Figure 5. Ln[ρ(B,T)/ρ(0,T)] vs M(B,T)^2/ T from 240 K to 350 K with magnetic field up to 5 T. The line is a guide to the eye.

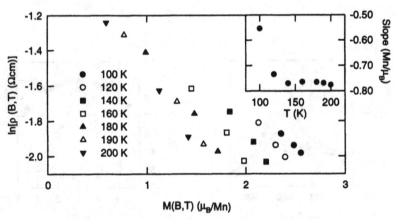

Figure 6. Ln[ρ(B,T)] vs M (B,T) from 100 K to 200 K with magnetic field up to 5 T. The inset is the fitted slope for different temperatures.

In this temperature region, magnetic field has its greatest influence on the resistivity. The magnetoresistance $\Delta\rho$(B) /ρ(0) = [ρ(B)- ρ(0)] / ρ(0) achieves a maximum value of about -60% with B=5 T at the transition temperature 210K. Above T_c, the entropy of disorder prevents the formation of magnetic polarons[11] and the electronic conduction is accomplished by carrier hopping from one center to another. As a consequence, the resistivity curve shows an activated behavior. The resistivity curves are fitted with $\rho(T) \sim \exp[E_p/k_B T]$, where E_p and k_B are the activation energy and the Boltzmann's constant respectively. The resulting activation energies are 47.5, 44.6, 43.9 and 43.0 meV for B=0, 1, 3, 5 T respectively. The decreasing values are consistent with the picture whereby activation energy is diminished when the moments are line up by a magnetic field[11]. The activation energies are relatively lower than those reported on the polycrystalline samples[12, 13].

The field dependences of magnetoresistance at 170 K (i.e., below T_c), at 210 K ($T = T_c$) and 270 K ($T > T_c$) are shown in Figure 4. The magnetoresistance has rather different curvature above and below T_c indicating different magnetic scattering mechanisms. This can further be gleaned from Fig. 5 and Fig. 6. Figure 5 shows the resistivity in the paramagnetic state ($T > T_c$) plotted according to Eq. 2.

$$\rho(B,T) = \rho(0,T) \exp[-M(B,T)^2/k_B T]. \tag{2}$$

The scaling behavior below T_c is quite different. In reference 12, Hundley et al. was the first to use the phenomenological scaling expression

$$\rho(B,T) = \rho_m \exp[-M(B,T)/M_0]. \tag{3}$$

to describe the correlation between the magnetic moment and the electronic conduction in the ferromagnetic region. Figure 6 shows the plot of $\ln[\rho(B,T)]$ plotted against $M(B,T)$. The insert is the slope $d\{\ln[\rho(B,T)]\}/d M(B,T)$ at different temperatures. The expression holds quite well from 140 K to 200 K except that ρ_m appears temperature dependent rather than being a constant. Deviations from this behavior at low temperatures were noted previously[6]. Both Fig. 5 and 6 show electronic conduction strongly modified by the magnetic state of the sample.

CONCLUSION

We studied the magnetic and electronic transport properties of a single crystalline sample of $La_{0.64}Pb_{0.36}MnO_3$. Magnetization curves along the three different orthorhombic crystal orientations are the same and we detect no hysteresis. A transition from a ferromagnetic to a paramagnetic state is found at 210 K. At T_c, a maximum negative magnetoresistance of about -60% is reached. The correlation between the magnetic moment and the electronic transport is examined both for the ferromagnetic state as well as for the paramagnetic state. At very low temperatures ($T < 15$ K) resistivity increases with decreasing temperature which might indicate the presence of a quantum tunneling mechanism. Investigations of thermal transport properties on this single crystal sample are in progress.

* Also: Physics & Physical Chemistry Department, General Motors Research and Development Center, Warren, MI, 48090.

ACKNOWLEDGEMENTS

The authors wish to thank Dr. D. T. Morelli for valuable discussions and for assistance with the magnetization measurements. JY wishes to thank General Motors Corp. for the DEGS Fellowship support. The research was supported in part by ONR Grant No. N00014-92-J-1335.

REFERENCES

1. G. H. Jonker and J. H. Van Santen, Physica (Utrecht) **16**, 337, 599 (1950).
2. C. Zener, Phys. Rev. **82**, 403 (1951).
3. P. W. Anderson and H. Hasegawa, Phys. Rev. **100**, 675 (1955).
4. P. G. deGennes, Phys. Rev. **118**, 141 (1960).
5. A. J. Millis, P. B. Littlewood and B. L. Shraiman, Phys. Rev. Lett. **74**, 5144 (1995).
6. B. Chen, A. G. Rojo, C. Uher, H. L. Ju and R. L. Greene, Phys. Rev. B **55**, 1 (1997).
7. A. H. Morrish, B. J. Evans, J. A. Eaton and L. K. Leung, Can. J. Phys. **47**, 2961 (1969).
8. H. L. Ju, J. Gopalakrishnan, J. L. Peng, G. C. X. Qi. Li, T. Venkatesan and R. L. Greene, Phys. Rev. B **51**, 6143 (1995).
9. M. Jaime, M. B. Salamon, K. Pettit, M. Rubinstein, R. E. Treece, J. S. Horwitz and D. B. Chrisey, Appl. Phys. Lett. **68**, 1576 (1996).
10. M. Jaime, M. B. Salamon, M. Rubinstein, R. E. Treece, J. S. Horwitz and D. B. Chrisey, Phys. Rev. B **54**, 11914 (1996).
11. N. F. Mott and E. A. Davis, *Electronic Processes in Non-crystalline Materials* (Oxford University Press, New York, 1979).
12. M. F. Hundley, M. Hawley, R. H. Heffner, Q. X. Jia, J. J. Neumeier, J. Tesmer, J. D. Thompson and X. D. Wu, Appl. Phys. Lett. **67**, 860 (1995).
13. B. Chen, C. Uher, D. T. Morelli, J. V. Mantese, A. M. Mance and A. L. Micheli, Phys. Rev. B **53**, 5094 (1996).

EFFECTS OF CHROMIUM ION IMPLANTATION ON THE MAGNETO-TRANSPORT PROPERTIES OF $La_{0.7}Ca_{0.3}MnO_3$ THIN FILMS

P. S. I. P. N. DE SILVA *, N. MALDE *, A. K. M. A. HOSSAIN *, L. F. COHEN *
K. A. THOMAS **, R. CHATER **, J. D. MACMANUS-DRISCOLL **, T.J.TATE [†],
N. D. MATHUR [‡], M. G. BLAMIRE [‡] and J. E. EVETTS [‡]
*Blackett Laboratory, Imperial College, Prince Consort Rd, London, SW7 2BZ, UK.
**Materials Department, Imperial College, Prince Consort Rd, London, SW7 2BP, UK.
[†]Depatment of Electrical Engineering, Imperial College, Prince Consort Rd, London, SW7 2BT, UK.
[‡]Materials Department, University of Cambridge, CB2 3QZ, UK.

ABSTRACT

Thin films of colossal magnetoresistance material $La_{0.7}Ca_{0.3}MnO_3$ were implanted with different fluence 200keV Cr ions. Resistivity measurements in zero and applied fields of up to 8T were made in order to determine the effects of the implanted magnetic ions on the magnetoresistance (MR). As the Cr fluence was increased, the resistivity increased and the metal-insulator transition (MI) temperature was suppressed to values below the experimentally accessible temperature range as a result of oxygen loss and the creation of defects. However, for the highest fluence of $5x10^{15}$ ions/cm^2, a *re-entrant* metal-insulator type transition was observed. Furthermore a significant improvement in the low field MR was observed for fields less than 500mT. These results are interpreted in terms of substitution of Cr ions onto Mn sites and the creation of a magnetically inhomogeneous material and the influence of oxygen deficiency.

INTRODUCTION

Recent investigations of colossal magneto-resistance (CMR) materials have been focussed towards applications. To this end, materials are sought with higher Curie temperatures (T_c), and greater MR change in the vicinity of T_c and in lower magnetic fields. Magnetoresistance effects can be separated into two types: high field effects associated with the intrinsic paramagnetic to ferromagnetic and simultaneous insulator to metal transitions[1]; and low field effects usually associated with grain boundaries[2].

Recent progress has been made towards understanding the origin of low field MR effects. Many reports have been made of polycrystalline thin films[3,4] where the low field MR effect increases as the sample is cooled away from T_c. In artificial grain boundary structures[5,6] it has been demonstrated that the low field effect disappears close to T_c. Of promise for applications, low field effects have also been reported in epitaxial thin films[7,8] where the effect is limited to temperatures around T_c, although this behaviour is not yet completely understood.

The influence of ion implantation on low and high field MR effects have not been widely studied. Chen *et al.*[9] implanted 200keV Ar$^+$ ions at fluence between 10^{11} and 10^{15} ions/cm^2, and found that the major influence of the implantation was to create defect sites and possibly oxygen loss, and thereby suppress T_c. In the present study $La_{0.67}Ca_{0.33}MnO_3$ thin films were implanted with Cr ions in order to study the influence on low and high field MR effects.

EXPERIMENT

The films were grown by pulsed laser deposition and patterned by ion milling. The thickness of the films was around 200nm as measured by profilometry. The thicknesses of the films are known only to within 20nm and for clarity, resistance not resistivity is plotted. Before implantation the films were annealed in 0.8 atm. O_2 for 4 hours at 900°C. These annealing conditions had previously been found to lead to saturation of film properties[10].

X-Ray (Bragg-Bretano and rocking curve) measurements were made on similar, saturated and unimplanted films. The films were (001) oriented with no peak splitting indicating distortion from an ideal cubic lattice, and the rocking curve width was ~0.2°. Note that low temperature low-field grain boundary effects are usually only seen in polycrystalline thin films, with larger rocking curve widths.

The films were then implanted with fluences of 10^{13} (Cr1), 10^{14} (Cr2), 5×10^{14} (Cr3) and 5×10^{15} (Cr4) ions/cm^2. A further sample which underwent a saturation anneal only but was not implanted, was used for comparison and was denoted as the virgin film (vir). The lowest fluence Cr dose was chosen to coincide with a fluence used in Ref. 9 in order that the effects of the two different types of ions could be compared.

The implantation modelled using the TRIM program[11] for a 200nm film, predicted high damage for the higher dose samples and amorphisation. Rutherford backscattering spectra on the Ar$^+$ implanted films[9] also showed almost complete disorder near the sample surface for the highest dose of 10^{15} ions/cm^2. Assuming a film density of 6.5 g/cm^3, the maximum in the Cr content was predicted to occur at 100nm, and if all the Cr substituted into the lattice this would imply for the highest dose, a 2.7% substitution per formula unit. The success of this was confirmed by Secondary Ion Mass Spectroscopy (SIMS) analysis which showed that for the highest implantation dose, the profile of the variation of Cr ions with depth, did indeed show a maximum around 100nm.

Resistance as a function of temperature, R(T), was measured between 20-300K using the standard 4 point technique and constant current. Magnetoresistance was measured in fields up to 8T using a superconducting magnet. After measurements were made on the implanted films, they were again subjected to a saturation anneal.

RESULTS

Figure 1 shows the resistance in 0T and 8T for the 5 films where it can be seen that the resistance generally increases as the dose of the Cr implantation is increased. For the lowest dose, the MI transition is shifted down to lower temperature. For the films implanted at fluences of 10^{14} ions/cm^2 and 5×10^{14} ions/cm^2, the resistivity behaviour becomes semiconductor-like at all temperature measured although significant MR is observed. The 10^{13} ions/cm^2 implantation qualitatively reproduced that shown in Ref. 9. However, in the present study it is found that increasing the fluence further to 5×10^{15} ions/cm^2 causes a peak in the resistance, reminiscent of a MI transition. This measurement was repeated two weeks later on the same film and although the details of the R(T) curve had changed, certain features such as the magnitude of the resistance at the peak temperature, and the temperature at which the peak occurs (T_p) were comparable. Most importantly the presence of a peak in the resistivity is still clearly visible.

The MR behaviour where MR(%) = 100 x $[\rho(0) - \rho(B)]/\rho(0)$ and B = 8T, is shown as a function of temperature in Figure 2. Even the lowest dose implantation significantly broadens the transition, suggestive of oxygen or other stoichiometric inhomogeneity. Figure 3 shows the MR(B) behaviour of the virgin, lowest dose and highest dose films at their respective T_p. Most

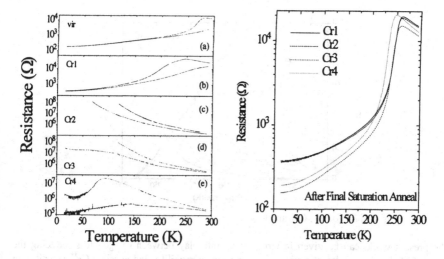

Fig 1. Left Panel: R(T) for all 5 films in 0T (upper curves in each frame) and in 8T (lower curves). Right Panel: R(T) for implanted films after saturation anneal.

striking is the enhancement of the low field behaviour in the implanted films. For the high dose film, the measurements made two weeks apart are shown and it can be seen that at this temperature no significant changes to the low field MR behaviour has occurred.

After the post-implantational saturation anneal, the films showed magneto-transport properties resembling the virgin sample except T_p was depressed systematically by an amount which increased to 25K for the highest dose film. This depression in T_p suggests that there are residual defects which impair the mobility. The enhanced low field MR behaviour at T_p also vanishes after the anneal. This behaviour is consistent with the previously reported Ar^+ implantation study, which also found that the ion-induced damage could be annealed by high temperature treatment[9].

The magnetic properties of the $LaMn_{1-x}Cr_xO_3$ system were studied by Jonker[12] in 1956. He reported that the $180°$ $Cr^{3+}(d^3)$ - O - $Mn^{3+}(d^4)$ interaction was positive i.e. ferromagnetic superexchange could occur between Cr^{3+} and Mn^{3+} but not double exchange as between Mn^{3+} and Mn^{4+} in $La_{1-x}Ca_xMnO_3$ because of the absence of carriers.

Recently, Gundakaram et al.[13] confirmed the Jonker result, by finding $LaMn_{1-x}Cr_xO_3$ to be ferromagnetic and semiconductor-like, although showing MR ~20% in 6T. Interestingly, in systems with localised carriers such as the charge ordered insulator $Pr_{0.5}Ca_{0.5}MnO_3$, Cr substitution was found to induce a MI transition[14] implying that the charge ordered symmetry was broken and/or the creation of a second channel for carriers to move through was made possible. The mechanism for this remarkable behaviour is not known although it was suggested that the mixed valence states of chromium were important in understanding the interplay between ferromagnetism and antiferromagnetism in these materials. However this result holds a possible key to the present study, where carriers are present but there is also a large amount of disorder causing localisation. Re-entrant behaviour of the MI transition has not been observed before in the $La_{0.67}Ca_{0.33}MnO_3$ system. A qualitative explanation of the R(T) behaviour discussed

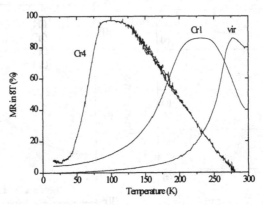

Fig 2. MR(%) vs T in 8T for vir, Cr1 and Cr4.

in the present work, can be given in terms of an interplay between oxygen loss reducing the number of Mn^{4+} ions and defect creation reducing carrier mobility, and possible Cr^{3+} substitution on Mn sites allowing a second channel for carriers. For the low fluence implantation, oxygen loss and defect creation dominate, resulting in an increased resistance and a depression of T_p, as observed in $La_{0.67}Ba_{0.33}MnO_{3-z}$ (Ref. 2) and Ar^+ implanted films[9]. It is also possible that cation vacancies will be created, becoming interstitials within the lattice[15]. At the highest dose, a second conduction channel possibly due to the substitution of Cr ions dominates and T_p reappears and the resistance is lowered.

The MR properties in the implanted thin films are of technological interest since the highest fluence sample shows a dramatic improvement of the low field MR at its T_p. The re-entrant MI transition confirms that the observations are not entirely due to oxygen loss and that some Cr might be substituting on the Mn sites, creating a second conduction channel for carriers. The enhancement of low field MR at T_p and a broadening of the high field MR(T) suggests that Cr^{3+} implantation causes the films to have a distribution of properties, perhaps arising from inhomogeneous Cr implantation. Indeed the SIMS analysis indicated that the Cr content changes as a function of depth. The possibility of ferromagnetic clusters imbedded in a non-ferromagnetic matrix, and the influence this might have on enhancing the low field MR behaviour, has been alluded to by a number of authors[16,17,18]. This behaviour may in some cases also be attributed to locally increased anisotropic strain.

However, it is important to consider whether oxygen loss alone could have created the enhanced low field MR properties. Consistent with the present observations, Mahendiran et al. showed that oxygen loss lowers T_p, broadens MR(T) and can separate T_p and T_c by as much as 100K[19]. Of particular interest is the study by Ju et al.[2], showing that oxygen loss enhances the low field R(B) slope at low temperatures. The low field enhancement witnessed in the present study would suggest enormous oxygen loss had taken place in the implanted films and it is not clear if this would occur in the grains or at the grain boundaries.

Unless the Cr was mobile during the anneal and diffused away from the Mn sites (possibly to the grain boundaries), the post anneal behaviour where the low field enhancement has disappeared suggests that the oxygen deficiency and defects played an important role. If Cr did not segregate to the grain boundaries after the anneal, then it appears that the second

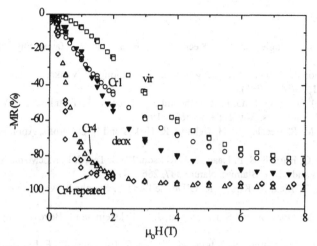

Fig 3. -MR(%) for vir, Cr1 and Cr4 at their T_p's of 275K, 245K and 95K respectively. Also shown are the data for a sample deoxygenated at a partial pressure of 10^{-7} atm. O_2 (deox).

conduction channel created by the implantation becomes insignificant after the anneal in comparison with the resurrected low vacancy, high mobility Mn^{4+} - O - Mn^{3+} path.

There is a possibility that the reentrant R(T) behaviour results from annealing of the films at the high implantation dose. Raman spectra of films which have been subjected to reduced oxygen atmospheres have been measured previously[20]. Various phonon modes are seen to shift systematically as oxygen is reduced from the films. Building on this result the Raman spectra of the Cr implanted films suggest that oxygen and defect damage is increased continuously across the doping range negating the high does annealing concept[21]. We do not rule out that the enhanced R(B) curve shown in Figure 3 for the heavily implanted sample could result from the extreme deoxygenation of the film as Figure 3 suggests, but we believe that the reentrant behaviour is in some way attributable to the presence of the Cr.

CONCLUSIONS

In summary, the transport properties of $La_{0.67}Ca_{0.33}MnO_3$ thin films implanted with 200keV Cr ions (10^{14} - 5×10^{15} ions/cm^2) were studied. A re-entrant behaviour in the MI properties were observed and interpreted as substitution of some Cr on Mn sites. The results can be understood by considering the effects of two competing effects. At low doses, magnetic ion implantation causes disorder and oxygen loss, reducing the Mn^{4+} content and reducing carrier mobility but at higher doses some Cr substitution occurs on the Mn sites allowing double exchange effects to occur through a second conduction channel. The dramatic improved low field R(B) properties are attributed to the creation of a magnetically inhomogeneous material, and the effects of oxygen deficiency. The exact role of oxygen deficiency in determining, if at all, the enhanced low field behaviour is currently being investigated. In a further attempt to rule out the possibility of annealing effects causing this behaviour, samples which have been implanted with a non-magnetic ion i.e. Al are being studied.

REFERENCES

[1] R. M. Kusters, J. Singleton, D. A. Keen, R. McGreery and W. Hayes, Physica B **155**, 362 (1989).

[2] H. L. Ju, J. Gopalakrishnan, J. L. Peng, Q. Li, G. C. Xiong, T. Venkatesan and R. L. Greene, Phys. Rev. B **51**, R6143 (1995).

[3] A.Gupta, G. Q. Gong, G. Xiao, P. R. Duncombe, P. Lecoeur, P. Trouilloud, Y. Y. Wang, V. P. Dravid and J. Z. Sun, Phys. Rev. B **54**, R15629 (1996).

[4] G. J. Snyder, M. R. Beasley, T. H. Geballe, R. Hiskes and S. DiCarolis, Appl. Phys. Lett. **69**, 4254 (1996).

[5] N. D. Mathur, G. Burnell, S. P. Isaac, T. J. Jackson, B. -S. Teo, J. L. MacManus-Driscoll, L. F. Cohen, J. Evetts and M. G. Blamire, Nature **387**, 266 (1997).

[6] K. Steenbeck, T. Eick, K. Kirsch, K. O'Donnell and E. Steinbeiß, Appl. Phys. Lett. **71**, 968 (1997).

[7] J. O'Donnell, M. Onellion, M. S. Rzchowski, J. N. Eckstein and I. Bozovic, Phys. Rev. B **55**, 5873 (1997).

[8] M. G. Blamire, N. D. Mathur, S. P. Isaac, B. -S. Teo, G. Burnell and J. E. Evetts, preprint.

[9] C. -H. Chen, V. Talyansky, C. Kwon, M. Rajeswari, R. P. Sharma, R. Ramesh, T. Venkatesan, J. Melngailis, Z. Zhang and W. K. Chu, Appl. Phys. Lett. **69**, 3089 (1996).

[10] K. A. Thomas, P. S. I. P. N. de Silva, L. F. Cohen, A. K. H. A. Hossain, M. Rajeswari, T. Venkatesan, R. Hiskes and J. L. MacManus-Driscoll, submitted to J. Mater. Res.

[11] J. Ziegler and P. Biersack, *The Stopping and Range of Ions in Solids* (Pergammon, New York, 1985).

[12] G. H. Jonker, Physica **XXII** 707 (1956).

[13] R. Gundakaram, A. Arulraj, P. V. Vanitha, C. N. Rao, N. Gayathri, A. K. Raychaudhuri and A. K. Cheetham, J. Solid State Chem. **127**, 354 (1996).

[14] B. Raveau, A. Maignan and C. mMartin, J. Solid State Chem. **130**, 162 (1997).

[15] P. S. I. P. N. de Silva, F. M. Richards, L. F. Cohen, J. A. Alonso, M. J. Martínez-Lope, M. T. Casais, K. A. Thomas and J. L. MacManus-Driscoll, submitted to J. Appl. Phys.

[16] Z. Guo, J. Zhang, N. Zhang, W. Ding, H. Huang and Y. Du, Appl. Phys. Lett. **70**, 1897 (1997).

[17] S. Jin, H. M. O'Brien, T. H. Tiefel, M. McCormack and W. W. Rhodes, Appl. Phys. Lett. **66**, 382 (1995).

[18] N. R. Washburn, A. M. Stacey and A. M. Portis, Appl. Phys. Lett. **70**, 1622 (1997).

[19] R. Mahendiran, R. Mahesh, A. K. Raychaudhuri and C. N. R. Rao, Solid State Comm. **99**, 149 (1996).

[20] N. Malde, P. S. I. P. N. de Silva, A. K. M. A. Hossain, L. F. Cohen, K. A. Thomas, J. L. MacManus-Driscoll, N. D. Mathur and M. G. Blamire, in press Solid. State Comm.

[21] N. Malde, to be published.

EVALUATION OF RAMAN SCATTERING IN La$_{1-x}$M$_x$MnO$_3$ SINGLE CRYSTALS DUE TO STRUCTURAL AND MAGNETIC TRANSITIONS

V.B. PODOBEDOV*,1, A. WEBER*, D.B. ROMERO**, J.P. RICE*, and H.D. DREW**
* Optical Technology Division, National Institute of Standards and Technology, Gaithersburg, MD 20899, USA.
** Department of Physics, University of Maryland, College Park, MD 20742, USA
1 Institute of Spectroscopy, Russian Academy of Science, Troitsk Moscow Region, 142092 Russia.

ABSTRACT

The study of doped and undoped La$_{1-x}$M$_x$MnO$_3$ single crystals ($x = 0$ to $x = 0.3$) in the temperature range from 5 K to 423 K is reported. The activity of phonon vibrational modes in optical spectra is analyzed and a comparison to experimental Raman data is presented. Different contributions from optical phonons to the Raman scattering process are discussed. It is shown that both the value of doping and temperature have a significant influence on Raman features, and these are strongly affected by structural and magnetic phase transitions. Due to the symmetry present in distorted lanthanum manganese compounds their Raman spectra were found to contain comparable contributions from first- and second-order phonon scattering. The common features related to both the reduction of the Jahn-Teller (JT) distortion and the temperature-induced phase transition were found in the Raman spectra of La$_{1-x}$M$_x$MnO$_3$ crystals. An anomalous behavior of the low-frequency Raman mode in doped systems was found and explained in terms of the temperature-dependent ionic radius of the La/Sr ion sites. It was found that the strong dependence of the A$_g$ Raman mode on the value of doping may serve as a useful tool for optical characterization of La$_{1-x}$M$_x$MnO$_3$ compounds, including the inhomogeneity of the surface.

INTRODUCTION

The variety of structural transitions related to cubic, rhombohedral, tetragonal, orthorhombic and monoclinic crystal symmetries as well as magnetic phases (para- and ferromagnetic insulators and metals, canted antiferromagnetic insulator) were intensively studied in lanthanum manganese compounds by different techniques [1-4]. A strong dependence of the main lattice parameters and bond lengths over a wide range of temperature and doping was demonstrated in several studies [1, 2]. In particular, the key parameters of the giant magnetoresistivity (GMR) materials, the Curie temperature T_c and magnetoresistance, strongly correlate with a microstructural Mn-O-Mn bond angle due to related changes in the lattice [5]. It was shown that an external magnetic field can control the structural phase transition in La$_{0.83}$Sr$_{0.17}$MnO$_3$ crystal as soon as the magnetic moments and charge carriers are coupled to the crystal structure [3]. A strong electron-phonon coupling related to JT splitting of the Mn^{3+} ion [4] was found to play an important role for the understanding of the metallic and electronic properties of La$_{1-x}$Sr$_x$MnO$_3$ system.

Optical spectra of phonons, including those in Raman scattering (RS), are known to be sensitive to many of the above lattice effects. However, in the perfect cubic perovskite structures the first-order RS is forbidden by the selection rules. Therefore, RS in the nearly-cubic structures of the doped lanthanum manganese materials is extremely weak. Previous Raman studies of the La$_{1-x}$Sr$_x$MnO$_3$ systems [6, 7], as well as closely related studies of perovskite-like ferroelectrics have indicated some difficulties in the interpretation of experimental data. Few known infrared (IR) studies were carried out with ceramic samples and with small spectral resolution. In view of the great attention given to this system, any information about optical phonons in the La$_{1-x}$M$_x$MnO$_3$ materials is of current interest.

Here we report a Raman study of the La$_{1-x}$M$_x$MnO$_3$ system. The experimental data presented are for Sr-doped single crystals, although samples doped by Ca or Pb and Nd$_{1-x}$Sr$_x$MnO$_3$ crystals were found to have some common spectroscopic features as well.

329

EXPERIMENT

The spectra were excited by the radiation of an Ar+ ion laser (λ_i = 514.5 nm).To avoid both possible damage of a sample and the uncertainty in its temperature due to heating, the excitation power was kept below 50 mW. Different polarization geometries were realized by means of a half-wave silica plate and wide-band polarizer. Raman spectra were obtained with a triple-stage multichannel spectrometer equipped with a CCD detector operating at 140 K. All spectra were detected in a back-scattering geometry with a spectral resolution of about 4 cm^{-1}. The millimeter-sized La$_{1-x}$Sr$_x$MnO$_3$ single crystals were prepared by the floating-zone technique. Other details of the experimental set-up have been published in [8].

RESULTS

The evaluation of Raman spectra of La$_{1-x}$M$_x$MnO$_3$ materials comprises the measurements of both intensity and position of selected peaks that are most representative for a particular structural or magnetic phase, as well as for a stoichiometry of the compound, doping value and so on. Due

Table I. Factor group analysis for orthorhombic (D_{2h}), rhombohedral (D_{3d}) and cubic (O_h) symmetries of the La$_{1-x}$M$_x$MnO$_3$ perovskite structures.

Factor group	Space group	Z*	predicted fundamental vibrational modes Raman-active modes	IR modes
D_{2h}	D_{2h}^{16} (mmm)	4	$7A_g + 5B_{1g} + 7B_{2g} + 5B_{3g}$	$8A_u + 9B_{1u} + 7B_{2u} + 9B_{3u}$
D_{3d}	D_{3d}^6 (R3c)	2	$A_{1g} + 4E_g$	$2A_{1u} + 3A_{2u} + 5E_u$
O_h	O_h^1 (m3m)	1	None	$3F_{1u} + F_{2u}$

* Z is the number of formula units per unit cell

to the properties of the RS tensor for different symmetries, the latter may be found from experimental polarized spectra. The activity of vibrational modes in optical spectra for the three most commonly reported cubic, rhombohedral, and orthorhombic symmetries of La$_{1-x}$M$_x$MnO$_3$ materials are summarized in the Table I.

Doped crystals (x > 0)

It was found that Raman spectra of La$_{1-x}$M$_x$MnO$_3$ single crystals have a complex structure that may be represented by three different components: first-order RS, second-order RS, and electronic RS [6, 9]. The first component follows the selection rules for vibrational excitations (Table I) and thus makes it useful for diagnostic purposes. The second component is due to the density of vibrational states and usually contains broad-band features. The electronic Raman scattering in La$_{1-x}$M$_x$MnO$_3$ materials observed together with phonon scattering is like a continuous background and was studied in [7]. Typical Raman spectra of doped La$_{1-x}$M$_x$MnO$_3$ single crystals are shown in Fig. 1. The spectra are normalized by the thermal factor $[1+n(\omega)]=[1-\exp(\hbar\omega/k_BT)]^{-1}$ and therefore describe the imaginary part of the Raman susceptibility, $Im\chi$. The above three components are present in each of the three spectra. It was found that sharp peaks located at the top of the wide band have different polarizations and may be separated in different scattering geometries [9]. One can note the intensity redistribution in the Raman spectra as well as the

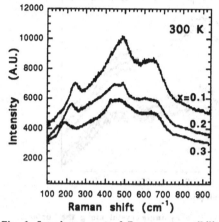

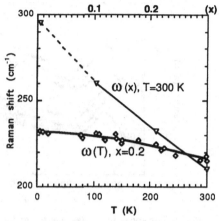

Fig. 1. Imaginary part of Raman susceptibility, $Im\chi$, for $La_{1-x}Sr_xMnO_3$ single crystals with different doping (x).

Fig. 2. Raman shift of the A_g mode in $La_{1-x}Sr_xMnO_3$ samples versus x ($T = 293$ K) and T ($x = 0.2$).

significant shift of the low-frequency mode with doping. Below we will consider the nature of these spectroscopic features as well as their possible applications.

The low-frequency portion of the vibrational spectrum in $La_{1-x}M_xMnO_3$ crystal is usually assigned to the external vibrations. When interpreting some previous IR studies of perovskite-like materials, one may suppose that in this frequency range the vibrational modes related to motions of both La and M ions with respect to the MnO_6 octahedra may appear. Such assignment in the present Raman spectra is also supported by a strong dependence of the low-frequency A_g mode on both the kind of the doping ions [6] and their concentration (Fig. 2).

From Fig. 2, upper curve, it follows that the above A_g mode near 200 cm-1 at $T = constant$ undergoes a strong (more than 50 cm-1) change in Raman shift as the doping increases from 0.1 to 0.3. Furthermore, the sign of this change was found to be opposite to that expected from the difference in the atomic weights of La(139) and Sr(87). Such behavior is unusual, for example, in terms of what is known for mixed crystals where the frequency change is driven mainly by the mass effect. We consider therefore that the reason for above anomalous change is due to the nature of interaction between La/Sr ions and MnO_6 octahedra that can overcompensate the mass effect. While the lanthanum or strontium ions occupy the sites between oxygen octahedra in $La_{1-x}Sr_xMnO_3$ compounds, it was shown in Ref. [10] that the ionic radius of this site, $<r_A>$, increases with Sr doping. Accounting for the overlap of the Mn-site d orbitals and oxygen p orbitals, the change of ionic radius of a La site was found to be strongly responsible for the significant wide-range dependence of the Curie temperature T_c [5]. We consider that the increase of the ionic radius $<r_A>$ strongly affects the force constant in the La/Sr - MnO_6 vibrational system and that this effect explains the corresponding change in the Raman shift of the related low-frequency mode.

The lower curve in Fig. 2 demonstrates the Raman shift of this band with temperature for a fixed value of doping, $x = 0.2$. The change in Raman shift of ~ 20 cm-1 was found to exceed significantly the value known for the usual anharmonic effect in the indicated temperature range. Coming back to the consideration of the effect of the ionic radius $<r_A>$ on the vibrational mode, one may believe that a temperature dependence of $<r_A>$ may contribute to the reduction of the Raman shift with temperature. In this case, both the anharmonic and the force constant effects have the same sign and thus provide the change in Raman shift of about 20 cm-1.

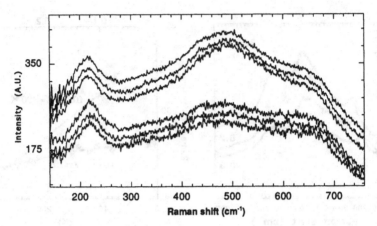

Fig. 3. Variations in Raman spectra of $La_{0.8}Sr_{0.2}MnO_3$ single crystal along the surface.

As a non-destructive optical technique, Raman spectroscopy has also an advantage that is often used for the spatially-resolved analysis. A laser beam may be focused into a diffraction-limited spot of a diameter $D = 4\lambda f/\pi d$. In our case, for $\lambda = 514.5$ nm, diameter of a laser beam $d = 3$ mm, and focal length of focussing lens $f = 100$ mm, D is approximately 0.02 mm. This value corresponds to an in-plane spatial resolution of the RS techniques. In Fig. 3 some Raman spectra from the same $La_{0.8}Sr_{0.2}MnO_3$ single crystal are shown as a function of position of the laser beam on the crystal. The upper and lower sets of spectra correspond to a displacement of ± 1.5 mm from an almost invisible boundary line on the surface of the crystal. All spectra have the same intensity scale but were shifted in the vertical direction within each top and bottom sets. The obvious difference between both sets is the intensity of the peak at about 500 cm⁻¹. It was found that this peak is usually stronger in the orthorhombic phase of $La_{1-x}Sr_xMnO_3$ system (see also Fig. 4). Accounting for the relatively small accumulation time (2 min.), this example demonstrates the principal possibilities of Raman spectroscopy as a potential tool for a fast diagnostic of the $La_{1-x}Sr_xMnO_3$ crystal surface or films.

Another practical application, fast non-destructive measurement of the doping value x in a $La_{1-x}Sr_xMnO_3$ system, is based on the observed change in the Raman shift, $\Delta\omega > 50$ cm⁻¹, of the band associated with the low-frequency A_g mode. For this purpose, the upper curve in Fig. 2 may be used as a calibration guide. The position of the A_g mode, at least over the range of x from 0.1 to 0.3, may be determined in practice to within an uncertainty of $\delta s = \pm 2$ cm⁻¹. Therefore, the estimated uncertainty in the experimental value of x is of the order of $\Delta x(\delta s/\Delta\omega) = 0.2$ (2/50), that is better than 8 %. To use this method for the determination of x it is necessary that the calibration curve be obtained from samples with well-defined x as well as of known stoichiometric composition. The latter parameter may affect the vibrational spectrum, and a study of this phenomenon is now in progress.

Undoped crystal ($x = 0$)

The assignment of vibrational modes in Raman spectra at $x = 0$ was done within the orthorhombic D_{2h}^{16} symmetry reported so far for undoped $LaMnO_3$ compounds (Table I, [6]).

Most of predicted modes were detected in the following scattering geometries: A_{1g} - $xx, yy,$ and zz; B_{1g} - xy; B_{2g} - xz; B_{3g} - yz. Unpolarized spectra of $LaMnO_3$ single crystal are shown in Fig. 4. As in the case of doped samples, at $x = 0$ the Raman spectra also exhibit significant transformation related to structural and magnetic transitions. The intensity of selected representative modes as well as ordinary and anomalous Raman shifts may be found from experimental spectra. This can be therefore evaluated in terms of qualitative and quantitative descriptions.

As indicated in Fig. 4, the intensity of the Raman spectrum of $LaMnO_3$ single crystal gradually decreases with increasing temperature. From neutron studies it is known that, at high temperatures, the $LaMnO_3$ compounds undergo an orthorhombic to rhombohedral structural transition. Therefore, the intensity of the bands at 493 cm-1 and 609 cm-1, related to the D_{2h}^{16} orthorhombic symmetry, decreases. It is interesting to note that doping of the crystal has the same effect on Raman spectrum (Fig. 4, the bottom spectrum). This behavior is in qualitative agreement with the decrease of the JT distortion with increasing doping. JT distortion is known to be responsible for the orthorhombic distortion of the nearly-cubic perovskite $La_{1-x}M_xMnO_3$ structure.

An irregularity in the Raman shift for some peaks was found at T close to $T_N \approx 140$ K. At this temperature the phase transition from paramagnetic insulator to canted-antiferromagnetic is known to occur [1, 2]. The position of peaks associated with the B_g modes above 600 cm-1 was found to be strongly dependent in the limited temperature range from $T = 100$ K to $T = 180$ K (Fig. 5). Accounting for previous IR [11] and Raman [6, 9] studies of related crystals, the high-frequency B_g mode in $La_{1-x}M_xMnO_3$ may be assigned to the internal vibration related to the mutual Mn-O motion within the oxygen octahedron. Due to both the location and the relatively strong intensity of the B_g mode close to 600 cm-1, this mode may be classified as a stretching Mn-O vibration. The variations of the lattice parameters, unit-cell volume, and sublattice magnetization of the orthorhombic $LaMnO_3$ phase reported in [12], have shown their irregular behavior in the temperature range from 100 K to 150 K. We have found that the Raman shift of the B_g modes near 600 cm-1 (Fig. 5) strongly correlates with the value of magnetization. Therefore, the observed irregular shift may be due to a spin-lattice interaction arising from the stabilization energy required to bring the MnO_6 octahedra to a particular structural configuration. As $LaMnO_3$ undergoes a paramagnetic to canted-antiferromagnetic transition, the MnO_6 octahedra change accordingly in response to the antiferromagnetic ordering of the Mn spins.

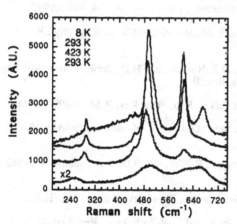

Fig. 4. Comparison of Raman spectra of doped (bottom spectrum) and undoped crystals at different temperatures.

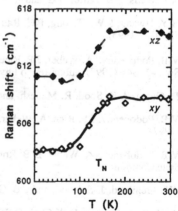

Fig. 5. Raman shift of the high- frequency B_g modes in undoped crystals versus temperature.

CONCLUSIONS

Raman spectroscopy is shown to be a useful tool for the study of the phase transitions in lanthanum manganese compounds. Experimental Raman spectra exhibit their strong dependence on the value and kind of doping. They are sensitive to structural and magnetic transitions driven by both doping and temperature. The phonon Raman scattering from doped crystals have demonstrated some anomalies that were explained by both the temperature and doping dependence of the ionic radius of a La/Sr-site. The dependence of the Raman shift of the low-frequency mode on the value of doping x may be used for diagnostic purposes. Raman spectra of undoped $LaMnO_3$ crystal were found to be in an agreement with an orthorhombic D_{2h}^{16} space group. An effect of the spin-lattice interaction was found in Raman spectra of undoped $LaMnO_3$. It is assigned to a paramagnetic-canted antiferromagnetic phase transition near $T_N = 140$ K.

ACKNOWLEDGEMENTS

The authors gratefully acknowledge Dr. Raju V. Datla for fruitful discussions and support of this work. We thank Professor S. Bhagat from University of Maryland for some samples of $La_{1-x}M_xMnO_3$ crystals and for interest in the present study, and Dr. J. Hougen for helpful discussions.

REFERENCES

1. H. Kawano, R. Kajimoto, M. Kubota, and Y. Yoshizawa, Phys. Rev. **B 53**, R14709 (1996).

2. P.G. Radaelli, D.E. Cox, M. Marezio, S-W. Cheong, P.E. Schiffer, and A.P. Ramirez, Phys. Rev. Lett. **75**, 4488 (1995).

3. A. Asamitsu, Y. Moritomo, Y. Tomioka, T. Arima, and Y. Tokura, Nature **373**, 407 (1995).

4. A.J. Millis, P.B. Littlewood, and B.I. Shraiman, Phys. Rev. Lett. **74**, 5144 (1995).

5. H.Y. Hwang, S.W. Cheong, P.G. Radaelli, M. Marezio, and B. Batlogg, Phys. Rev. Lett. **75**, 914 (1995).

6. V.B. Podobedov, A. Weber, J.P. Rice, D.B. Romero, and H.D. Drew, Bulletin Am. Phys. Soc. **42**, 341 (1997); submitted to Phys. Rev. **B**.

7. R. Gupta, A.K. Sood, R. Mahesh, and C.N.R. Rao, Phys. Rev. **B 54**, 14899 (1996).

8. V.B. Podobedov, J.P. Rice, A. Weber, and H.D. Drew, J. of Superconductivity **10**, 205 (1997).

9. V.B. Podobedov, A. Weber, D.B. Romero, J.P. Rice, and H.D. Drew, submitted to Solid State Commun.

10. W. Archibald, J.-S. Zhou, and J.B. Goodenough, Phys. Rev. **B 53**, 14445 (1995).

11. K.H. Kim, J.Y. Gu, H.S. Choi, G.W. Park, and T.W. Noh, Phys. Rev. Lett. **77**, 1877 (1996).

12. Q. Huang, A. Santoro, J.W. Lynn, R.W. Erwin, J.A. Borchers, J.L. Peng, and R.L. Green, Phys. Rev **B 55**, 14987 (1997).

PRESSURE AND ISOTOPE EFFECTS IN THE MANGANESE-OXIDE PEROVSKITES

J. B. Goodenough and J.-S. Zhou, Center for Materials Science & Engineering, ETC 9.102, University of Texas at Austin, Austin, TX 78712-1063

ABSTRACT

Measurements of the temperature dependence of the resistivity $\rho(T)$ and thermoelectric power $\alpha(T)$ under several hydrostatic pressures on $^{18}O/^{16}O$ isotope-exchanged $(La_{1-x}Nd_x)_{0.7}Ca_{0.3}MnO_3$ polycrystalline samples spanning the structural O'-O orthorhombic transition have demonstrated that the perovskite tolerance factor t increases with pressure, signaling an unusually compressible Mn-O bond. They have also indicated a change at the O'-O transition from static to dynamic cooperative Jahn-Teller deformations, from a second-order to a first-order magnetic transition, from Mn(IV) to two-Mn polarons in the paramagnetic region, from Mn(IV) polarons to a vibronic state below T_c, and a phase segregation in the O phase above T_c that traps out mobile polarons into ferromagnetic Mn(IV)-rich clusters within a Mn(IV)-poor matrix. Specific heat data show a transfer of spin entropy to configurational entropy on cooling through T_c in the O phase, and a single-crystal study of $La_{1-x}Sr_xMnO_3$, x = 0.12 and 0.15, has demonstrated a transition from polaronic to itinerant e electrons below T_c within the O phase. Magnetic-susceptibility measurements in low fields confirm the phase segregation above T_c and the existence of a ferromagnetic glass below T_c in the O phase. The intrinsic "colossal" magnetoresistance (CMR) is attributed to the growth to their percolation threshold of the ferromagnetic clusters existing above T_c, and the dramatic rise in T_c with increasing tolerance factor t in the O-orthorhombic phase to an increase in the density, and also the mobility, of the untrapped polarons above T_c.

INTRODUCTION

An intrinsic "colossal" magnetoresistance (CMR) has been observed in the perovskites $Ln_{0.7}A_{0.3}MnO_3$, where Ln is one or a combination of rare-earth atoms and A = Ca, Sr, or Ba. By varying the mean ionic radius $<r_A>$ of the $Ln_{0.7}A_{0.3}$ atoms, it is possible to vary the geometrical perovskite tolerance factor

$$t \equiv (A-O)/\sqrt{2}\,(Mn-O) \qquad (1)$$

where A-O and Mn-O are equilibrium metal-oxygen bond lengths for, respectively, twelvefold and sixfold oxygen coordination. Different thermal-expansion coefficients and compressibilities for the A-O and Mn-O bonds make t = t(T,P). For ambient pressure and temperature, t may be calculated from the sums of empirical ionic radii available in Tables [1].

Fig. 1 is a T-t phase diagram for the $Ln_{0.7}A_{0.3}MnO_3$ family of compounds. The CMR is a maximum near T_c in the O-orthorhombic phase at the cross-over from the O'-orthorhombic phase; it decreases with increasing T_c in the range 0.96 < t < 0.97. The O-orthorhombic structure is generated from the ideal cubic perovskite by a cooperative rotation of the corner-shared MnO_6 octahedra around a cubic [110] axis, which becomes the b axis of the orthorhombic phase; it is characterized by a $c/a > \sqrt{2}$. In the O' structure, a cooperative Jahn-Teller ordering of occupied σ-bonding e orbitals at high-spin Mn(III) ions into (001) planes makes $c/a < \sqrt{2}$. The O'-O transition marks an order-disorder transition for the occupied e orbitals. The R-rhombohedral phase is generated from the cubic perovskite structure by a cooperative rotation of the MnO_6 octahedra about a [111] axis. Rhombohedral symmetry does not remove the twofold e-orbital degeneracy and is incompatible with a cooperative Jahn-Teller ordering, either static or dynamic.

In the R phase, the Mn(III) localized e electrons of the O' phase are transformed into itinerant σ^*-band states of e-orbital parentage. Of particular interest are the electronic properties of the O phase where the e electrons undergo a transition from localized to itinerant behavior.

In the O' phase, the magnetic interactions are described by superexchange (SE) theory [2] and the long-range magnetic order below T_c has Type A antiferromagnetic order [3], *i.e.* ferromagnetic (001) planes coupled antiparallel to one another with a canted-spin ferromagnetic component along the c-axis, *i.e.* perpendicular to the Dzialoshinskii vector $D \| b$. The O' phase is a canted-spin ferromagnetic insulator (CFI) in which the mobile holes are small-polaron Mn(IV) species. In the R phase, the magnetic interactions are dominated by the de Gennes [4] double-exchange (DE) coupling of the localized t^3 configuration (S = 3/2) by itinerant σ^* electrons of e-orbital parentage. Below T_c, the R phase is a ferromagnetic metal (FM). The PI-PM transition (paramagnetic insulator (polaron) - paramagnetic metal) occurs within the R phase.

We have measured the temperature dependence of the resistivity $\rho(T)$ and of the thermoelectric power $\alpha(T)$ under different hydrostatic pressures and upon ^{16}O to ^{18}O isotope exchange on samples that span the O' to O critical calculated tolerance factor $t_c \approx 0.953$. We have also measured molar magnetic susceptibilities $\chi_m(T)$ and specific heats $c_p(T)$. From our data, we argue for the following evolution of electronic properties with increasing t across the O phase:

(1) At the O'-O transition, the cooperative Jahn-Teller electron-lattice coupling changes from static to dynamic and the ferromagnetic e-electron SE interactions change from anisotropic to isotropic.

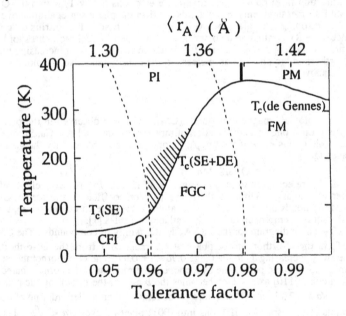

Fig. 1.   Temperature-tolerance factor phase diagram for the $Ln_{0.7}A_{0.3}MnO_3$ family of compounds. Shaded area corresponds to domain of CMR.

(2) A discontinuous change in the kinetic energy of the electrons at the O'-O transition induces a first-order change in the equilibrium Mn-O bond length and therefore a double-well potential that is manifest not only in a highly compressible Mn-O bond, but also in a phase segregation below a temperature $T_s > T_c$ into Mn(IV)-rich clusters within a Mn(IV)-poor matrix.

(3) Above T_s, polarons change with increasing t from Mn(IV) species to two-manganese clusters; with decreasing temperature in the interval $T_c < T < T_s$, the mobile polarons become progressively trapped in Mn(IV)-rich clusters, and those remaining in the Mn(IV)-poor matrix may become progressively converted back from two-manganese clusters to Mn(IV) species.

(4) In the interval $T_c < T < T_s$, the volume of the Mn(IV)-rich clusters grows with an applied field H and/or hydrostatic pressure, and the CMR occurs where the ferromagnetic clusters percolate.

(5) The Curie temperature T_c is determined by the sum of SE and Zener [5] DE ferromagnetic interactions in the Mn(IV)-poor matrix, and the stronger Zener DE component increases with the density and mobility of the two-manganese polarons in the Mn(IV)-poor matrix.

(6) A discontinuous change in the kinetic energy of the electrons on lowering the temperature through T_c induces a first-order phase change at T_c, the volume change at T_c decreasing with increasing t.

(7) Below T_c, the Mn(IV)-rich clusters grow beyond the percolation limit, and the trapped polarons are progressively released to a vibronic state with the exchange of spin entropy for configurational entropy; the resulting ferromagnetic glass is a conductor (FGC), but not a metal, even though the resistivity exhibits a metallic temperature dependence below T_c.

INTERATOMIC EXCHANGE

In an octahedral site, the high-spin Mn(III) configuration t^3e^1 contains a single σ-bonding e electron in a twofold orbital degeneracy, which makes Mn(III) a strong Jahn-Teller ion; the Mn(IV) configuration t^3e^0 contains only half-filled, π-bonding t orbitals that have their spin degeneracy removed by a strong intraatomic Hund exchange field to give S = 3/2. The t^3 configurations remain localized on both Mn(III) and Mn(IV) ions, and the $(180° - \phi)$ t^3 - $O:2p_\pi$ - t^3 superexchange interactions are everywhere antiferromagnetic. However, the interatomic interactions via σ-bonding e electrons are stronger and dominate where they are operative.

In the O' phase, ordering of the occupied e orbitals by a cooperative, static Jahn-Teller deformation shifts the oxygen atoms away from one Mn neighbor toward the other to create ferromagnetic $e^1 \cdots O:2p_\sigma$ - e^0 SE interactions in an (001) plane whether the e^0 orbitals are at a Mn(III) or a Mn(IV) ion; polaronic Mn(IV) species do not allow DE real electron transfer between Mn(III) and Mn(IV) ions in a time τ_h short relative to a spin-relaxation time τ_s. Orbital ordering restricts the ferromagnetic $e^1 \cdots O:2p_\sigma$-e^0 interactions to the (001) basal planes [6]. However, these anisotropic SE interactions become isotropic where the Jahn-Teller deformations become dynamic, as has been demonstrated in the system $LaMn_{1-x}Ga_xO_3$ [7].

The ferromagnetic DE interaction was first postulated by Zener [5] for a Mn(III):t^3e^1 - $O:2p_\sigma$ - Mn(IV):t^3e^0 cluster in which an α-spin electron e_α displaces the α-spin $2p_\sigma$ electron on the oxygen atom to the empty e orbital on a Mn(IV) ion preferentially if the spin of the t^3 configuration of the Mn(IV) ion is parallel to the transferred spin; the preference is due to the strong intraatomic Hund exchange field. Implicit in this mechanism is not only a $\tau_h < \omega_0^{-1}$, where ω_0^{-1} is the period of the oxygen-atom vibration between the Mn atoms, but also a polaron transfer time $\tau_p < \tau_s$ if the coupling is to be global. Zener assumed a diffusional motion with no activation energy, $\mu_p = eD_o/kT$, to obtain a polaron resistivity $\rho \sim T$ below T_c and a short τ_p.

In an alternate model, de Gennes [4] assumed that the e electrons may be described with a tight-binding band model in which the electron-energy transfer integral is spin-dependent, *i.e.* $t_{ij} = b_{ij}\cos(\theta_{ij}/2)$ where θ_{ij} is the angle between spins on adjacent cations and b_{ij} is the usual spin-independent energy transfer integral. With the assumption of a homogeneous electronic system, the total exchange energy along the c axis of the O' phase, which was not distinguished from the O phase by de Gennes, would be $\Delta\varepsilon_{ex} \sim |J|S^2\cos\theta - xb\cos(\theta/2)$ where J is the antiferromagnetic $t^3 - 2p_\pi - t^3$ superexchange energy parameter. Optimization of the angle θ between canted spins of the ferromagnetic (001) planes gave

$$\cos(\theta_o/2) = xb/4|J|S^2 \qquad (2)$$

In this model, θ_o decreases smoothly with x until full ferromagnetic order ($\theta_o = 0$) is achieved. The de Gennes model appears to be applicable for the FM phase R where $\theta_o = 0$, but it is not applicable to an O phase containing strong electron-lattice coupling to dynamic Jahn-Teller deformations.

EXPERIMENTAL RESULTS AND DISCUSSION

We investigated five compositions in the system $(La_{1-x}Nd_x)_{0.7}Ca_{0.3}MnO_3$: x = 0.2, 0.4, 0.55, 0.75, and 1.0 with t = 0.964, 0.960, 0.957, 0.952, and 0.946, respectively; they span the critical tolerance factor t_c for the O' - O transition at room temperature and ambient pressure. Fig. 2 shows $\alpha(T)$ for each composition at ambient pressure; the inset shows the variation with pressure for x = 0.4 of the temperature T_{max} of the maximum value of the $\alpha(T)$ curve [8]. For each x, the maximum in $\alpha(T)$ falls a little above T_c. The resistances of the samples x = 0.75 and 1.0 were too high to be measured below T_c, indicating a $0.952 < t_c < 0.957$ at room temperature. A higher thermal expansion of the A-O relative to the Mn-O bond makes t increase with temperature, and at temperatures T > 300 K we find a $t_c < 0.952$. Four conclusions can be drawn from these data:

(1)   A temperature-independent $\alpha(T) = \alpha_o$ at high temperatures signals polaronic conduction with

$$\alpha_o \approx (k/e)\ln[(1-c)/c] \qquad (3)$$

Since a strong Hund's intraatomic exchange field constrains electron transfer from a Mn(III) ion to a polaronic neighbor, the spin-degeneracy factor is taken as $\beta = 1$ in Equation (3). If the polarons were randomly distributed Mn(IV) species, they would have a fractional site occupancy $c = xN/N = x$, where N is the number of Mn atoms, and we should expect an $\alpha_o \approx +38$ µV/K for x = 0.3. On the other hand, two-Mn clusters would have $c = xN/(N/2) = 2x$, which would give $\alpha_o = -20$ µV/K corresponding to the measured values for $t < t_c$ above 500 K. Since iodometric titration gave 2.98 ± 0.01 oxygen atoms per formula unit, we conclude that the polarons are two - Mn clusters, *i.e.* Zener pairs, at high temperatures. A somewhat higher α_o for the x = 1.0 sample with $t < t_c$ appears to signal the coexistence of Mn(IV) species and Zener pairs at high temperatures in this composition.

(2)   Pressure increases T_{max} in the same way as an increase in t, which signals a dt/dP > 0 and therefore a more compressible Mn-O bond. This unusual situation has only been found at a cross-over from localized to itinerant electronic behavior; it would signal a double-well potential for the Mn-O bond. From the Virial theorem for central-force fields, which states: $2 <T> + <V> = 0$, an increase in the mean kinetic energy $<T>$ of a system of antibonding

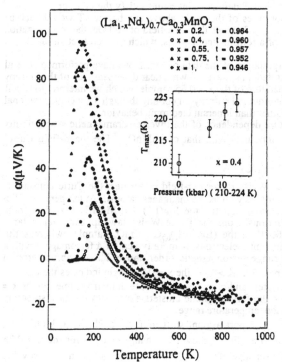

Fig. 2. Thermoelectric power versus temperature for compositions x = 0.2, 0.4, 0.55, 0.75, and 1.0 of $(La_{1-x}Nd_x)_{0.7}Ca_{0.3}MnO_3$. Inset: pressure dependence of T_{max} for x = 0.4.

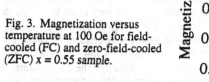

Fig. 3. Magnetization versus temperature at 100 Oe for field-cooled (FC) and zero-field-cooled (ZFC) x = 0.55 sample.

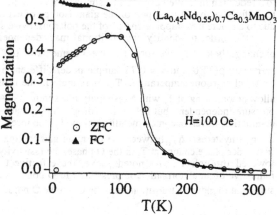

it follows that the data require either an $\varepsilon = \varepsilon_F$ over an extended temperature range or an electronic state in which $\sigma(\varepsilon)$ for energies $(\varepsilon-\varepsilon_F)$ is identical to that for $(\varepsilon_F-\varepsilon)$. A vibronic state with strong electron coupling to dynamic J-T deformations could satisfy the latter criterion over a range of temperatures since the mobilities of the states above and below kT would each be determined by the J-T phonon modes. Fig. 3 shows that, in a field of 100 Oe, the magnetization data below T_c for x = 0.55 is typical of a ferromagnetic glass, which indicates retention of two magnetic phases below T_c.

To test further the above interpretation of the data of Fig. 2, we have performed several additional experiments. Specific-heat data [11] have shown a near disappearance of the entropy change at the magnetic-ordering temperature of the x = 0.55 sample, which we interpret to signal an exchange of spin entropy for configurational entropy on cooling through T_c; a configurational entropy below T_c requires polaronic rather than itinerant electronic behavior.

Fig. 4 compares the temperature dependence of the inverse paramagnetic susceptibility $\chi_m^{-1}(T)$ for several $Ln_{0.7}A_{0.3}MnO_3$ compounds with that of $LaMnO_3$. The mean-field Curie-Weiss law

$$\chi_m^{-1} = (T-\theta_p)/C = (T/C) - W \qquad (4)$$

gives $T_c = \theta_p = CW$, where W is the Weiss molecular-field parameter; the Curie constant C contains contributions from the rare-earth as well as the manganese atoms. At temperatures T > 500 K, the Zener pairs are not superparamagnetic; the Mn(IV) and Mn(III) ions contribute individually to C. However, deviations from the Curie-Weiss law appear over a large temperature interval $T_c < T < 500$ K in the $(La_{1-x}Nd_x)_{0.7}Ca_{0.3}MnO_3$ samples whereas the $La_{0.7}Ba_{0.3}MnO_3$ sample, which has itinerant e electrons, is more typical of a ferromagnet with a smaller temperature range of short-range ferromagnetic order above T_c. An abrupt drop in $\chi_m^{-1}(T)$ with decreasing temperature near 275 K > T_c in the x = 0.2 sample indicates the onset of ferromagnetic order in Mn(IV)-rich clusters and/or an abrupt increase in their volume. In the x = 0.55 and 0.75 samples, the volume of the ferromagnetic clusters appears to increase smoothly with decreasing temperature over a wide temperature range.

Zhao et al [12] were the first to report a giant shift in T_c on exchanging $^{18}O/^{16}O$ in $La_{0.8}Ca_{0.2}MnO_3$, which has the O structure. Fig. 5 shows the change in $\alpha(T)$ for the x = 0.75 compound at ambient pressure on the exchange of $^{18}O/^{16}O$. The increase in the maximum value of $\alpha(T)$ means a reduction in the concentration of mobile polarons just above T_c and is equivalent to a reduction in t. We have found a similar isotope effect in samples with smaller x, i.e. with t > t_c. It follows that at least part of the giant isotope effect on T_c reported by Zhao et al [12] was due to an increased trapping energy of the polarons above T_c with an increased oxygen mass M_O.

In order to identify any additional mass dependence of the interatomic ferromagnetic exchange below T_c, we turned to measurements of $\rho(T)$ under pressure. Fig. 6 shows the resistivity $\rho(T)$ of the x = 0.75 sample under different hydrostatic pressures; this sample has t < t_c below room temperature. The maximum in the resistivity occurs very close to T_c, which allows monitoring of T_c with a resistivity measurement. Above T_c, the $\rho(T)$ curve is essentially pressure-independent, but below T_c it drops sharply with decreasing temperature; however, the conduction never becomes metallic below T_c, see inset. At T_c, pressure introduces a marked thermal hysteresis ΔT_c between the warming and cooling curves. Although the ΔT_c indicates a first-order phase change at T_c in the O phase, the resistivity drops smoothly as the temperature is lowered through T_c. As noted above, the effective t increases with hydrostatic pressure, indicative of an abnormally compressible Mn-O bond; in the x = 0.75 sample, a P = 1.6 kbar is sufficient to induce a transition from the O' to the O phase at T_c.

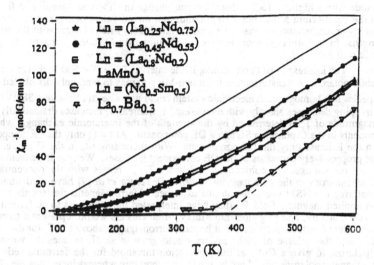

Fig. 4. Inverse molar paramagnetic susceptibility versus temperature for several $Ln_{0.7}A_{0.3}MnO_3$ samples compared to that for $LaMnO_3$ (solid line).

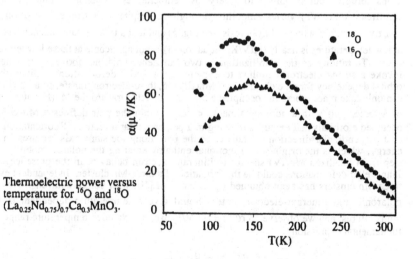

Fig. 5. Thermoelectric power versus temperature for ^{16}O and ^{18}O $(La_{0.25}Nd_{0.75})_{0.7}Ca_{0.3}MnO_3$.

electrons will be compensated by a decrease in the mean potential energy <V> and hence in the equilibrium M-O bond length. A double-well potential for the Mn-O bond would make pressure favor the state with a higher <T>. A discontinuous change in <T> would result in a first-order change in the equilibrium Mn-O bond length and, in a mixed-valent perovskite system, could stabilize at lower temperatures a phase segregation into Mn(IV)-rich clusters with shorter Mn-O bond lengths in a Mn(IV)-poor matrix of longer Mn-O bond lengths at a ratio Mn (IV)/Mn = 0.3.

(3)   The increase in $\alpha(T)$ on cooling in the interval $T_{max} < T < 500$ K signals a trapping out of mobile polarons; a transformation from two-Mn polarons to randomly distributed mobile Mn(IV) species in a homogeneous electronic system could not raise $\alpha(T)$ above + 38 μV/K. The extent of trapping decreases sharply with increasing t-t_c whereas T_c increases dramatically with t-t_c. The magnitude of T_c is determined by the strength of the interatomic exchange, which has two components in the O phase: a SE and a DE component. At $t = t_c$, only the SE component is present; in the R phase only the DE component. With increasing t-t_c in the O phase, the SE component progressively decreases as the DE component increases. We may infer from the data of Fig. 2 that the stronger DE component, and hence T_c, increases with the concentration of mobile Zener polarons in the paramagnetic matrix. DeTeresa et al [9] have used small-angle neutron scattering (SANS) to demonstrate the existence of ferromagnetic clusters above T_c that grow in an applied magnetic field **H**; they have interpreted these clusters to be ferromagnetic polarons. Alternatively, a segregation into Mn(IV) -rich clusters would represent a trapping of polarons into more conductive clusters that become ferromagnetic above the T_c for the Mn(IV)-poor matrix, and the volume of such clusters would grow in an **H,** as does the volume of a magnetic polaron, to give a CMR at the percolation threshold for the ferromagnetic phase. Moreover, magnetic polarons would not be sensitive to pressure whereas increasing the effective t with pressure would lower the polaron trapping energy and therefore increase the T_c of the matrix as is observed, see inset of Fig. 2. The onset temperature $T_s < 500$ K for polaron trapping is poorly defined.

(4)   Below T_c, the charge carriers are not trapped in ferromagnetic clusters and $\alpha(T)$ drops abruptly, but smoothly, to a very low temperature-independent value. However, it is significant that $\alpha(T)$ varies smoothly through T_c despite a first-order transition at T_c (the resistivity exhibits a thermal hysteresis, see Fig. 5) and that a temperature-independent $\alpha(T) \approx \alpha_o$ at low temperatures is nearly 0 μV/K. Equation (3) is not applicable at these low temperatures.

To rationalize the stabilization of two-Mn Zener polarons above T_c, it is necessary to invoke a strong electron coupling to dynamic Jahn-Teller deformations. Given the twofold orbital degeneracy for an e electron at a Mn(III) ion, the e-electron transfer to a Mn(IV) neighbor can only take place where the occupied e orbital is directed toward the Mn(IV) atom. Formation of a Zener pair with a hopping time $\tau_h < \omega_0^{-1}$ within the pair defines a Mn-O-Mn axis of occupied e orbitals that excludes hopping in a perpendicular direction. Reorientation of the axis to a perpendicular direction or transfer of the pair along the same axis requires an activation energy, which is the origin of the motional enthalpy entering the polaron mobility. A second step from localized Mn(IV) species to itinerant-electron behavior in the presence of dynamic Jahn-Teller deformations could be the formation of a three-Mn cluster. Independent evidence for three-Mn clusters has been obtained by Louca et al [10] with pulsed neutrons. However, both polaronic and itinerant-electron states should give a temperature-dependent $\alpha(T)$ at low temperatures, and we observe an $\alpha(T) = 0$ μV/K over an extended temperature range. From the fundamental expression

$$\alpha(T) = \frac{k}{e} \int \frac{(\varepsilon - \varepsilon_F)}{kT} \frac{\sigma(\varepsilon)}{\sigma} d\varepsilon$$

342

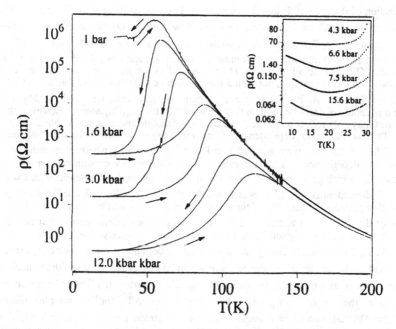

Fig. 6. Resistivity versus temperature for x = 0.75 sample for different hydrostatic pressures.

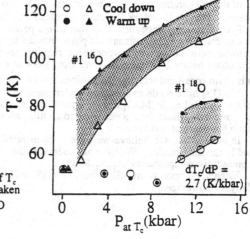

Fig. 7. The pressure dependencies of T_c (warming) and T_c (cooling) taken from $d\rho/dt = 0$ for ^{16}O and ^{18}O of sample x = 0.75.

Since the maximum in $\rho(T)$ is located near T_c, this type of measurement allows investigation of the shift of T_c and the change in ΔT_c on exchanging $^{18}O/^{16}O$. Fig. 7 shows the variation of T_c and ΔT_c with pressure for x = 0.75 with $^{16}O/^{18}O$. In the O' phase, there is no thermal hysteresis at T_c; the magnetic transition appears to be second-order. A ΔT_c sets in abruptly at the O'- O transition at t_c as pressure increases t, and t_c is shifted from a little below 2 kbar in the ^{16}O sample to a little below 11 kbar in the ^{18}O sample. It follows that the ^{18}O sample is equivalent to the ^{16}O sample, but with a lower effective tolerance factor t as was also deduced from the $\alpha(T)$ data. At P = 11 kbar > P_c, a giant isotope coefficient $d\ln T_c/d\ln M_O = 4.9$ was found; it is about six times larger than that reported [12] for $La_{0.8}Ca_{0.2}MnO_3$. However, this giant coefficient may only reflect a change in the density of mobile charge carriers above T_c.

To test the mass dependence of the ferromagnetic state below T_c, we note that $\Delta T_c \approx 27$ K at P_c for ^{16}O decreases to 19 K at P_c for ^{18}O. To clarify whether this reduction is due to a pressure dependence or a mass dependence, we did pressure experiments on a ^{16}O sample with x = 0.85, which gave an O'- O transition at the same P_c as the ^{18}O sample with x = 0.75. A pressure-independent $\Delta T_c \approx 27$ K was found for the ^{16}O samples; the coefficient $dT_c/dP = 4$ K/kbar, though reduced from that of the x = 0.75 sample, remained higher than the $dT_c/dP = 2.7$ K/kbar of the ^{18}O sample at the same P_c. These experiments demonstrate that both ΔT_c and dT_c/dP found in the ferromagnetic O perovskite manganites are mass-dependent.

At T_c, the Gibbs free energy ΔG is identical for both the ferromagnetic and paramagnetic phases, and the hysteresis ΔT_c is due to a surface strain energy associated with the nucleation of the second phase as a result of the volume change ΔV occurring at a first-order transition. The larger ΔV, the larger ΔT_c; therefore, the larger ΔT_c for ^{16}O samples indicates a $\Delta V(^{16}O) > \Delta V(^{18}O)$, a conclusion that is verified by the greater pressure sensitivity dT_c/dP for the ^{16}O compared to the ^{18}O sample. If, as is generally assumed, a polaronic to itinerant electronic transition occurs on cooling through T_c, then the mass dependence of ΔT_c is due to a mass dependence of the paramagnetic-state volume since there would be no mass dependence of the volume of an itinerant-electron state. But from the Virial theorem, the lower polaron kinetic energy of the ^{18}O paramagnetic state would make $\Delta V(^{18}O) > \Delta V(^{16}O)$, which is opposite to what is observed. We are therefore forced to conclude that the volume of the ferromagnetic state below T_c is more mass dependent than that of the paramagnetic state, a higher mobility of the charge carriers in the ^{16}O ferromagnetic phase introducing a smaller volume than is found in the ^{18}O ferromagnetic phase. This conclusion requires retention of some form of polaronic or vibronic phase below T_c at the cross-over from static to dynamic Jahn-Teller site deformations, which is again consistent with our conclusions from the $\alpha(T)$ and $c_p(T)$ data.

We [13] have made experiments under pressure on single-crystal samples x = 0.12 and 0.15 of the system $La_{1-x}Sr_xMnO_3$ that have provided independent evidence of a phase segregation above T_c and a transition from a polaronic to an itinerant electronic behavior below T_c in the O orthorhombic phase of this system.

In conclusion, we believe we have demonstrated the basic features of Fig. 1 that we outlined in the Introduction. We have not addressed the phenomenon of charge ordering, which is not present for compositions $Ln_{0.7}A_{0.3}MnO_3$.

Financial support from NSF Grant DMR 9528826 and the Robert A. Welch Foundation is gratefully acknowledged.

REFERENCES

1. R.D. Shannon and C.T. Prewitt, Acta Crystallogr. B25, 725(1969); 26, 1046 (1970)

2. J.B. Goodenough, Prog. Solid State Chem. 5, 145 (1972)

3. E.O. Wollan and W.C. Koehler, Phys. Rev. 100, 545 (1955)

4. P.-G. de Gennes, Phys. Rev. 118, 141 (1960)

5. C. Zener, Phys. Rev. 81, 440 (1951)

6. J.B. Goodenough, Phys. Rev. 100, 564 (1955)

7. J.B. Goodenough, A. Wold, R.J. Arnott, and N. Menyuk, Phys. Rev. 124, 373 (1961)

8. W. Archibald, J.-S. Zhou, and J.B. Goodenough, Phys Rev. B53 14445 (1996)

9. J.M. DeTeresa, M.R. Ibarra, P.A. Algarabel, C. Ritter, C. Marquina, J. Blasco, J. García, A. del Moral, and Z. Arnold, Nature 386, 256 (1997)

10. D. Louca, T. Egami, E.L. Brosha, M. Röder, and A.R. Bishop, Phys. Rev. B56, R8475 (1997)

11. J.B. Goodenough, J. Appl. Phys. 81, 5330 (1997)

12. G.M. Zhao, K. Konder, H. Keller, and K. A. Müller, Nature 381, 676 (1996)

13. J.-S. Zhou, J.B. Goodenough, A. Asamitsu, and Y. Tokura, Phys. Rev. Lett. 79, 3234 (1997)

MAGNETO-TRANSPORT PROPERTIES IN LAYERED MANGANITE CRYSTALS

T. KIMURA,* Y. TOMIOKA,* T. OKUDA,* H. KUWAHARA,* A. ASAMITSU,* and Y. TOKURA*,**

* Joint Research Center for Atom Technology (JRCAT), Tsukuba 305, Japan
** Department of Applied Physics, University of Tokyo, Tokyo 113, Japan

ABSTRACT

Anisotropic charge transport and magnetic properties have been investigated for single crystals of the layered manganite, $La_{2-2x}Sr_{1+2x}Mn_2O_7$ ($0.3 \leq x \leq 0.5$). Remarkable variations in the magnetic structure as well as in the charge-transport properties are observed with changing doping-level x. A crystal with $x=0.3$ behaves like a 2-dimensional ferromagnetic metal in the temperature region between ~ 90 K and ~ 270 K, and shows the interplane tunneling magnetoresistance at lower temperatures. These characteristic charge-transport properties are attributed to the interplane magnetic coupling between the adjacent MnO_2 bilayers, and are strongly affected by the application of pressure as well as low magnetic fields through the change in magnetic structure. With increase of the carrier concentration toward $x=0.5$, the charge-ordered phase is stabilized and dominates the charge transport and magnetic properties.

INTRODUCTION

The recent observations of the large negative magnetoresistance (MR) effect have shed renewed light on the study of perovskite manganites, producing a great deal of interest in underlying physics. The perovskite manganites offer a unique opportunity for the study of novel aspects of the strongly correlated system. One of the most attractive and puzzling issues in the research of these systems is the interplay of the charge, spin and orbital degrees of freedom. The basis for the theoretical understanding of these systems is the notion of "double-exchange interaction" related to the strong (Hund's rule) coupling between itenerant electrons and localized spins. It has lately been argued, however, that the structural, electronic and magnetic properties are governed by not only the double-exchange interaction but also the superexchange interaction, (collective) Jahn-Teller distortion, charge and orbital ordering phenomena, etc.

Recently, extensive studies [1-6] have also been performed in the so-called Ruddlesden-Popper (RP) structure series for manganese oxides which are characterized by the chemical formula $(RE, AE)_{n+1}Mn_nO_{3n+1}$ (RE and AE being trivalent rare earth or divalent alkaline earth ions, respectively). The basic structure in this homologous series is based on alternate stacking of rock-salt-type block layers $(RE, AE)_2O_2$ and n MnO_2-sheets along the c-axis, as shown in the insets of Fig. 2. For the $n=1$ $RE_{1-x}AE_{1+x}MnO_4$ compound, the structure is basically the same as that of famous high-T_c cuprate $La_{2-x}Sr_xCuO_4$. The $n=\infty$ $RE_{1-x}AE_xMnO_3$ compound is the well-known colossal magnetoresistive materials with the provskite structure.

The $n=2$ member of the RP series, $RE_{2-2x}AE_{1+2x}Mn_2O_7$, also shows a wide variety of physical properties like the $n=\infty$ (perovskite) analogs, including large MR [2] and magnetostriction [4, 7] effects related to paramagnetic insulator to ferromagnetic metal transition. The charge-ordering transition has also been observed in the bilayered manganite with the doping level $x=0.5$ [8]. One of the most distinctive features for the bilayered manganite is its anisotropic characters in charge transport (the resistivity ratio $\rho_c/\rho_{ab} > 10^2$) and in magnetic interaction (the exchange interaction ratio $|J_{ab}/J_c| > 10^2$) due to the layered structure.

347

Another remarkable feature for the bilayered manganite is the interplane tunneling magnetoresistance (TMR) effect which may provide a novel approach to the large MR attainable at low magnetic fields [9]. Several kinds of the ferromagnetic tunneling junctions with use of perovskite manganites have so far been investigated, such as trilayer junctions [10], grain-boundary junctions in polycrystalline samples [12] and in thin-films on bicrystal substrates [13]. The bilayered compound is composed of ferromagnetic-metallic (FM) MnO_2 bilayers with intervening insulating (I) $(La,Sr)_2O_2$ blocks. In other words, the bilayered manganite intrinsically contains the infinite arrays of FM/I/FM tunneling junctions in its crystal structure. In such a quasi-two-dimensional (quasi-2D) FM, the interplane as well as inplane charge dynamics (and hence the MR characteristics) is expected to critically depend on the interlayer magnetic coupling between the FM MnO_2 bilayers.

Here we show results of systematic study on the magneto-transport properties of single crystals of the bilayered manganite, $La_{2-2x}Sr_{1+2x}Mn_2O_7$, with various carrier concentrations. The carrier doping, magnetic fields, and pressure effects on the charge transport of the layered manganite are extensively investigated, and compared with other RP phases ($n=1, \infty$).

EXPERIMENTAL

A series of $La_{2-2x}Sr_{1+2x}Mn_2O_7$ single crystals with various doping levels ($0.3 \leq x \leq 0.5$) was grown by the floating zone method as reported previously [3]. The grown crystals were characterized by the powder X-ray, the four-circle single crystal X-ray, and high-resolution transmission electron diffraction measurements to ensure single phase of the bilayer structure. The grown crystals were oriented using X-ray back-reflection Laue technique, and cut out into rectangular slab specimens. Sample dimensions were typically $1 \times 1 \times 0.1$ mm³ with rectangular wide faces parallel to the ab-plane($\parallel MnO_2$ bilayers). Figure 1 exemplifies the x-ray diffraction pattern of a obtained slab specimen with $x=0.3$. Only the $(0\ 0\ l)$ peaks of bilayered 327 structure with the c-axis length of 20.35 Å are observed. Resistivity measurements were made by the conventional four-probe technique with current parallel (ρ_{ab}) and perpendicular (ρ_c) to MnO_2 bilayers, under magnetic fields up to 7 T and quasi-hydrostatic pressures up to 1.1 GPa. The voltage and current electrodes were formed by silver paste with heat treatment at 550°C. The magnetic field H was provided by a split-type superconducting magnet. The quasi-hydrostatic pressure was obtained with a clamp type pressure cell using Fluorinert as a pressure-transmitting medium.

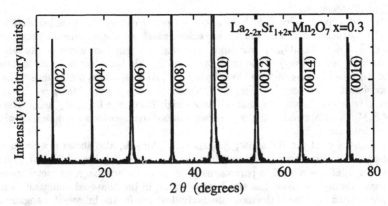

Fig. 1: $\theta - 2\theta$ X-ray diffraction scan of a $La_{2-2x}Sr_{1+2x}Mn_2O_7$ ($x=0.3$) crystal.

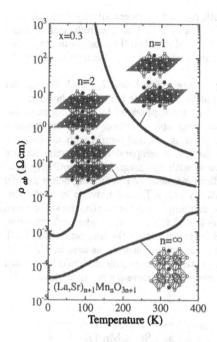

Fig. 2: Temperature dependence of in-plane resistivity (ρ_{ab}) for $n=1$, 2, and ∞ members of Ruddlesden-Popper phase $(La,Sr)_{n+1}Mn_nO_{3n+1}$. The nominal hole concentration is fixed at $x=0.3$. Inset: The crystal structure of $(La,Sr)_{n+1}Mn_nO_{3n+1}$ ($n=1$, 2, and ∞). Shaded planes represent MnO_2 layers.

Figure 2 shows the temperature dependence of the ρ_{ab} in the series of RP phases ($n=1,2$, and ∞) with the nominal hole concentration fixed at $x=0.3$. The ρ_{ab} as well as its temperature dependence changes from metallic to semiconducting, with decreasing the number (n) of MnO_2 sheets. In the $n=\infty$ compound, the steep drop of the resistivity at ~360 K corresponds to the onset of ferromagnetic ordering which is driven by the double exchange interaction. The $n=1$ phase with an isolated MnO_2 sheet, on the other hand, does not undergo the ferromagnetic-metallic transition and remains insulating down to the lowest temperature. In the $n=2$ compound, ρ_{ab} shows a semiconducting temperature dependence above the room temperature. With decreasing temperature, ρ_{ab} shows a broad maximum around ~270 K, and then shows a metallic temperature dependence in contrast to a semi-conducting behavior of ρ_c. This result implies that the short-range ferromagnetic correlation which extends only within MnO_2 bilayers evolves with decrease of temperature below ~270 K [9]. With further decreasing temperature the long-range spin-ordering takes place around $T_c \approx 90$ K. Below T_c the spin correlation extends over the adjacent MnO_2 bilayers, and reduces the spin scattering of the conduction electron in the transport process along the c-axis as well as the ab-plane, which is reflected in steep drops of ρ_{ab} and ρ_c around T_c.

RESULTS AND DISCUSSION

Pressure-enhanced interplane TMR in $La_{2-2x}Sr_{1+2x}Mn_2O_7$ ($x=0.3$)

Let us focus on the pressure and magnetic field effects on the resistivity in $La_{2-2x}Sr_{1+2x}Mn_2O_7$ ($x=0.3$). Figures 3(a) and (b) show the temperature profiles of ρ_{ab} and ρ_c at several pressures under magnetic fields of 0 and 3 T, respectively. All these data were taken in the warming run. In the temperature range above T_c, the both ρ_{ab} and ρ_c monotonically decrease with increasing pressure under the magnetic fields of 0 T and 3 T. A similar reduction of resistivity has also been observed in $n=\infty$ phase, $La_{1-x}Sr_xMnO_3$ [14]. The results indicate that the application of pressure enhances the transfer interaction t for e_g electrons as far as the temperature-region above T_c is concerned.

The most striking feature is found in the temperature region below T_c. Both the ρ_{ab} and ρ_c are remarkably increased by applying pressure at zero magnetic field, as displayed in Fig. 3(a). In other words, the steep drops of ρ_{ab} and ρ_c are suppressed by applying pressure, which signals the pressure-suppression of the long-range spin-ordering. Under the pressure of 1.1 GPa the ρ_{ab} shows no drop due to the spin-ordering but dull temperature dependence characteristic of the 2D metallic state down to the lowest temperature. A similar feature is observed in the pressure effect in ρ_c. Compared with the magnetic susceptibility data suggestive of the evolution of the interplane antiferromagnetic (AF) coupling below $T_N \approx 60$ K [20], the pressure-induced change in the resistivity below T_c may have an intimate connection to the change of the interplane magnetic coupling. In other words, the short-range 2D spin ordering appears to extend down to the lowest temperature under high pressure, which sensitively affects the charge transport. By contrast, under the magnetic field of 3 T the pressure effect on ρ_{ab} and ρ_c becomes considerably small below T_c, as seen in Fig. 3(b). By applying high magnetic fields, the 3D ferromagnetic spin-arrangement should be realized even at high pressures. This may explain the rather weak pressure-dependence of resistivity at high magnetic fields.

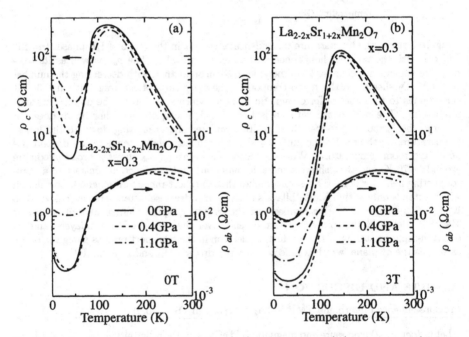

Fig. 3: Temperature dependence of inplane (ρ_{ab}) and interplane (ρ_c) resistivity under magnetic fields of 0 T (a) and 3 T (b) with $H\|c$ at pressures of 0, 0.4, and 1.1 GPa in the $La_{2-2x}Sr_{1+2x}Mn_2O_7$ ($x=0.3$) crystal.

For an origin of the pressure effect on the interlayer coupling, we may need to consider the orbital degrees of freedom of the e_g-like conduction electrons. Since the orbital strongly couples with lattice, the structural modification by the application of pressure may reflect on the change in the orbital state. In $La_{2-2x}Sr_{1+2x}Mn_2O_7$ ($x=0.4$), Argyriou et al. [16] observed a larger pressure-induced change in the Mn-O apical bond length of the MnO_6 octahedron than that in the Mn-O inplane bond length. In other words, the application of hydrostatic pressure reduces the Jahn-Teller distortion in the MnO_6 octahedron, which may cause the crystal field to change and the orbital state of the e_g electron to shift from the $3d_{3z^2-r^2}$ orbital to the $3d_{x^2-y^2}$ orbital. Recently, Ishihara et al. [17] have studied theoretically the pressure effect on the spin and orbital states in the bilayered manganite, taking account of the anisotropic change in the Mn-O bond length. Considering the energy difference between $3d_{x^2-y^2}$ and $3d_{3z^2-r^2}$ orbitals in the pressure-deformed lattice, they showed that the applied pressure can weaken the interlayer charge and spin couplings through the stabilization of the $3d_{x^2-y^2}$ orbital. The change in the orbital character may thus be responsible for the observed pressure effect on the interlayer coupling in the bilayered manganite.

The field dependence of ρ_c have been measured with $H\|c$ at 4.2 K under the pressures of 0 and 0.7 GPa, as displayed in Fig. 4. For these measurements, the crystals were slowly cooled from room temperature to 4.2 K at 7 T, and then magnetic fields were swept cyclically. At ambient pressure the ρ_c rapidly decreases with increasing magnetic fields, and becomes nearly constant above the saturation field $H_{sat}\approx0.2$ T. The interplane MR, $[\rho_c(0 \text{ T})-\rho_c(0.8 \text{ T})]/\rho_c(0.8 \text{ T})\sim490$ %, is much larger than the inplane MR, $[\rho_{ab}(0 \text{ T})-\rho_{ab}(0.8 \text{ T})]/\rho_{ab}(0.8 \text{ T})\sim25$ % [20]. This has been ascribed to the presence of the interplane tunneling MR process for the current-perpendicular-to-plane (CPP) configuration. Namely, the c-axis transport of the spin-polarized electron is blocked at the insulating block due to the AF-type coupling between the adjacent MnO_2 bilayers, but the magnetization process removes such AF-coupling boundaries and allows the interplane tunneling of spin-polarized electrons. In addition, we have observed the non-linear $I-V$ characteristics in ρ_c, suggestive of a FM/I/FM tunneling process [18, 19].

As noted in Fig. 4, the magnitude of the interplane MR is drastically enhanced up to $\sim4,000$ % by applying pressure. This is because the ρ_c value at zero field remarkably increases by high pressure while remaining nearly pressure-independent at magnetic fields more than H_{sat} [see and compare Fig. 3(a) with (b)]. H_{sat} is increased slightly with the increase of pressure. Such an enhanced MR effect can be closely related to the reduction of the interplane magnetic coupling. The weak pressure-dependence of ρ_c at $H \geq H_{sat}$ suggests that the pressure effect is small in the absence of the magnetic domain boundaries. By contrast, at low temperatures ($\leq T_c$) ρ_c shows strong pressure-enhancement at zero field. This indicates that conduction electrons moving along the c-axis in the pressure-induced 2D FM state suffer from even stronger scattering than those in the weakly AF-coupled state at ambient pressure.

In the real material, magnetic domains must be present within each ferromagnetic MnO_2 bilayer below T_N at ambient pressure. The spin arrangement for this concept are illustrated in the insets of Fig. 4. The regions of the interplane AF alignment are therefore incorporated with local regions of the interplane F alignment [inset (a) of Fig. 4]. The regions of the interplane F alignment which remain as defects in the globally AF state, are hence likely to reduce ρ_c and lead to the relative decrease in the interplane MR. (If this is the case, the microscopic current path for the measurement of ρ_c must be highly inhomogeneous.) The suppression of the interplane coupling by pressure may decouple these defects of the F alignment as well as the global AF-coupled state, and cause the increase of the resistivity at zero field [inset (b) of Fig. 4]. At $H \geq H_{sat}$ the decoupled spin domain of each MnO_2 bilayer is aligned along the field direction, which can be viewed as the transition from the 2D ferromagnetic (paramagnetic along the c-axis) to the 3D ferromagnetic state [inset (c) of Fig. 4]. This arises from the fact that the 2D FM state is a highly diffuse metal as seen in $\rho_{ab}-T$ curves at 1.1 GPa of Fig. 3(a). The pressure-enhanced inplane MR thus reflects the deconfinement transition of the spin-polarized carriers, being reminiscent of the spin-valve effect [21].

Possible enhancement of MR due to AF fluctuations

The metallic regime in ρ_{ab} above T_c disappears by degrees with increasing the nominal hole concentration. We display in Fig. 5(a) and (b) the temperature profiles of the ρ_{ab} in x=0.3 and 0.4 crystals under several magnetic fields. In the x=0.4 crystal, the both ρ_{ab} and ρ_c show a steep increase toward T_c with a large activation energy ($\approx$30 meV), which contrasts with the metallic ρ_{ab} in the x=0.3 crystal. The MR ratio in the x=0.4 crystal immediately

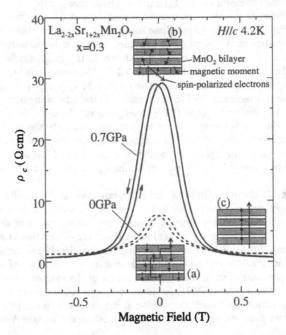

Fig. 4: Normalized interplane resistivity ρ_c as a function of a magnetic field parallel to the c axis at 4.2 K in a $La_{2-x}Sr_{1+2x}Mn_2O_7$ (x=0.3) crystal. Inset shows schematic spin arrangement; (a) At ambient pressure, the MnO_2 bilayers exhibit the interlayer antiferromagnetic ordering, but include the ferromagnetic one as defects. The spin-polarized electrons may move along the c-axis in an inhomogeneous path, through the ferromagnetically-coupled region, although the antiferromagnetically-ordered region serves as a barrier. (b) The application of pressure suppresses the interlayer magnetic coupling, and make the conducting passes in the ferromagnetically-ordered region extinguish. (c) Applications of a magnetic field aligns all of the magnetic moments irrespective of pressure. Spin-polarized electrons can hop along the c-axis over the whole region of the specimen.

above T_c [ρ_{ab}(0 T)-ρ_{ab}(7 T)] /ρ_{ab}(7 T)≈2×10^4 %] is much larger than that in the x=0.3 crystal [ρ_{ab}(0 T)-ρ_{ab}(7 T)] /ρ_{ab}(7 T)≈3×10^2 %]. One of the possible origins for this anomalous behavior of the x=0.4 crystal is related to the charge-/orbital-ordering which takes place in specimens with the hole concentration near x=0.5, as described in the next section. The semiconducting resistivity above T_c has been observed also in pseudo-cubic perovskite manganites at a hole concentration near x=0.5 with controlled one-electron band width [23]. In these systems, the AF charge-ordering instability competes with the ferromagnetic double-exchange interaction near above T_c, and may be relevant to the semiconducting behavior. The competition gives rise to the phase instability and the extremely large MR at temperatures immediately above T_c due to the field suppression of AF fluctuation. For bilayered compounds, recent neutron scattering measurements revealed that the long-lived AF clusters within bilayers coexist with ferromagnetic critical fluctuations above T_c for the x=0.4 crystal [22]. The evolution of the semiconducting behavior and the enhanced MR effect with increasing the hole concentration arises from the localization effect due to the inplane AF correlation, that is perhaps related to the charge-/orbital-ordering.

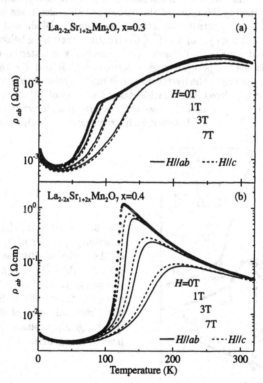

Fig. 5: Temperature dependence of ρ_{ab} under various magnetic fields with different field orientations ($H\|c$ and $H\perp c$), in La$_{2-2x}$Sr$_{1+2x}$Mn$_2$O$_7$ [x=0.3 (a) and 0.4 (b)] crystals. The appreciable differences near T_c between the field directions parallel to the c-axis and the ab-plane are related to the easy axis of magnetization ($H\|c$ for x=0.3 crystal, $H\|ab$ for x=0.4 crystal).

Charge-ordering phenomena

With further increasing the nominal hole concentration toward $x=0.5$, the ferromagnetic-metallic state disappears in bilayered manganites, which is closely related to the stabilization of the charge-ordered state at a commensurate value ($x=0.5$) of the hole concentration. The charge-ordering accompanies, in general, a remarkable change in lattice parameters as well as magnetic and transport properties. Figure 6 displays the temperature profiles of the inplane resistivity in $(La,Sr)_{n+1}Mn_nO_{3n+1}$ ($n=1$, 2, ∞) with $x=0.5$. The system becomes more insulating with the increase of n also at this carrier concentration. The charge-ordering phenomena in $n=\infty$ [24, 25] and $n=1$ [2, 26, 27] phases have been studied extensively so far. In case of the $n=\infty$ phase, the charge-ordered state becomes stable with decreasing the one-electron band width by choice of smaller A-site cations [$Nd_{0.5}Sr_{0.5}$; $T_{CO}\approx158$ K (dashed lines in Fig. 6), $Pr_{0.5}Ca_{0.5}$; $T_{CO}\approx230$ K, ...] although the charge ordering does not show up in $La_{0.5}Sr_{0.5}MnO_3$ with a maximal band width (Fig. 6). In the $n=1$ phase $La_{0.5}Sr_{1.5}MnO_4$, the ρ_{ab} steeply increases around $\sim$220 K, which corresponds to the onset of charge-ordering. Recent resonant X-ray diffraction measurements demonstrated that the charge-ordering accompanies the orbital-ordering as shown in the inset of Fig. 6 [28]. For $n=2$ phase $LaSr_2Mn_2O_7$, the resistive jump signaling the onset of the charge-ordering also takes place at $\sim$220 K as in the $n=1$ phase. Recent electron diffraction measurements [8] showed the appearance of superlattice spots indicating ($\frac{1}{4}$ $\frac{1}{4}$ 0) and ($\frac{3}{4}$ $\frac{3}{4}$ 0) structural scattering below T_{CO}, which is the same pattern observed for the $n=1$ phase. The hysteresis was observed during the cooling and warming runs at the temperature region between $\sim$50 K and $T \sim$160 K. Although we have not understood the origin the hysteresis yet, it may be related to the first-order melting transition of the charge-ordered state. Incidentally, Battle et $al.$ [5] reported on the basis of the neutron powder diffraction that the low-temperature phase for $x=0.5$ is the layered antiferromagnet with the AF coupling between each ferromagnetic MnO_2 layer within a bilayer.

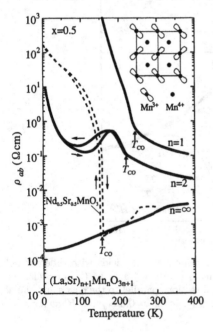

Fig. 6: Temperature dependence of inplane resistivity (ρ_{ab}) for $n=1$, 2, and ∞ members of the series of Ruddlesden-Popper phases $(La,Sr)_{n+1}Mn_nO_{3n+1}$. The nominal hole concentration is fixed at $x=0.5$. Dashed lines represent the results of $Nd_{0.5}Sr_{0.5}MnO_3$. Inset: Schematic view of (0 0 1) projection of the charge and orbital ordering pattern in the layered manganite.

SUMMARY

We have performed systematic measurements of the inplane and interplane magneto-transport properties for bilayered manganite crystals, $La_{2-2x}Sr_{1+2x}Mn_2O_7$ ($0.3 \leq x \leq 0.5$). Most of characteristic features observed in the pseudo-cubic perovskite manganite are reproduced in the bilayered manganite, such as a large MR effect due to AF fluctuations ($x=0.4$), charge-ordering ($x=0.5$) by the control of the carrier concentration. In addition, novel phenomena derived from the 2D nature have been found in the bilayered manganite which is viewed as composed of ferromagnetic-metallic MnO_2 bilayers with intervening non-magnetic insulating $(La,Sr)_2O_2$ blocks. The $x=0.3$ crystal appears to behave like a 2D ferromagnetic metal in the temperature range immediately above T_c. At low temperatures below T_c under ambient pressure, the interplane antiferromagnetic coupling is dominant, which is attributed to the origin of the large interplane MR effect observed under a low magnetic field. The MR effect is related to the interplane tunneling of the fully spin-polarized electrons which is hindered at the interplane magnetic domain boundaries on the insulating $(La,Sr)_2O_2$ layers but is restored during the magnetization process. Furthermore, the application of pressure weakens the interplane magnetic coupling, and makes the charge-transport more 2D-like. As a result, a huge interplane MR is observed up to ~ 4000 % at ≈ 1 GPa and 4.2 K.

ACKNOWLEDGMENTS

We would like to thank R. Kumai, T. G. Perring, G. Aeppli, S. Ishihara, M. Tamura, J. Q. Li, Y. Matsui, Y. Moritomo, and T. Arima for helpful discussions. This work, supported in part by NEDO, was performed in JRCAT under the joint research agreement between NAIR and ATP.

REFERENCES

[1] R. A. M. Ram, P. Ganguly, and C. N. R. Rao, J. Solid State Chem. 70, 82 (1987).

[2] Y. Moritomo, Y. Tomioka, A. Asamitsu, and Y. Tokura, Y. Matsui, Phys. Rev. B 51, 3297 (1995).

[3] Y. Moritomo, A. Asamitsu, H. Kuwahara, and Y. Tokura, Nature 380, 141 (1996).

[4] J. F. Mitchell, D. N. Argyriou, J. D. Jorgensen, D. G. Hinks, K. D. Potter, and S. D. Bader, Phys. Rev. B 55, 63 (1997).

[5] P. D. Battle, M. A. Green, N. S. Lasky, J. E. Millburn, M. J. Rosseinsky, S. P. Sullivan, and J. F. Vente, Chem. Commun., 767 (1996); P. D. Battle, M. A. Green, N. S. Lasky, J. E. Millburn, P. G. Radaelli, M. J. Rosseinsky, S. P. Sullivan, and J. F. Vente, Phys. Rev. B 54, 15967 (1996).

[6] H. Asano, J. Hayakawa, and M. Matsui, Appl. Phys. Lett. 71, 844 (1997).

[7] D. N. Argyriou, J. F. Mitchell, C. D. Potter, S. D. Bader, R. Kleb, and J. D. Jorgensen, Phys. Rev. B 55, R11965 (1997).

[8] J. Q. Li, Y. Matsui, T. Kimura, and Y. Tokura, Phys. Rev. B, (to be published).

[9] T. Kimura, Y. Tomioka, H. Kuwahara, A. Asamitsu, M. Tamura, and Y. Tokura, Science 274, 1698 (1996).

[10] Yu Lu, X. W. Li, G. Q. Gong, Gang Xiao, A. Gupta, P. Lecoeur, J. Z. Sun, Y. Y. Wang, and V. P. Dravid, Phys. Rev. B **54**, R8357 (1996).

[11] J. Z. Sun, W, J. Gallagher, P. R. Duncombe, L. Krusin-Elbaum, R. A. Altman, A. Gupta, Yu Lu, G. Q. Gong, and Gang Xiao, Appl. Phys. Lett. **69**, 3266 (1996); J. Z. Sun, L. Krusin-Elbaum, P. R. Duncombe, A. Gupta, and R. B. Laibowitz, Appl. Phys. Lett. **70**, 1769 (1997).

[12] H. Y. Hwang, S-W. Cheong, N. P. Ong, and B. Batlogg, Phys. Rev. Lett. **77**, 2041 (1996).

[13] N. D. Mathur, G. Burnell, S. P. Isaac, T. J. Jackson, B.-S. Teo, O. J. L. MacManus-Driscoll, L. F. Cohen, J. E. Evetts, and M. G. Blamire, Nature **387**, 266 (1997).

[14] Y. Moritomo, A. Asamitsu, and Y. Tokura, Phys. Rev. B **51**, 16491 (1995).

[15] T. Kimura, A. Asamitsu, Y. Tomioka, and Y. Tokura, U. Phys. Rev. Lett. **79**, 3720 (1997).

[16] D. N. Argyriou, J. F. Mitchell, J. B. Goodenough, O. Chmaissem, E. S. Short, and J. D. Jorgensen, Phys. Rev. Lett. **78**, 1568 (1997).

[17] S. Ishihara, S. Okamoto, and S. Maekawa, J. Phys. Soc. Jpn. **66**, 2965 (1997).

[18] M. Julliere, Phys. Lett. A **54**, 225 (1975).

[19] S. Maekawa and U. Gäfvert, IEEE Trans. Magn. **MAG-18**, 707 (1982).

[20] T. Kimura, Y. Tomioka, A. Asamitsu, and Y. Tokura, Phys. Rev. Lett. **79**, 3720 (1997).

[21] B. Dieny, V. S. Speriosu, S. S. P. Parkin, B. A. Gurney, D. R. Wilhoit, and D. Mauri, Phys. Rev. B **43**, 1297 (1991).

[22] T. G. Perring, G. Aeppli, Y. Moritomo, and Y. Tokura, Phys. Rev. Lett. **78**, 3197 (1997).

[23] H. Kuwahara, Y. Tomioka, Y. Moritomo, A. Asamitsu, M. Kasai, R. Kumai, and Y. Tokura, Science **272**, 80 (1996); S. Tokura, H. Kuwahara, Y. Moritomo, Y. Tomioka, and A. Asamitsu, Phys. Rev. Lett. **76**, 3184 (1996); H. Kuwahara, Y. Moritomo, Y. Tomioka, A. Asamitsu, M. Kasai, R. Kumai, Y. Tokura, Phys. Rev. B **56**, 9386 (1997).

[24] K. Knizek, Z. Jirak, E. Pollert, F. Zounova, and S. Vratislav, J. Solid State Chem. **100**, 292 (1992); Y. Tomioka, A. Asamitsu, Y. Moritomo, H. Kuwahara, and Y. Tokura, Phys. Rev. Lett. **74**, 5108 (1995).

[25] H. Kuwahara, Y. Tomioka, A. Asamitsu, Y. Moritomo, and Y. Tokura, Science **270**, 961 (1995).

[26] B. J. Sternlieb, J. P. Hill, and U. C. Wildgruber, G. M. Luke, C. Nachumi, Y. Moritomo, and Y. Tokura, Phys. Rev. Lett. **76**, 2169 (1996).

[27] Y. Moritomo, A. Nakamura, S. Mori, N. Yamamoto, K. Ohoyama, and M. Ohashi, Phys. Rev. B , (to be published).

[28] Y. Murakami, H. Kawada, H. Kawata, M. Tanaka, T. Arima, Y. Moritomo, and Y. Tokura, Phys. Rev. Lett. (to be published).

AUTHOR INDEX

357

SUBJECT INDEX

Printed in the United States
By Bookmasters